AF844513

Sustainability and the City

Ecocritical Theory and Practice

Series Editor
Douglas A. Vakoch, California Institute of Integral Studies, USA

Advisory Board:
Bruce Allen, Seisen University, Japan; Hannes Bergthaller, National Chung-Hsing University, Taiwan; Zélia Bora, Federal University of Paraíba, Brazil; Izabel Brandão, Federal University of Alagoas, Brazil; Byron Caminero-Santangelo, University of Kansas, USA; Simão Farias Almeida, Federal University of Roraima, Brazil; George Handley, Brigham Young University, USA; Isabel Hoving, Leiden University, The Netherlands; Idom Thomas Inyabri, University of Calabar, Nigeria; Serenella Iovino, University of Turin, Italy; Daniela Kato, Kyoto Institute of Technology, Japan; Petr Kopecký, University of Ostrava, Czech Republic; Serpil Oppermann, Hacettepe University, Turkey; Christian Schmitt-Kilb, University of Rostock, Germany; Heike Schwarz, University of Augsburg, Germany; Murali Sivaramakrishnan, Pondicherry University, India; Scott Slovic, University of Idaho, USA; J. Etienne Terblanche, North-West University, South Africa; Julia Tofantšuk, Tallinn University, Estonia; Cheng Xiangzhan, Shandong University, China; Hubert Zapf, University of Augsburg, Germany

Ecocritical Theory and Practice highlights innovative scholarship at the interface of literary/cultural studies and the environment, seeking to foster an ongoing dialogue between academics and environmental activists.

Recent Titles

Sustainability and the City: Urban Poetics and Politics edited by Lauren Curtright and Doris Bremm
Ecocultural Ethics: Critical Essays edited by Rayson K. Alex, S. Susan Deborah, Reena Cheruvalath, and Gyan Prakash
Critical Ecofeminism by Greta Gaard
Writing the Earth, Darkly: Globalization, Ecocriticism, and Desire by Isabel Hoving
Ecological Entanglements in the Anthropocene edited by Nicholas Holm and Sy Taffel
Ecocriticism, Ecology, and the Cultures of Antiquity edited by Christopher Schliephake
Ecotheology and Nonhuman Ethics in Society: A Community of Compassion edited by Melissa Brotton
The Ethics and Rhetoric of Invasion Ecology edited by James Stanescu and Kevin Cummings
Dark Nature: Anti-Pastoral Essays in American Literature and Culture edited by Richard J. Schneider

Sustainability and the City

Urban Poetics and Politics

Edited by
Lauren Curtright and Doris Bremm

LEXINGTON BOOKS
Lanham • Boulder • New York • London

Published by Lexington Books
An imprint of The Rowman & Littlefield Publishing Group, Inc.
4501 Forbes Boulevard, Suite 200, Lanham, Maryland 20706
www.rowman.com

Unit A, Whitacre Mews, 26-34 Stannary Street, London SE11 4AB

Copyright © 2017 by Lexington Books

All rights reserved. No part of this book may be reproduced in any form or by any electronic or mechanical means, including information storage and retrieval systems, without written permission from the publisher, except by a reviewer who may quote passages in a review.

British Library Cataloguing in Publication Information Available

Library of Congress Cataloging-in-Publication Data Available

ISBN 978-1-4985-3659-2 (cloth: alk. paper)
ISBN 978-1-4985-3660-8 (electronic)

Printed in the United States of America

Contents

List of Figures

Acknowledgments

We thank, at Lexington Books, Douglas Vakoch for encouraging us to edit this collection, as well as Lindsey Porambo and Nick Johns for their patience and help with its production. We are grateful to Rebecca Weaver for her feedback on the introduction and to administrators who made possible the travel that benefited our international collaboration. The genesis of this collection was a panel that we co-organized at the 86th Annual Convention of the South Atlantic Modern Language Association. An earlier iteration of the section on Thomas Cole and Edgar Allan Poe in the introduction was a paper that Lauren Curtright presented at the Interdisciplinary Nineteenth-Century Studies Conference hosted by the Georgia Institute of Technology in 2015. We thank the organizers, panelists, and audiences of both conferences for their insights into, and enthusiasm for, our topic. Last but not least, our gratitude goes to our friends, our families, and all of the collection's contributors for their support of this project.

Introduction

The Role of the Humanities in Urban Sustainability

Lauren Curtright

On the same day as homemade bombs were detonated in New Jersey and Manhattan in September 2016, the *New York Times* ran an article in its Real Estate section about the shift toward verisimilitude in representations of New York City apartments. The author, Ronda Kaysen, argues that living quarters depicted in contemporary television shows set in Manhattan, Brooklyn, and Queens reflect the reality of a generation who "earned about 20 percent less in real wages in 2014 than they would have earned in 2000."[1] These fictional apartments, Kaysen observes, contrast starkly to the plethora of fantastical urban dwellings on television screens in the 1990s, including character Carrie Bradshaw's "Upper East Side brownstone" with its conspicuous "walk-in closet" in HBO's *Sex and the City*. Despite Kaysen's attention to important issues of sustainable development, including wage stagnation, rent control, and displacement by gentrification, her article took a backseat to news of terrorism. Urban residents' real or perceived vulnerabilities to terrorist attacks overshadowed, at least temporarily, Kaysen's discussion of long overdue popular cultural attention to a veritable housing crisis.

In reality, the imminence of terrorism in cities is no more threatening than the skyrocketing costs of urban living. The fire on December 2, 2016, that killed thirty-six people in "The Ghost Ship," an illegally tenanted warehouse in Oakland, California, is grim proof of the fatal consequences of unaffordable housing. Nevertheless, in the contemporary popular imagination, terrorism looms as a preeminent urban threat. The September bombings in New York City and the more recent murders at a Christmas market in Berlin and, on New Year's Eve, a nightclub in Istanbul—among other terrorist attacks in urban centers—seem to affirm the belief that democratic cities invite terrorism. Charles M. Blow predicts in his *New York Times* editorial column that "[s]o long as free people live in free societies and increasingly cluster in urban areas, they will be soft and easy targets for those who have apocalyptic dreams of watching the world burn."[2] However much it resonates with common fears about cities, this conclusion is no less of a cultural construct than the various depictions of urban housing on television. Kaysen and Blow's writings touch on larger narratives—of excess, struggle, and disaster—that demonstrate the difficulty of conceiving of urban sustainability despite cities' allure.

In contrast to *Sex and the City*, and in comparison to current television shows that break away from illusions of equitable urban access, *Sustainability and the City: Urban Poetics and Politics* aims to "get real" about cities' abilities and potentials to foster sustainable development. Paradoxically, *Sustainability and the City* maintains that any hope for achieving urban sustainability will require taking seriously the ways in which cities are imagined. Whether they offer comfort or warning, representations of cities' prospects for sustainability must be understood as key sites of reflection, innovation, and advocacy. Therefore, the essays in this collection analyze myriad creations of, as well as obstructions to, conceptual links between urbanism and sustainability. Rethinking cities' relationships to sustainable development from an international, cultural studies perspective with social justice as its goal, *Sustainability and the City* troubles any easy convergence of its eponymous terms. Still, its contributors share an optimism that cities can achieve sustainability, but they insist that cities will if, and only if, participation in the effort is inclusive of all groups and dimensions of humanity.

The "themes of urbanization and sustainability" are "relative newcomers to the environmental studies field," write the cofounders of the University of Kentucky's Center for Sustainable Cities.[3] In its short life, the term "urban sustainability" has been used interchangeably with "sustainable development," which gained momentum with the *Brundtland Report*, "the publication of the work of the Brundtland Commission" of the United Nations General Assembly, titled *Our Common Future*, in 1987.[4] Emphasizing "integrative thinking," *Our Common Future* urged conservationists to adapt to, rather than resist, urbanization and globalization.[5] It thus proposed unity between traditionally polarized concepts in environmentalist representations, including humanity versus nature and city versus country. It also enfolded an equity agenda into the theory and practice of sustainability, thereby challenging environmentalists to define sustainability comprehensively. If the UN's goal of comprehensive sustainability requires critical thinking about culture, then its enactors must make room for the humanities to join the discussion. Indeed, *Our Common Future* implicitly puts the onus of sustainable development on humanistic endeavors—specifically, on inquiries into what sustainability means (poetics) and the social conditions that make it possible (politics).

CITIES GOING DOWN IN HISTORY

Scholars in the humanities are uniquely positioned to show that, despite the newness of "urban sustainability" as a concept, contradictory views on the capacity of cities to sustain humanity—and the capacity of humans to sustain cities—have a long history. Blow's expression of the city-terrorism equation, for example, is one recent iteration in a long line of associating cities with violence that the freedoms and burdens of capitalist democracies presumably incite. Others, nearly two centuries old, include a series of oil paintings titled *The Course of Empire* by Thomas Cole, the famed American painter credited with founding the Hudson River School, and Edgar Allan Poe's poem "The City in the Sea." The history of *The Course of Empire*'s production is especially revealing of the long-standing conceptual weight of urban sustainability. In the middle of working on what was his first multi-canvas series, Cole took a break, not to rest but, rather, to embark on a distraction that both the artist and the American public favored at the time. In his essay on Cole published in the Smithsonian's journal *American Art*, David Bjelajac describes the artist's situation as follows:

> When strapped for cash in 1836 and in need of a respite from his five-panel Course of Empire Series, an allegorical cycle of a mythical empire's rise and decline, Cole wrote his sympathetic patron Luman Reed that he was about to paint over a large canvas originally intended for the *Consummation* panel [...] to create instead on that canvas a quick, easily marketable American landscape.[6]

The resulting work, shown in figure 0.1, titled *View from Mount Holyoke, Northampton, Massachusetts, after a Thunderstorm*, more commonly known as *The Oxbow*, represents Cole's literal retreat from the exhaustion brought on by imagining the future of cities. Unlike *Consummation*, *The Oxbow* uses what Alan Wallach calls "a traditional landscape vocabulary, evocative of the beautiful, sublime, and picturesque."[7] Cole stepped back from *Consummation* to recuperate his spirit and refill his purse by assuming an un-urban prospect, ranked at the time as "[p]robably the richest view in America."[8] The relative ease of linking a natural scene to sustainability comes across when Cole writes, "in looking over the yet uncultivated scene, the mind's eye may see far into futurity[,] [w]here [...] mighty deeds shall be done in the now pathless wilderness."[9]

In contrast, the vantage point of *Consummation*, shown in figure 0.2, is located in the thick of what Bjelajac terms "overheated, overbuilt urban decadence," as represented by multimodal thoroughfares.[10] While humans seem to traverse the frame of *Consummation*, Cole draws the viewer's gaze to the center of the painting and limits one's perspective to the backs of most

Figure 0.1 Thomas Cole's *View from Mount Holyoke, Northampton, Massachusetts, after a Thunderstorm (The Oxbow)*, 1836. *Source*: Held in the Metropolitan Museum of Art, New York City.

of its figures. If a prospect like that in *The Oxbow* fosters, in Cole's words, "freedom's offspring—peace, security, and happiness," then Cole's shift to a more proximate point of view in *Consummation* creates a paradoxical sense of being simultaneously locked into and out of the spoils of civilization.[11] Given *Consummation*'s effect of constricting viewers to forebode them of the subsequent paintings in the series, *Destruction* and *Desolation*, it is no wonder that Cole hesitated and backed away from *The Course of Empire* before delving into its last three panels. Nonetheless, the consequence is fascinating: the story behind, and the canvas beneath, *The Oxbow* show, respectively, a man and an empire on the verge of exhaustion.

For reasons related to both the artist's personal life and his professional goals, *Consummation* was exceedingly difficult for Cole. Months after completing the first two paintings in *The Course of Empire*, Cole complained in his journal:

> I have scarcely done anything, not finished a single picture. I have made an outline of the 3rd picture in the series, but it is unsatisfactory & I shall commence another; indeed my mind has been occupied with so many cares & anxieties[,] sickness of my Mother & Father &tc., & so many interruptions that it has not been in proper tone for pursuing my profession [. . .] My best days are passing away without being able to apply talents I possess so entirely to my Art as I should wish.[12]

Figure 0.2 Thomas Cole's *The Course of Empire: The Consummation of Empire*, 1836.
Source: Held in the Collection of the New-York Historical Society.

Nearly a year later, in spite of giving another "painstaking" go at *Consummation*, which, according to Ellwood Parry, "occupied most of his time during the winter of 1835–1836," Cole was still struggling with the painting.[13] Parry writes that Cole's letters to Reed from this period indicate that *Consummation* caused the artist "profound anxiety and despair" and "intense dissatisfaction."[14] From sketches underneath *The Oxbow*'s layers of paint as revealed by infrared photography, Parry diagnoses "Cole's problems" with *Consummation* as the artist's difficulty with realizing an upward-looking "perspective system" and the "spatial compression" of "a built environment [...] involving impossibly long colonnades and other structures, piled up layer on layer."[15] Even though Cole would ultimately achieve this visual scheme for conveying urban precariousness in his final *Consummation*, the natural prospect in *The Oxbow* gave him a temporary reprieve from the effort. After returning to *The Course of Empire*, he managed to pull off his desired close-in view of the series' civilization in its glorious heyday and to sustain that exhausting vantage point through the city's violent destruction and serene decay.

Cole's contemporary and compatriot Poe also used optics of exhaustion to represent cities. Initially, in 1831, "The City in the Sea" was titled "The Doomed City" and appeared as "The City of Sin" in the *Southern Literary Messenger* in 1836—the same year as Cole put his final touches on *The Course of Empire*. Poe's opening word of the poem is "Lo!," which the speaker repeats in the first line of the last stanza. Based on how the word appears in Middle English poetry, the *Oxford English Dictionary* defines *lo* in "two distinct" ways: as (1) "an exclamation indicating surprise, grief, or joy, and also used (like O!) with vocatives," and as (2) "the imperative of Look," in which case, the word is "Used to direct attention to the presence or approach of something." That Poe had the latter meaning in mind is arguable in "The Doomed City" but becomes increasingly supportable by his revisions of the poem, in which he alters or omits "O! no—O! no," removing any suggestion by the first version that the *los* in Poe's first and final stanzas are variants of an expression of surprise rather than commands for readers to behold visible evidence that "Death has reared himself a throne" and "a stir is in the air."[16] Drawing readers into submission, these orders have a similar effect as Cole's shift from a prospect in *The Arcadian or Pastoral State* to a closer view of the scene in *Consummation.*

As much as by collapsing distance, Cole and Poe signify exhaustion through their representations of time. Adding an abstract component to *The Oxbow*'s identity as a palimpsest, Parry argues that Cole's 1836 pastoral echoes *Consummation*'s focus on time because *The Oxbow* "was meant to incorporate the principles of the moving panorama."[17] *The Course of Empire* series tracks parts of the earth's daily rotation on its axis and its yearly cycle around the sun: as Parry points out, the setting of *The Savage State* is sunrise in the spring while *Desolation*, shown in figure 0.3, depicts twilight in autumn.[18] Cole's diagram for installing the series in the third-floor art gallery of Reed's home in Manhattan reinforces the sense of the passage of time by three additional panels along the series' top border, which suggest, respectively, the sun's climb from the eastern horizon, its position directly overhead, and its descent through the western half of the sky. Although *Consummation* and *Destruction* visually invoke ancient Greece and Rome and narrate nineteenth-century theories on the demise of classical antiquity, Cole's contemporaries understood *The Course of Empire* as an implicit commentary on their present day—that is, on Jacksonian America. Regarding *Consummation*, they asked themselves, "What does this work say about the prospects of our Republic?," Carl Pfluger surmises.[19] The same question applies to Poe's "The City in the Sea." In particular, Cole and Poe respond to one of the greatest anxieties of their age: the ascendancy of timekeeping. While Cole avoided placing explicit anachronisms in *The Course of Empire*, the ovals patterned in *Consummation*—by hanging wreaths that indicate victory—and in *Destruction*—by broken and submerged shields that indicate defeat—hint at clocks. These implicit timepieces enhance the natural signs of the series' major theme. As Clayton Marsh explains, whether free or enslaved, the population of the United States in the 1830s "suddenly found itself rushing to meet the demands of a new economic order that rewarded

Figure 0.3 Thomas Cole's *The Course of Empire: Desolation,* 1836. *Source*: Held in the Collection of the New-York Historical Society.

speed and punctuality."[20] Marsh argues that two of Poe's short stories from the late 1830s—"The Devil in the Belfry" and "The Man That Was Used Up"—"provide the most insightful literary responses to this historical situation."[21] Produced during the same decade, Cole's series and Poe's poem about the demise of civilization also point to a widespread fear of the social ramifications of "a mechanical conception of time."[22]

The natural measures of time in *The Course of Empire* indicate that, just as surely as the sun will rise again, the series' succession of eras will repeat itself; however, the total absence of human life in *Desolation* diminishes confidence in any chance of reproduction. Pfluger describes the last panel in Cole's series as a "view of the natural world quietly reclaiming its territory from the fallen city."[23] However, in *Desolation*, one striking vestige of humanity remains: a full moon hangs in the sky like a clock on a wall. Moreover, the moon's reflection on the water reinforces its symbolism of mechanical time—an artifice that seems to mirror nature. Likewise, "The City in the Sea" may give hope to the reader when Poe's speaker casts doubt on the impression that the city is locked in "eternal rest" by noticing the ways in which movements of light "So blend the turrets and shadows there/That all seem pendulous in air."[24] This suggests either that the city's suspension is merely a trick of light or that a force is moving the city like a clock. Nevertheless, visual cues that belie the city's stasis, which culminate with "But lo, a stir is in the air!/The wave—there is a movement there!," only indicate that its towers are "sinking."[25] Dwayne Thorpe calls "The City in the Sea" Poe's "nihilistic vision" of "the triumph of time."[26] Like Poe's, Cole's assurance of his civilization's mobility does not guarantee its revitalization but, rather, signals the impossibility of its recovery from exhaustion. Nearly two centuries later, movement, time, and their quotient—speed—continue to shape cities, and to what extent urbanity may be constant or conservationist remains as much a question of culture as of natural law.

On the one hand, in their representations of the inevitability of urban demise, Cole's *The Course of Empire* and Poe's "A City in the Sea" evince that antebellum America was fascinated with the seeming futility of cities' sustainability. On the other hand, in light of Bernard Rosenthal's argument that the notion of nineteenth-century Americans' "deep antipathy to the city" is a "modern myth" and, in fact, "the great vision of the age was preeminently urban," one is likely to discern symbioses between human and nonhuman elements in that period's other meditations on sustainability and the city.[27] Explorations of competing historical perspectives on urban sustainability are needed as part of the emergent urban humanities. Past representations of cities map a history of uncertainty and disagreement about the possibility of urban sustainability and illuminate beliefs about sustainability with which urbanists must still contend.

Niall Ferguson gives the name "cyclical history" to Cole and his contemporaries' assumption that great civilizations fall, and he argues that it carries on in scholarship, as well as common knowledge, today.[28] Ferguson observes that, despite, or perhaps because of, its foreboding, this model gives societies "little incentive to address problems that are unlikely to manifest themselves for a hundred years or more."[29] Moreover, this assumption maintains a paralyzing conceptual and temporal division between sustainability as defined for the "more-than-human"—a term that both Christopher Schliephake and Lisa FitzGerald (citing Stacy Alaimo) use in their essays in this collection—versus for society. In result, Ferguson writes, "rhetorical pleas to save the planet for future generations are insufficient to overcome the conflicts over economic distribution between rich and poor countries that exist in the here and now."[30] Ferguson's critique of cyclical history raises the possibility and significance, for sustainability, of identifying and evaluating alternative models—a task taken up by the contributors to this collection.

CITIES ON THE RISE

Today, sustainability is linked by necessity to cities. In his talk at the *TedCity2.0* conference in Manhattan in 2013, Enrique Peñalosa emphasized the unprecedented scale of humanity's ongoing urbanization by noting that the number of cities in the world is projected to double over the next forty to fifty years.[31] Pointing to infrastructural improvements in Bogotá, Colombia (sidewalks, bikeways, and dedicated lanes for Bus Rapid Transit), which he oversaw as that city's mayor, and proposing "hundreds of kilometers of greenways crisscrossing cities in all directions," Peñalosa argues for "democratic" designs of cities of the future to increase the health, happiness, and mobility of the working poor and thereby promote urban equity.[32] Because Peñalosa's proposals would not only increase equity but also reduce human dependence on fossil fuels and lower carbon emissions that are warming the planet, they exemplify comprehensive sustainability. While traditional deployments of the term "sustainability" have emphasized "avoiding the long-term depletion of natural resources," as the *Oxford English Dictionary* defines this word, the nascent concept of comprehensive sustainability considers the so-called natural environment and human politics in equal measure and means not only reconciling preservation and development but also focusing on social justice and equality.

This altered definition of sustainability has coincided with a shift in perceptions of the city from the antithesis of environmentalism to the new "promised land" for the human species. However, it remains uncertain what it means that the world is urbanizing at the same time as climate change is becoming visible. One reason cities are now recognized as key sites of sustainability is that, as environmentalists have come to focus on the relationship between energy consumption and climate change, many believe cities hold the advantage. For example, echoing David Owen's *Green Metropolis: Why Living Smaller, Living Closer, and Driving Less Are the Keys to Sustainability*, architect and author Ellen Dunham-Jones points out, "the average urban dweller in the U.S. has about one-third the carbon footprint of the average suburban dweller"; thus, she concludes, "from a climate change perspective, the cities are already relatively 'green.'"[33] Although city lifestyles are generally more energy efficient than their suburban counterparts, Pamela Robinson and Ann Dale remind us that urban dwellers as a whole "consume some 75 percent of the world's energy and produce 80 percent of its greenhouse gases."[34] Moreover, in stark contrast to Dunham-Jones, Harm de Blij writes in his "Forward" to *Urban Sustainability: A Global Perspective*, "the demands of urban dwellers, ranging from shelter to diets, are far greater per capita than those of rural residents, so that consumption of construction materials and meat products grows exponentially as urbanization proceeds" around the world.[35] Dunham-Jones's qualifier "relatively" is important: in her praise of cities, she is comparing urban cores to suburbs in the United States, in particular. However, from a global perspective, as lands of plenty, cities paradoxically mean exhaustion and unequal distribution of resources.

Perhaps cities are increasingly figured as sites where comprehensive sustainability may occur simply because cities are where the future of humanity is predicted to lay. As Gary Hustwit's 2011 documentary film *Urbanized* highlights, "over half the world's population now lives in an urban area, and 75 percent will call a city home by 2050."[36] Moreover, the rate of urbanization is staggering—altogether, cities gain "5.5 million people every month"—and points to the urgency of the UN's "New Urban Agenda," the subject of its Habitat III conference in Quito, Ecuador, in October 2016.[37] Continuing what Mauricio Rodas, Mayor of Quito, characterized as the "much deeper" and "much more conceptual" way of discussing urban issues at Habitat III than in the past calls for urban poetics, including interpretations of what draws people to cities in the first place.[38] A list of contributing factors may start with cities' relatively greater and more varied number of practical opportunities to access the means to produce and consume the determined necessities of human life, but explanations must also include the cultural aspects that characterize urban environments. Cities are seen as places to find, in high concentration, distinguished forms of human expression, exciting venues of entertainment, progressive politics, and dynamic physical landscapes. Nonetheless, cities serve as harbors of destitution, exploitation, insecurity, and instability, and critiques of cities' ills—whether seen as causes or effects of cities' seeming offers to escape desperation and "make it big"—rage on. Do humans see their own image in cities? Does human fascination with cities stem from beliefs in and about humanity itself in spite of, or perhaps because of, human blemishes and sins? That would help to explain one dominant story of urbanization, in which the city, in the end, triumphs against all odds. Do cities, as the quintessential embodiment of human efforts, appeal to hubris and narcissism in their promise to save humanity from itself? Better (or worse) yet, are cities simply the last stops in the human species' death drive? The variety and complexity of answers to these questions make clear that deciphering the meanings of cities and resisting simple equivalences of cities and sustainability—tasks for which cultural and literary theorists are particularly qualified—are just as important projects as scientific endeavors for comprehensive sustainability.

The central questions of *Sustainability and the City* concern whether, why, and for whom it may be desirable, or even possible, to make cities sustainable. This collection considers whether *de rigueur* positing of cities as engines of sustainability falls in line with old habits of positivist, wishful thinking, which, arguably, are largely responsible for the ongoing crisis—for humans and nonhumans alike—known as climate change. In other words, it asks whether coalescing the terms "sustainability" and "city" may actually obstruct human action to combat what, from some angles, seems impending, self-imposed apocalypse. To answer these questions and others about the urban turn, the authors in this collection attend to culture—to art, cinema, critical theory, literature, pedagogy, philosophy, and quotidian experience. For example, in his essay in this collection, Karim Wagih Fawzi Youssef criticizes the "positivistic approach" to making cities sustainable by physically conforming them to one perceived ideal. Instead, he argues, urban sustainability requires a foundation in, and openness to, various representations of the city. Youssef stresses that the physical and social dimensions of cities are constantly reshaping each other. Along similar lines, Schliephake writes in his essay that "[s]tatistics and numbers have to be interpreted and are narratively framed, so that they always mirror the cultural place and context in which they are created by various means." And Joseph Donica observes in his essay that "physical models for sustainability are only as strong as the assumptions behind them and the consensus of the public for whom sustainable cities should be built." Simply put, the sustainability of cities depends as much on conceptual and political work as technical practice. The challenge for the urban humanities is to produce frameworks that equitably sustain people and their environments rather than encourage, at best, naive optimism or, at worst, appropriation, exploitation, and destruction. Inclusivity is at the heart of this challenge. Therefore, with chapters written by scholars working within and across a variety of fields, *Sustainability and the City* addresses representations of the relationship between sustainability and the city in diverse locales.

OVERVIEW OF THIS COLLECTION

Sustainability and the City uniquely contributes to "third-generation" discourse on "sustainable development" by considering meanings and manifestations of efforts toward sustainability in a wide range of urban cultures. Moreover, it demonstrates that urbanization is inextricable from discussions on sustainability because, although cities' populations and nations' gross domestic products differ, all cities are primary locations of: the circulation of excess capital; socioeconomic divisions; political resistance; friction between human and nonhuman worlds; and the confluence of art, policy, and identity formation in placemaking. The essays in this collection thus offer lenses for testing out in, and reconfiguring in order to adapt to, other urban settings. *Sustainability and the City* especially insists that, in any context, the ongoing conversation on urban sustainability must not only bridge "the traditional separation of the natural and social sciences" but also fully incorporate the humanities.[39] This collection maintains that critical endeavors and creative expressions that fall within the purview of the humanities are vital to closing the conceptual gulf, as well as the practical gap, between human and nonhuman conservation. As Schliephake writes in his essay, "culture has to be seen as an integral part of sustainability, not only because urban planning relies on narratives and is rooted in long cultural traditions, but also because our cities are cultural products themselves." At stake in bringing "the cultural dimension" of cities to the fore, he argues, is enabling "the environmental impacts of cities [to] be reconciled with the social topics of equality, justice, and prosperity." The essays gathered here share this aim.

Sustainability and the City is divided into three parts. "Part I: Conceptualizing Urban Sustainability" consists of four essays that approach urban sustainability from various disciplinary vantage points to emphasize history, ideology, pedagogy, and critical theory. Together, the authors in "Part I" argue that the ways in which one talks, teaches, and thinks about sustainability in cities can make all the difference in implicating a multitude of participants and perspectives in its realization. Next, the four essays in "Part II: *In Situ* Sustainable Urbanism" foreground, as vital elements for fostering comprehensive sustainability, less tangible aspects of human actions and experiences than data captures. In their analyses of urban commons on four different continents, the authors in "Part II" underscore the influences of fear, identity, memory, myth, tranquility, trust, and cultural values, as well as the production, exhibition, and even destruction of art—and other forms of what Claudia Mantovan refers to as "socio-cultural animation"—on dynamics of public spaces. They point to the need for governing bodies to recognize and respond to these factors in public decision making, especially in urban communities strained by conflict, disaster, gentrification, globalization, renewal, poverty, and/or violence. Finally, to further spotlight the role of the humanities in both theorizing and actualizing urban sustainability, this collection moves from the diverse set of interpretations of on-the-ground urban phenomena in "Part II" to a compilation of five readings of sustainability in different media and genres—sound art, drama, fiction, and film—set in, or evocative of, cities in "Part III: Representations of Sustainability and the City." The authors in "Part III" convince us that sustainable development is impossible without the production, reception, and analysis of creative works because art not only expresses, but also breathes life into, humans' relationships with their environment and all of the hopes, joys, anxieties, suffering, and ambivalence that those relationships entail. Their essays carve out a place for artists and critics to help realize social justice in cities, which generate remarkable power, but power that is too often and too easily used destructively, unfairly, and wastefully despite cities' unique capacities to inspire and sustain humanity.

In "The Imaginative Fabrics of Urban Sustainability: From Metaphor to Matter and from Management to Culture in Eco(logical) City (Re-)Design," Schliephake marshals the term "urban ecology" against the rigidly administered, top-down arrangements of "so-called eco-cities." Arguing that "[m]etaphor and matter are mutually dependent on one another in concepts of urban ecology," Schliephake discusses New Songdo, South Korea, in particular, to bolster criticism of eco-cities for omitting "the cultural aspects of urban ecology and communal identity" from their visions of, and designs for, sustainability. The purpose of this critique is to both reveal and counter the use of "sustainability" as an empty "label" or, worse yet, a means of perpetuating or exacerbating inequality. "What is needed is not so much a look ahead to shiny urban projects, but rather a look back to the environmental history of our cities," as well as attention to residents' own stories and creativity, he argues.

Looking to narratives of the past to rethink urban sustainability in the present, Anirban Adhya and Philip D. Plowright challenge conventional paradigms for understanding the city in relation to wilderness in their essay "Wilderness as City: Conflicting Values in Framing Sustainable Urbanism." They theorize that, in the United States, the concept of wilderness developed from a Judeo-Christian view of "uncivilized" space into an "extended issue of urbanity." Rather than standing for its polar opposite, as is often assumed, wilderness actually links to the construct of the city in American ideology. "Wilderness is not, in fact, about nature but a projection of urban values centered on private ownership, human benefit, and human control," they contend. Adhya and Plowright thus deconstruct the American conglomerate of the individual's right to property and right to self that continues to obstruct key projects of urban sustainability, such as dense housing, mixed-use zoning, and public transit. They explain that, in the twentieth century, property came to hold "psychosocial," as well as economic, value for Americans. The resulting paradox is that movements for public spaces that rejuvenate the individual, such as that led by famed landscape architect Frederick Law Olmsted, "reinforced urban space as an extension of one's needs for privacy." Developed in tandem with this urban construct, rather than "conceived as a non-human space, wilderness is more correctly considered as a place for *a single isolated* human" (original emphasis). Wilderness is thus a "privileged construction," which efforts for comprehensive sustainability must confront.

Youssef argues that as the humanities finds its footing in urban sustainability studies, there must occur "an ontological questioning of the city-sustainability nexus," in "Unraveling the Poles of Suburb and City." Youssef updates a modernist image of the city as hierarchically structured to a postmodernist conception of the city along the lines of Gilles Deleuze and Félix Guattari's theory of the rhizome, "characterized by a horizontal ontology and an instantaneous network paradigm," and of Alan Walks's "paradigm of interpenetration of the urban and the suburban." Youssef thereby challenges dominant models of the city that separate its core from its periphery. Especially vital to his shift in paradigm in representing urban space is Youssef's term "pole," which he describes as having greater complexity, extensity, and relationality than Kevin Lynch's modernist "node" and, therefore, as more conducive to comprehensive sustainability.

While Youssef offers a different conceptual approach for professional designers and planners to take to help make cities sustainable, Iuliu Ratiu considers how college administrators, faculty, and students choose to incorporate sustainability into higher education—the training ground for those who may one day take charge of diagnosing and trying to solve cities' problems. In his essay "The Rhetoric of Sustainability: Teaching Writing to (Mostly) Engineering Students in an Urban Setting," Ratiu exposes "greenwashing," that is, "efforts made by corporations to fund eco-friendly research in order to atone for their negative impacts on the environment." He advances rhetoric and communication toward the forefront of urban sustainability initiatives by demonstrating ways in which, from courses in these fields, students acquire skills "to see through the smokescreen of greenwashing"—skills which, Ratiu argues, are just as vital to urban sustainability as future engineers' "technical innovations to solve some of the problems caused by industries employing unsustainable practices." Providing guidance for instructors on teaching sustainability-themed "ecocomposition," Ratiu discusses implementation of a "solution-based" pedagogy in his first-year writing courses, in which students developed multimedia interpretations of waste in Atlanta, the so-called capital of the New South (in the United States), at this region's most highly ranked university for engineering.

Moving from the proverbial ivory tower to the streets, "Part II" presents specific practices that exemplify successes and challenges of comprehensive sustainability in cities. In "In a Fragile Context: Exploring the Challenges of Informal Sector Urban Gardening in Rio de Janeiro, Brazil," Lea Rekow discusses community gardens in the Rocinha and Manguinhos favelas, including "the largest urban food garden in South America," to reveal "the challenges of operating in internal armed conflicts." These urban agricultural projects represent comprehensive sustainability because they not only mitigate ecological degradation but also improve residents' agency, security, and well-being. However, the projects become vulnerable in the "absence of strong community leadership" and with the lack of "intra-community trust" that can result from the complex "perceptions of control over land" that develop from urban poverty. They also face serious challenges from pollution, inadequate healthcare and sanitation services, setbacks in government funding, and homicidal violence by the police. If such obstacles may be overcome—which will require "consistent and adequate institutional support in both hard and soft infrastructure, including [...] accountable policing procedures that protect the informal sector, rather than terrifying it"—then, Rekow argues, "these gardens could flourish as vibrant, participatory, and sustainable enterprises managed by the people living in favelas."

In "Back to the City: Sustainability and Creativity as a New Approach to Revive the City from the Inside," Heide Imai expands the meaning of "urban commons" by studying varied, everyday practices of residents of the Yanaka district in Tokyo. She thereby rethinks the term "sustainability" in light of rapid transformations to "everyday practices of in/exclusion" and "urban identities" spurred by ongoing globalization, urban renewal projects, and the "commodification of marginal urban places." Her focus is on ways in which "creative practices and small, urban places shape each other and the contemporary urban landscape of Tokyo to be sustainable in times of urban change." Imai's interviewees run small shops that are sustainable because they intersect their own personal and professional lives—which Imai characterizes as "alternative" because they are often relatively simple and slow-paced, as well as based on producing handcrafted goods, other forms of creativity, and interests in preservation, including nostalgia. Even if they don't participate in organized community-building or collective decision-making, and even if they don't regularly communicate with their neighbors, her interviewees have, in effect, created urban commons in Yanaka by "using the creative and productive potential of a derelict, ordinary place to revive the place for all."

In "Coexistence, Conflict, and Governance in Multi-Ethnic Districts: Two Case Studies in the Municipalities of Padua and Venice, Italy," Mantovan looks at the challenges to urban life posed, increasingly, by criminalization of public behavior and by segregation of the poor and the rich in city centers—particularly, near railway stations—due to globalization, which has made it "no longer possible to confine otherness to faraway places." Usually, people whom established residents perceive as threats to security are recent immigrants or migrants. To appease autochthonous populations, city administrators enact regulations that either are merely "symbolic" or actually work against everyone's safety and best interests. Mantovan shows that not only inclusive cultural activities but also alternative narratives—such as exhibits of photography "showing old and new citizens" alike—created by residents and supported and guided by local authorities become "a crucial area of intervention" against "counterproductive" responses to perceptions of criminality in cities.

In terms of criminalization and segregation of populations, the situations that Mantovan describes in Italian cities are strikingly similar to ones in the United States. However, some reversals occur between them. On the one hand, Donald Trump won the 2016 U.S. Presidential election on a platform of exclusion—specifically, of restoring "law and order" to "inner cities" and deporting and barring immigrants from the country—that appealed predominantly to rural voters. On the other hand, in many urban communities that Hillary Clinton carried in the election, such as the eastside of in-town Atlanta, where I live, "migrants" are largely gentrifying middle-class whites who, using online forums like *Nextdoor*, aptly touted as "the *private* social network for your neighborhood" (my emphasis), reproduce narratives of criminality to exclude working-class African Americans whose families have resided in the neighborhood for a half century—since earlier white residents fled to the suburbs in the wake of the Civil Rights Movement. Demonstrating the value of internationalizing the discussion of comprehensive sustainability, Americans' current public discourse on both the Left and the Right could be vastly improved by adopting Mantovan's conclusion that criminalization of immigrants and the poor is a misdirected reaction to a "decline of the traditional channels of social and political representation and mediation." Cities everywhere require public officials' support of inclusive organizations, activities, and events to make urban commons genuinely secure for all.

Like Mantovan, Donica shows the significance of urban residents in peril having control over narratives in order to counter those of the dominant media and influence governance in "The Erosion of the Cultural Commons and the Possibilities of Participatory Urbanism: Public Art in New Orleans, Detroit, and Port-au-Prince." "Part II" concludes with his analysis of public art that memorializes recent natural or economic disasters in New Orleans and Detroit in the United States and in Port-au-Prince, Haiti. Donica interprets specific recent public art projects in these cities as "powerful expressions" of "participatory urbanism," which impact "spaces and surfaces," as well as residents' identities. These projects, he argues, challenge the status quo under neo-capitalism, wherein "[t]he lack of participation by the public in competitions for public art mirrors the lack of representation that many communities have in the urban spaces that they use." The Ninth Ward Katrina Memorial, the Heidelberg Project, and the Tele Ghetto art project represent "a different type of memorial [that] has emerged in the past decade that works against concealment and toward repair" of devastated urban communities; "[t]hey offer a set of strategies for thinking about the city that does not reach out for assumed publics but seeks participation from potential publics—publics who are seeking a space for participation."

Readers may link Donica's alternative models to other public art projects that, while not memorials per se, also reclaim and give continuity to endangered narratives. For example, they compare to work by Rick Lowe, whom the hosts of the podcast series *Equity Speaks* describe as "a world renowned artist and social sculptor."[40] Lowe strives to create projects that are, in his words, "both poetic and symbolic" and that also "really address the issues" of oppressed communities "in a direct way."[41] Lowe's aim with organizations like Project Row Houses in Houston, Texas, is for artists, led by the experiences and needs of urban residents, "to renovate and recontextualize something that was considered the worst in the neighborhood to something that symbolically represents the aspiration of [the] community."[42]

By way of Donica's focus on public art, this collection transitions from social realms to texts. The contributors to "Part III" orient different fields in the humanities—music, English and European literatures, and Francophone film studies—to critical studies of urban sustainability. In the first two chapters of "Part III," Alexander Kleinschrodt and Lisa FitzGerald consider how different kinds of urban performances—musical and dramatic, respectively—dissociate cities from apocalypse. Kleinschrodt makes a case for, in general, the role of artists in environmentalism and, in particular, the impact of sound art on sustainable development in "The Ambivalence of Noise: *Requiem for Fossil Fuels* by Bruce Odland and Sam Auinger." By capturing "the noisy reality of New York City," Kleinschrodt argues, the eponymous piece uses the traditional form of the Catholic requiem mass to congregate "a world society, jointly but not without grief, disbanding itself from fossil fuels, motivated by the belief that it is possible to abandon the route leading to a truly apocalyptic scenario." In "Urban Ecology: Envisioning and Aestheticizing Modernity in Samuel Beckett's *Not I*," FitzGerald argues that Beckett's 1972 play simulates a city on the theater stage, where Mouth "makes manifest the physiological impact of living in a modern landscape." Mouth's "world as it is represented on stage is the ceaseless, surface white noise that is associated with living in the modern urban environment." Nevertheless, through Mouth's engagement with this environment, "the energy released in the course of the play is suggestive of a sustainable, ecological, and progressive model for living in urban space," thereby offering an alternative to the normative, albeit "waning," parity of urban density and apocalypse.

In "Sustainable Urban Development and Literature: A Central-European View," Katarzyna Szalewska analyzes textual representations of changes that have occurred in Polish cities since 1989. Her aim is to reflect on the relations between the assumptions of the so-called New Urbanism and urban poetics in Polish literature. She interprets the ways in which Polish novels by Sylwia Chutnik, Władysław Reymont, Joanna Bator, and Paweł Huelle negotiate competing visions of cities from the "anti-urbanist" view of the city "as Babylon, a monster, a space where apocalypse occurs" to "[u]rbanistic heroism, which is characteristic of the literature of the rising modernity (a belief that a city can be 'taken over', that it may serve for the purpose of upward mobility and personal success)" to the "critical urbanism (a criticism of a city as a space that is hostile to an individual and prevailing economic and social relations)" of postmodernism in Poland. She argues that the "lie" that fiction tells about a city "is essential for the city to function not only as a geographic construct but also as a symbolic space." Therefore, it will take new stories about their present spaces to enable cities like Gdańsk to work through their traumas of war and sociopolitical upheaval; only then will sustainable development in these cities become possible. Like Szalewska, Mehdi Kochbati, considers how memory figures into urban sustainability in "Exploring the American City's Environment: Friction between Fluctuating Urban Sustainability, Modern Myths of Space, and Memory Places in the Works of Paul Auster." Wandering through urban landscapes creates, destroys, doubles, and renews Auster's characters in *The New York Trilogy* and *Moon Palace*. Kochbati terms their desires and efforts to belong to the city as "memorial sustainability," which requires the characters to connect urban spaces to individual memories—theirs and Auster's own—and to the American collective memory, including myths associated with the western frontier.

Continuing the theme of the interrelatedness of human experiences, identities, and the environment in cities, Caitlin Yocco-Locascio wraps up *Sustainability and the City* with "Trash that Connects: The Everyday Rhizome of Cédric Klapisch's *Paris*." She argues that, Klapisch's 2008 film models comprehensive sustainability in drawing attention to, and making inextricable from each other, issues of urban waste and social justice. In *Paris*, viewers see recycling not only of literal trash but also of body organs (by transplants) and the movie camera's recordings (by digital film editing). Moreover, Yocco-Locascio theorizes that Klapisch (re)uses "trash" to represent Paris as a rhizome, a networked structure without hierarchy, and to thereby advocate for a more egalitarian city.

GOING FORWARD

While the network in Klapisch's *Paris* includes Cameroon, other than in Yocco-Locascio's essay, this collection lacks discussion of Africa, as well as other parts of the world where the need for, and challenges of, comprehensive sustainability are now most acutely felt. Thus, even with its pronounced internationalism, *Sustainability and the City* has limitations in scope. The twenty-first century's highest rates of urbanization are in African countries, and cities in Africa are "the world's least equal."[43] The United Nations' *World Urbanization Prospects: The 2014 Revision* projects that the percentage of the population living in urban areas in Africa will increase from 40 percent in 2014 to 50 percent in 2050 and that, during this period, 90 percent of urbanization worldwide will take place in Africa and Asia.[44] Imai discusses sustainable practices in Tokyo, which, despite its recent decline in growth, the United Nations projects will remain "the world's largest city" through 2030. The only other treatment of Asia in *Sustainability and the City* is Schliephake's of New Songdo. Thus, this collection does not touch on Asian cities in the Global South—the portion of Earth with the largest concentration of "megacities and large cities."[45] Indeed, only two chapters in *Sustainability and the City* focus on practices in urban centers of the Global South: Rekow looks at gardening in favelas of Rio de Janeiro (a megacity) and Donica at public art in Port-au-Prince (a medium-sized city).

However, contributors to this collection, including Donica, Kleinschrodt, Kochbati, Ratiu, and Yocco-Locascio, discuss cities of the Global North that diasporas from the Global South have significantly shaped, and cultural elements of which imperialism and globalization have transplanted worldwide: New Orleans and Detroit; New York City; Atlanta; and Paris, respectively. Indeed, from the contributors' vantage points, the lines between "developed" and "developing" parts of the world sometimes blur and other times harden. Szalewska, for example, calls attention to the chasm between meanings of urban sustainability in Western Europe and the so-called "Second World." Furthermore, vast differences exist within cities. Echoing Youssef, Szalewska characterizes Gdańsk as "a rhizome-city" with "many centers, thus many competing macro-narratives." Demonstrating Rio de Janeiro's exorbitant internal disparity in wealth, Rekow notes that Rocinha, "generally considered the largest favela in Brazil," is situated "between two of the city's wealthiest suburbs." Finally, recognized as a hub of not only global media, commerce, and travel but also human trafficking, Atlanta has the greatest income inequality of cities in the United States, making it a crucial site for discussing urban sustainability. There is a dire need to extend Ratiu's pedagogy beyond the Georgia Institute of Technology, especially to campuses in the Atlanta Metro with high percentages of racial minority, low-income, and/or immigrant students.[46]

The overall aim of *Sustainability and the City* is to spur others working in the humanities to test, in other contexts, the claims, methods, and questions put forth by the contributors to this collection. The editors hope that this collection inspires further humanistic inquiry into urban sustainability that not only looks to the past but also focuses on many more cities at present and in the future. No matter the setting, analyzing narratives, sounds, and visions of urban apocalypse, such as those offered by Cole and Poe, and endeavoring to create or find alternatives to them seem essential to achieving comprehensive sustainability; still, we might do well to keep representations of apocalypse in mind.

NOTES

1. Ronda Kaysen, "Fictional New York City Apartments Get Real," *New York Times*, September 16, 2016, http://nyti.ms/2cevKR9.

2. Charles Blow, "Dangerous World, Serious Leaders," *New York Times*, March 24, 2016, http://nyti.ms/1Riwxgo.

3. Ernest J. Yanarella and Richard S. Levine, *The City as Fulcrum of Global Sustainability* (London: Anthem Press, 2011), xvii.

4. Ann Dale, "Introduction," in *Urban Sustainability: Reconnecting Space and Place*, ed. Ann Dale, William T. Dushenko, and Pamela Robinson (Toronto: University of Toronto Press, 2012), 3.

5. Ibid., 3.

6. David Bjelajac, "Thomas Cole's *Oxbow* and the American Zion Divided," *American Art* 20, no. 1 (2006): 65, accessed April 1, 2017, http://www.jstor.org/stable/10.1086/504062.

7. Alan Wallach, "Thomas Cole's *River in the Catskills* as Antipastoral," *Art Bulletin* 84, no. 2 (2002): 343, accessed April 1, 2017, http://www.jstor.org/stable/3177272.

8. Nathaniel Parker Willis, *American Scenery; or, Land, Lake, and River: Illustrations of Transatlantic Nature, from Drawings by W. H. Bartlett, Engraved in the First Style of the Art*, vol. 1 (London: G. Virtue, 1840), 10, quoted in Ellwood C. Parry, "Overlooking the Oxbow: Thomas Cole's 'View from Mount Holyoke' Revisited," *American Art Journal* 34–35 (2003–2004): 14, accessed April 1, 2017, http://www.jstor.org/stable/3451056.

9. Thomas Cole, "Essay on American Scenery," *American Monthly Magazine* 1 (January 1836), posted on Thomas Andrews, "Cole," *California State University Northridge*, accessed April 1, 2017, https://www.csun.edu/~ta3584/Cole.htm.

10. Bjelajac, "Thomas Cole's *Oxbow*," 74

11. Cole, "Essay on American Scenery."

12. Thomas Cole, "Thoughts and Occurrences," in *Thomas Cole: The Collected Essays and Prose Sketches*, ed. Marshall Tymn (St. Paul, MN: J. Colet Press, 1980), 126, quoted in Parry, "Overlooking the Oxbow," 24–25.

13. Parry, "Overlooking the Oxbow," 30.

14. Ibid., 26.

15. Ibid., 30–31.

16. Edgar Allan Poe, "The City in the Sea," lines 1 and 42, *Edgar Allan Poe Society of Baltimore*, last modified October 15, 2015, http://www.eapoe.org/works/poems/cityseae.htm.

17. Parry, "Overlooking the Oxbow," 55.

18. Ibid., 50.

19. Carl Pfluger, "The Views and Visions of Thomas Cole," *The Hudson Review* 47, no. 4 (1995): 634, accessed April 1, 2017, http://www.jstor.org/stable/3851729.

20. Clayton Marsh, "Stealing Time: Poe's Confidence Men and the '*Rush* of the Age,'" *American Literature* 77, no. 2 (2005): 260, doi:10.1215/00029831-77-2-259.

21. Ibid., 260.

22. Ibid., 261.

23. Pfluger, "Views and Visions," 634.

24. Poe, "City in the Sea," lines 5 and 26–27.

25. Ibid., lines 42–43, 45.

26. Dwayne Thorpe, "Poe's 'The City in the Sea': Source and Interpretation," *American Literature* 51, no. 3 (1979): 395, accessed April 1, 2017, http://www.jstor.org/stable/2925393.

27. Bernard Rosenthal, "The Urban Garden: Nineteenth-Century American Views of the City," *Texas Studies in Literature and Language* 20, no. 1 (1978): 119, accessed April 1, 2017, http://www.jstor.org/stable/40754525.

28. Niall Ferguson, "Complexity and Collapse: Empires on the Edge of Chaos," *Foreign Affairs* 89, no. 2 (2010): 19, accessed April 1, 2017, http://www.jstor.org/stable/20699848.

29. Ibid., 20–21.

30. Ibid., 21.

31. Enrique Peñalosa, "Why Buses Represent Democracy in Action," TED video, 14:17, September 2013, http://www.ted.com/talks/enrique_penalosa_why_buses_represent_democracy_in_action.

32. Ibid.

33. Ellen Dunham-Jones, "Retrofitting Suburbia," TED video, 19:23, January 2010, http://www.ted.com/talks/ellen_dunham_jones_retrofitting_suburbia.

34. Pamela Robinson and Ann Dale, "Generational Responses: Why a Third?," in *Urban Sustainability: Reconnecting Space and Place*, eds. Ann Dale, William T. Dushenko, and Pamela Robinson (Toronto: University of Toronto Press, 2012), 13.

35. Harm de Blij, "Forward," in *Urban Sustainability: A Global Perspective*, ed. Igor Vojnovic (East Lansing: Michigan State University Press, 2012), xii.

36. *Urbanized*, directed by Gary Hustwit (London: Swiss Dots, 2011), film.

37. Next City, "Moment to Get Cities Right: Inside Habitat III, the Urbanization Summit of a Generation," YouTube video, 5:41, published December 12, 2016. https://youtu.be/hMGDYI9hdyg.

38. Ibid.

39. Robinson and Dale, "Generational Responses," 18.

40. PolicyLink, *A Conversation with Rick Lowe and Angela Glover Blackwell*, podcast audio, Equity Speaks, SoundCloud, 30:11, accessed April 1, 2017. https://soundcloud.com/policylink-1/es-rick-lowe-ep02.

41. Ibid.

42. Ibid.

43. Gavin Silber, "African Urbanism & Human Rights: A Cape Town Case Study," *Open Society Foundations* PowerPoint presentation, April 25, 2014, https://www.opensocietyfoundations.org/sites/default/files/african-urbanism-and-human-rights-20140507_0.pdf.

44. United Nations Department of Economic and Social Affairs, Population Division, *World Urbanization Prospects: The 2014 Revision* (New York: United Nations, 2015), xxi, 1, http://esa.un.org/unpd/wup/Publications/Files/WUP2014-Report.pdf.

45. Ibid., 16, xxi, 19. *World Urbanization Prospects* defines megacities, large cities, and medium-sized cities as having populations of 10 million or more, 5–10 million, and 1–5 million, respectively, in Figure I.10 (18).

46. Alan Berube and Natalie Holmes, "Some Cities Are Still More Unequal Than Others—An Update," *Brookings Institution*, March 17, 2015, http://www.brookings.edu/research/reports2/2015/03/city-inequality-berube-holmes.

BIBLIOGRAPHY

Berube, Alan, and Natalie Holmes. "Some Cities Are Still More Unequal Than Others—An Update." *Brookings Institution.* March 17, 2015. http://www.brookings.edu/research/reports2/2015/03/city-inequality-berube-holmes.

Bjelajac, David. "Thomas Cole's *Oxbow* and the American Zion Divided." *American Art* 20, no. 1 (2006): 60–83. Accessed April 1, 2017. http://www.jstor.org/stable/10.1086/504062.

Blow, Charles. "Dangerous World, Serious Leaders." *New York Times*, March 24, 2016. http://nyti.ms/1Riwxgo.

Cole, Thomas. "Essay on American Scenery." *American Monthly Magazine* 1 (January 1836): 1–12. Posted on Andrews, Thomas. "Cole." *California State University Northridge.* Accessed April 1, 2017. https://www.csun.edu/~ta3584/Cole.htm.

Dale, Ann. "Introduction." In *Urban Sustainability: Reconnecting Space and Place*, edited by Ann Dale, William T. Dushenko, and Pamela Robinson, 3–12. Toronto: University of Toronto Press, 2012.

de Blij, Harm. "Forward." In *Urban Sustainability: A Global Perspective*, edited by Igor Vojnovic, xi–xiv. East Lansing: Michigan State University Press, 2012.

Dunham-Jones, Ellen. "Retrofitting Suburbia." TED video, 19:23. January 2010. http://www.ted.com/talks/ellen_dunham_jones_retrofitting_suburbia.

Ferguson, Niall. "Complexity and Collapse: Empires on the Edge of Chaos." *Foreign Affairs* 89, no. 2 (2010): 18–32. Accessed April 1, 2017. http://www.jstor.org/stable/20699848.

Kaysen, Ronda. "Fictional New York City Apartments Get Real." *New York Times.* September 16, 2016. http://nyti.ms/2cevKR9.

Marsh, Clayton. "Stealing Time: Poe's Confidence Men and the '*Rush* of the Age.'" *American Literature* 77, no. 2 (2005): 259–89. doi:10.1215/00029831-77-2-259.

Next City. "The Moment to Get Cities Right: Inside Habitat III, the Urbanization Summit of a Generation." *YouTube.* December 12, 2016. https://youtu.be/hMGDYI9hdyg.

Parry, Ellwood C. "Overlooking the Oxbow: Thomas Cole's 'View from Mount Holyoke' Revisited." *American Art Journal* 34–35 (2003–2004): 6–61. Accessed April 1, 2017. http://www.jstor.org/stable/3451056.

Peñalosa, Enrique. "Why Buses Represent Democracy in Action." TED video, 14:17. September 2013. http://www.ted.com/talks/enrique_penalosa_why_buses_represent_democracy_in_action.

Pfluger, Carl. "The Views and Visions of Thomas Cole." *The Hudson Review* 47, no. 4 (1995): 629–35. Accessed April 1, 2017. http://www.jstor.org/stable/3851729.

Poe, Edgar Allan. "The City in the Sea." *Edgar Allan Poe Society of Baltimore.* Last modified October 15, 2015. http://www.eapoe.org/works/poems/cityseae.htm.

PolicyLink. *A Conversation with Rick Lowe and Angela Glover Blackwell.* Podcast audio. Equity Speaks. SoundCloud, 30:11. Accessed April 1, 2017. https://soundcloud.com/policylink-1/es-rick-lowe-ep02.

Robinson, Pamela, and Ann Dale. "Generational Responses: Why a Third?" In *Urban Sustainability: Reconnecting Space and Place*, edited by Ann Dale, William T. Dushenko, and Pamela Robinson, 13–28. Toronto: University of Toronto Press, 2012.

Rosenthal, Bernard. "The Urban Garden: Nineteenth-Century American Views of the City." *Texas Studies in Literature and Language* 20, no. 1 (1978): 119–38. Accessed April 1, 2017. http://www.jstor.org/stable/40754525.

Silber, Gavin. "African Urbanism & Human Rights: A Cape Town Case Study." *Open Society Foundations* PowerPoint presentation. April 25, 2014. https://www.opensocietyfoundations.org/sites/default/files/african-urbanism-and-human-rights-20140507_0.pdf.

Thomas Cole National Historic Site. *Explore Thomas Cole*. Last modified 2010. http://explorethomascole.org/.

Thorpe, Dwayne. "Poe's 'The City in the Sea': Source and Interpretation." *American Literature* 51, no. 3 (1979): 394–99. Accessed April 1, 2017. http://www.jstor.org/stable/2925393.

United Nations Department of Economic and Social Affairs, Population Division. *World Urbanization Prospects: The 2014 Revision*. New York: United Nations, 2015. http://esa.un.org/unpd/wup/Publications/Files/WUP2014-Report.pdf.

Urbanized. Directed by Gary Hustwit. London: Swiss Dots, 2011. Film.

Wallach, Alan. "Thomas Cole's *River in the Catskills* as Antipastoral." *Art Bulletin* 84, no. 2 (2002): 334–50. Accessed April 1, 2017. http://www.jstor.org/stable/3177272.

Yanarella, Ernest J., and Richard S. Levine. *The City as Fulcrum of Global Sustainability*. London: Anthem Press, 2011.

Part I

CONCEPTUALIZING URBAN SUSTAINABILITY

Chapter 1

The Imaginative Fabrics of Urban Sustainability

From Metaphor to Matter and from Management to Culture in Eco(logical) City (Re-)Design

Christopher Schliephake

"We have entered the urban millennium."[1] Kofi Annan's words, spoken in the year 2000, when he was Secretary-General of the United Nations, still sound like a self-fulfilling prophecy. The last decades have indeed witnessed a great increase in the urban population, as well as the growth of megacities and of far-reaching city regions with urban dwellers outnumbering rural dwellers for the first time in the history of humankind. And, as the latest UN report on the subject, issued in 2014, shows, we have not seen the end of it: by 2050, about 75% of the world's population is expected to live in cities.[2] This dramatic increase in city space and urban habitats will fundamentally alter our relationship to the planet, although nobody can predict what the exact consequences will be. Statistics and numbers have to be interpreted and are narratively framed, so that they always mirror the cultural place and context in which they are created by various means.[3] Annan, for instance, underlined the socioeconomic impacts connected to urbanization, their benefits, as well as their dangers. In his view, "cities are engines of growth and incubators of civilization," but can also turn into "places of exploitation, disease, violent crime, unemployment, and extreme poverty."[4] His comments echoed his concern for social topics of human rights and equality, issues whose relevance and immediacy he had witnessed firsthand due to his upbringing in a developing country. They also drew on a long tradition of thinking about cities in terms of their civilizational and cultural worth, especially with regard to the historical evolution of city-states, progress, and prosperity—a view that modernity has incorporated from antiquity. That the beneficial effects of cities are tied to public concern and policy is also implicit in this vision, and

Annan's call to "do more to make our cities safe and livable places for all" still rings true today.[5]

However, the parameters have shifted, and while the socioeconomic frameworks of urban lifestyles and quality of life continue to be pressing concerns, the environmental impacts of cities have been brought swiftly into focus in the wake of anthropogenic climate change and the depletion of resources on a global scale. Against this background, a different image seems to arise, which questions whether unbridled city growth can really be said to be limitless, and which suggests that large-scale urbanization could create more problems than it can actually solve. As Herbert Girardet notes in his 2008 book *Cities, People, Planet*, "the dependence of cities on the land beyond their periphery" and the impacts of urban living on those areas demand "new ways of looking at" urbanization.[6] His book is one example in a long line of contemporary publications and cultural media that study urban development in relation to climate change and that call for new ways of imagining the place of cities on a globalizing planet, as well as for innovative and interactive principles for situating cities in their respective environments. In this context, sustainability, resilience, and self-sufficiency have become key terms used by urban theorists, environmental activists, urban planners, and politicians alike. In the following, I want to trace the cultural roots and imaginations connected to these terms, and I want to sketch out the shift from metaphor to matter that co-occurs with the development of so-called eco-cities. While the general ideas behind eco-cities can be welcomed for their all-encompassing approach to what urban sustainability could entail, the socioeconomic fabrics and political ideologies are to be criticized for their top-down modeled management of urban environments. I will argue that culture has to be seen as an integral part of sustainability, because urban planning relies on narratives and is rooted in long cultural traditions. Moreover, our cities are (...). Only when the cultural dimension of sustainability is taken into account will the environmental impacts of cities be reconciled with the social topics of equality, justice, and prosperity that Annan evoked at the dawn of the millennium.

What will, in the following, be referred to as a "turn from metaphor to matter and from management to culture" is no totalizing or all-explanatory approach to the subject of urban sustainability, but rather an explorative perspective that tries to correlate two general patterns of development in recent years. There is no question that matter has come to the fore in environmental philosophy, as well as practice in shaping humanity's relationship to the nonhuman world. The "material turn" is both theory and practice, but in terms of urban planning and design it has its roots in cultural narratives and projections that have long imagined cities as circuits or metabolisms and that have themselves transferred principles as observed in nature to human-built environments. In taking the metaphorical conceptualizations of the urban literally and in perceiving the city as a scene of material exchange processes that can be administered, managed, and imposed, some of the implications ingrained in the original cultural imaginations (like, for instance, aspects of communal engagement with urban space) are in danger of being superseded by mechanistic concepts of eco-city planning where the human element is often missing or marginalized. The "cultural turn," on the other hand, has paved the way for a revaluation of the role of symbolization, language, and representation in society, replacing—or at least modifying—scientistic, often positivistic, and economic explanations of the "social." This is not to say that the "material" and "cultural" turns as described in this essay are opposites, but rather that they need to complement one another in sustainable (eco-) city (re)design. Where a materialistic view of city space leads to top-down management, a cultural view needs to highlight the social, historical, and imaginative dimensions that engage with the urban from the bottom-up. Built-from-scratch cities are ideal for showing these perceived shifts in urban design, not only because they take their impetus from cultural projections and stories, but also because they give clear prevalence to material dimensions of economic growth and environmental management. Eco-cities themselves are cultural fictions that need to be seen against their sociocultural background; this cultural aspect is in need of revaluation. Conceiving built-from-scratch cities as cultural narratives-turned-reality enables us, as I aim to show, to uncover the different dimensions of sustainability now prevalent in urban design—and to explore where the story of the urban future is in need of retelling.

FROM METAPHOR TO MATTER: URBAN SUSTAINABILITY DISCOURSE

Since the time of the Industrial Revolution, the decline of nature and the parallel ascendancy of an urbanized world have featured prominently in environmental and scientific debates. Because cities contribute massively to greenhouse gas emissions, demand huge amounts of energy, and produce heaps of waste, they are, it is claimed, places of pollution and hindrances on the path to a greener future. While I have dealt with these narrative templates elsewhere,[7] I have only marginally invoked the other side of the story, which sees cities as part of the solution to environmental degradation and as pioneers in progressive ecological policies and leadership. This could, for instance, be seen after the failure of a number of international climate summits, which repeatedly dashed high expectations by their unwillingness (or sheer incapability and ignorance) to reach binding emission-reduction targets and to impose transnational green initiatives. In the Climate Communiqué of the Large Cities Climate Leadership Group (C40) issued at Copenhagen, the mayors of the biggest cities involved in extensive decarbonization programs gave a wake-up call "that the future of our globe will be won or lost in the cities of the world."[8] "Aside from demonstrating that city governance was more progressive than policy-making at the state level," as Andrew Ross comments on their initiative, "the mayors' statement reflected a growing consensus that only in dense urban environments could efficient, low-carbon living be achieved on a mass scale."[9] This argument entails one of the key points that urban commentators make in their endorsement of city life over other forms of social organization, namely that population density and the compactness of urban spaces save large areas of nonhuman nature from occupation and create conditions enabling relatively short distances between people and their institutions, public transportation, and so on. However, this does not, of course, automatically engender sustainable urbanism per se, but rather depends on the willingness of administration and policy makers to implement programs that mitigate emission levels in urban spaces, reduce costs of—or create other incentives for—using public transportation services, and involve the most vulnerable and marginal groups in environmental justice initiatives. There are thus many variables involved in urban sustainability discourses that encompass social, material, and cultural dimensions.

The latter dimension has to do with how we imagine our cities and their place and function within a globalized world. As Susanna Hagan puts it, "the cultural context is one in which, over the last generation, a new metanarrative has risen out of the ashes of postmodernist relativism—'ecologism.'"[10] Hagan points to the increasing coalition between ecologist and urban thinking in her concept of "Ecological Urbanism," which "foregrounds a view of the city as literal and metaphorical ecosystem."[11] In architectural and planning terms, this entails focusing on the "metabolic processes" and "creat[ing] some version of 'artificial ecosystem': cities that achieve the same interdependent efficiencies and life-preserving redundancies as natural ecosystems."[12] In fact, much literature on urban development now stresses the ties between ecological thinking and urban practice. According to Tai-Chee Wong and Belinda Yuen, this also entails shifting from an "anthropogenic (human-centered)" perspective to one that highlights "the interdependency of human and non-human species and the 'rights' and 'intrinsic values' of non-human species in our pursuit for a sustainable ecosystem."[13] In their view, "ecological planning involves conceptual thinking in environmental urban sustainability, land use allocations, spatially designed and distribution patterns that contribute and lead to achieving such objectives of ecological balance."[14] Wong and Yuen's comment is interesting in so far as it merges the non-anthropocentric outlook of ecological justice advocates with a spatial focus on the city, which suggests that the achievement of environmental objectives depends on the human management and renegotiation of land use patterns and resources. As we will see, this is a paradox implicit in ecological models of urbanism in general, which stems from the aim of merging "ecology's characteristic scalar thinking" that "addresses the interrelationships between human culture and the biophysical environment [...] and the physical city and its bio-environment."[15]

The theoretical issue of scale, which is connected to many current environmental discussions about such concepts as deep time, the Anthropocene, or environmental footprints, has been incorporated into the scientific study of urban ecology since its beginning and concerns the identification of population-based spatial patterns and their relationships to ecological phenomena. More than a scientific collection of data on emission rates or population statistics, this also involves reflecting on how human meaning-making systems render their relationship to a more-than-human environment. While this is often done in a metaphorical way that conceptualizes anthropogenic city space as an ecosystem in its own right, it also means looking at the specific material conditions and physical, place-based contexts in which human-nature interactions take place. Metaphor and matter are mutually dependent on one another in concepts of urban ecology. As I want to show in the remainder of my essay, it is between these poles of metaphor and matter that urban sustainability arises as a concept at present, one in which the cultural imagination plays a key role for the formation and formulation of solutions to our environmental crisis.

Let us begin with the metaphorical aspect. To render cities as ecosystems has a long tradition in cultural history, and while "this simile was, and still is, used to describe a cluster of economic or social relations [...], the environmental sciences have made [it] literal," perceiving the city as a spatial area with its own metabolism and feedback systems.[16] In cultural-historic terms, the metaphor goes back as far as antiquity, when philosophers of the Greek city-state referred to cities as organisms—an image that has been taken up repeatedly by modern city planners, visionaries, and artists.[17] Included in this vision is an understanding of the underlying complexity of urban life, including the interaction and interplay of the various parts that make up the whole. To imagine the city as a body rather than a system of any kind has the advantage that the city is seen as a living entity that depends on the functioning of all parts for its survival, whereas the image of an ecosystem has the benefit of bringing other-than-human entities into focus as an integral part of a living landscape. The widespread tendency to visualize cities as biological occurrences is, on the one hand, tied to the scientific impulse to outline the biophysical foundation of cities; on the other hand, it may be seen as an attempt to discursively overcome a dualistic rhetoric that opposes cities and country, or culture and nature. Cities, these metaphors make clear, are also part of nature and of an interactive biosphere. Yet, as Richard Ingersoll reminds us in his critical overview, "the appearance of ecology within the discourse of architecture hinges on a historic paradox: every act of building betrays the environment, as it requires the displacement of 'natural' relationships."[18] Built environments are superimposed upon natural ones, feeding on natural resources and altering the landscape's morphology. Since cities merge "multi-layered biological and technical systems," Girardet has coined the term "eco-technical systems"[19] to underline the hybrid characteristic of cities. His term is useful because it avoids an anthropogenic perspective by describing the city as an *oikos* in which human and other-than-human entities are likewise involved without neglecting the controlled interventions in these relationships through technology and other means. Navigating these poles and realizing the paradox engrained in urban ecology are vital to any truly sustainable approaches to the city.

Urban ecology is, in this context, another way of framing the interrelatedness of human actions and the more-than-human environment within an urban space. Initially used in a metaphorical way by social scientists who studied the interaction between social groups (and their respective areas) in cities, urban ecology is now increasingly applied by urban planners, as well as natural scientists.[20] Common to their use of the term is an increasing tendency to render the material processes and metabolisms involved in urbanism. In Hagan's terms, urban ecology is a "way of seeing," especially the "material relationships between the built and the physical site," as well as their "socio-economic implications."[21] It is also a way of seeing what nature has to teach us with regard to adaption, construction, and resilience. Accordingly, natural processes are used as a model by many architects and planners who have begun to integrate concepts based on biomimicry.[22] While this model concerns the ways in which our built environments have to be reconstructed in order to withstand environmental impacts and mitigate emissions and energy consumption, it is based on the belief "that cities would be well advised to model themselves on the functioning of ecosystems to assure their long-term viability."[23] Inherent in this statement is a somewhat idealist image of ecosystems as beneficial (and balanced) for the continuity of the whole. While this does not hold in scientific terms because ecosystems are themselves marked by dominant or invasive species that lead to inconsistencies and periods of imbalance followed by regeneration,[24] it is still interesting to note the underlying impulse: Girardet argues that cities should create a metabolism in which every output is used as an input in order to minimize pollution and to maximize resource use.[25] Popular notions of ecosystems are thus turned into foils upon which new interactive processes within cities can be modeled. This entails, for instance, a rethinking of how waste, garbage, and sewage are treated—rather than something to be disposed of, they become raw materials that could be used for energy production. It also means re-visioning the ways in which our cities interact as place-based entities with their hinterlands and the global flows of travel and trade. Most of the products sold in cities are imported from elsewhere, and, in a time when agriculture has been relegated to areas far removed from urban centers, loads of "food miles" are piled up that lead to high emission rates for transport alone.[26] In order to render the impact of our cities on a global scale, William Rees and Mathis Wackernagel came up with the concept of the "ecological footprint," which they define as "the areas required to supply cities with food and forest products to absorb their output of wastes, and particularly their output of carbon dioxide."[27] Like the metaphor of an urban ecosystem, the ecological footprint analysis is a "seeing" device (in the sense of Hagan's use of the term): it helps us to visualize the material impact of cities on the planet, as well as their mutual interrelationship.[28] And it is from the realization of the material imbalance inherent in this connection that debates about urban sustainability stem in our modern age.

There are myriad ways of defining sustainability. In urban terms, sustainability is a slogan that unveils as much as it covers up. Usually it combines aspects of social equity, the political management of resources, as well as a consideration of the long-term effects of urban design on humans (and sometimes nonhumans) in the framework of a circumscribed city area.[29] As Girardet defines the term, "a 'sustainable city' enables all its citizens to meet their own needs and to enhance their well-being, without degrading the natural world or the lives of other people, now or in the future."[30] Girardet's anthropocentric and socially centered perspective is central to definitions of sustainability formulated by policy makers and underlines the necessity of shifting the focus on how to maintain resources, rather than deplete them, to attend to their just distribution.[31] This socioeconomic outlook is often supplemented by a technological one that addresses the problem of how to use the material by-products of urban life in order to reduce environmental pollution and mitigate climate impacts. Since not all technological interventions are environmentally friendly, and because economic interests do not always coincide with environmental justice, there are tensions inherent in the term,[32] which can undermine the formulated aims and efforts from the beginning. One example is the problem of the new versus the renewed:[33] is it more ecological to renovate already existing buildings with regard to insulation and energy consumption or to erect new buildings instead? What will become of the people living there? And who benefits, economically speaking, from grand-scale construction projects? Another aspect to consider is that, although cities can manage resources in a certain way, "they can never become fully self-sufficient. Therefore, individual cities cannot be considered 'sustainable' without acknowledging and accounting for their dependence on resources and populations from other regions around the world."[34] To put it differently, sustainability is a concept that focuses on engendering stability and maintaining balance within an environment whose manifold actors and agents are anything but stable or balanced and include areas far removed from the narrow outskirts of an individual city.[35] Sustainability means negotiating the parameters upon which the far-reaching material interrelationships and socioecological interactions rest and involves a process of reflection about which standards—and what exactly—should be maintained. Unfortunately, current approaches to the term often fall short of advocating these issues so that it appears as an empty catchword without meaning. Negotiating meaning and interpreting concepts is exactly what the humanities do, and it is time to shift the focus to what sustainability means from a cultural perspective—an issue that is still undertheorized at the moment and that is addressed by the essays in this collection. In the following, I want to tackle this question by looking at ecological redesign, as well as the construction and planning of so-called eco-cities. As I will show, eco-cities are a prime example of the ways in which sustainability is used as a cultural foil for making implicit claims about issues of management, matter, and social processes of participation. I will argue that the cultural level is central to any approach that self-labels its intentions as "sustainable": while urban planners use storytelling and other cultural media to communicate their visions, they should also look to the culture of a city as a resource to be used and from which to learn.

FROM MANAGEMENT TO CULTURE: NEGOTIATING THE NARRATIVE FABRICS OF ECO-CITIES AND SUSTAINABILITY

Sustainability evolved into a key term of urban planning and architecture over the last decades of the twentieth century. However, it has been an increasing sensibility to the possible interrelations between human resource consumption and degrading environmental conditions, as well as the realization of the long-term consequences of consumption and economic growth ideologies in the wake of the 2008 collapse, that has given sustainability a new urgency in the twenty-first century.[36] What can be witnessed all over the world are extensive construction projects that aim for a new type of city, one that incorporates ecological ideals in both a metaphorical and a material way. Bringing about (...) of creating cities as self-reliant systems that do not exclude natural forces but rather incorporate them in order to channel their energy and impact.[37] In Wong and Yuen's words, "this implies that cities should be conceptualized as ecosystems where there is an inherent circularity of physical processes of resources, [and] activities and residuals that must be managed effectively if the city's environmental quality is to be maintained."[38] This quotation incorporates all of the variables of ecological urbanism illustrated above and makes clear that the metaphor becomes a design principle in its own right, one that has to be "managed" and administered by officials and planners. As Hagan has shown, these principles are already ingrained in the UN Agenda 21, first formulated after the so-called Rio Earth Summit in 1992. Although very much centered on social justice in their formulation of "settlement objectives," the authors also proposed all-encompassing strategies for land use patterns and sustainable infrastructures.[39] In Hagan's analysis of this text, "it is a collection of ethically driven objectives, content without form or embodiment" with "objectives" that "don't lend themselves to imagery."[40] Thus the fabrics from which the eco-city concept sprung were vague, to say the least, and this still shows in contemporary design concepts, which have made clever use of the ethical objectives as labels for their far-reaching initiatives—initiatives that are, as we are about to see, neither always ecologically nor socially minded.

The trouble begins with the problem of how to accurately define the term eco-city. There is a plethora of definitions and characteristics attributed to it.[41] It can be seen as an "umbrella concept,"[42] which merges economic, social, and environmental aspects to reenvision human habitation in a globalized world. Richard Register first coined the term in his 1987 book *Ecocity Berkeley: Building Cities for a Healthy Future*, which put emphasis "on urban re-formation modelled on the circular consumption found in natural ecosystems, and achieved through environmental and social engineering."[43] Next to a focus on renewable energy production, this also includes aspects of social density and the avoidance of car traffic within the city area.[44] In order to guarantee and boost the development of eco-cities, the World Bank has incited the Eco2Cities program, which promotes financial support for cities and countries adopting green strategies in their city planning.[45] This has led developers to enter a race for financial support as more and more policy makers see the urgency of creating sustainable cities while also sensing the chance of participating in a new, profitable market at the same time. The latter aspect becomes apparent when one looks at the eco-city projects that have been implemented so far: Masdar City (Abu Dhabi), King Abdullah Economic City (KAEC, Saudi Arabia), and New Songdo (South Korea) are all designed as business hubs in a global network of travel trade. Either giant port cities (KAEC) or centers of air traffic (New Songdo), they are designed to meet the demands of a globalized economy. Moreover, all of them are master-planned cities, built from scratch on blank slates, without already existing historical ties to the environment or social infrastructures. Let us look at the example of New Songdo.[46]

With the help of international star architects and architectural consortia, one of the first eco-cities of the world has been built on Incheon, near Seoul. Situated in the Incheon Free Trade Zone, New Songdo could easily become the template for a new urban design focused on both ecological and economic functionality as a breeding ground for a new, relatively homogeneous middle class. This mega project is expected to fulfill the vision of bringing in modern technology and diversifying the country's economy by focusing growth on energy, transportation, and knowledge-based industries. Moreover, a design based on compactness should ensure that all its citizens are in close contact to green spaces. Active high technology in buildings and on streets regulates the metabolic exchange and flow of substances and people within the city, while reducing emissions to a bare minimum and making use of renewable energy. The animations by architects and design groups involved in the planning of Songdo promise a bright future for the city—one in which the inhabitants can live healthily and prosperously.[47] However, in these animations, the city is the star, not the people themselves. The people resemble nice decorations for the city's shiny urban fronts, but the city seems eerily devoid of life. The animations invite the viewer to visit the city via long tracking shots and aerial views that provide a dynamic overview over the different parts of the city. But we never actually touch the ground; there is nothing of the messiness, the hectic and noisy everyday life that comes along with thousands of people living in a dense urban environment. To be sure, an animation could never adequately capture this aspect of the urban experience, but it is a common denominator of eco-city designs that they seem to highlight order and structure as opposed to other factors that actual engagement with urban space creates.

It is interesting to note the utopian vision and political determinacy included in the planning of this large-scale urban project—while it holds great promise for a future based on mutual prosperity and well-being, there have been setbacks, as well. Initially planned for 70,000 inhabitants, the city now holds about half that number of people. The implementation of the project has been more costly than initially calculated, and the city primarily hosts a population whose mobility is constrained by various factors: I do not mean this in the literal, spatial sense of the word, but in both social and cultural terms. Whereas the public transportation system is excellent, and walking distances have been reduced to a bare minimum within the city space, Songdo is planned *for*, not *by* the inhabitants. Career paths and job opportunities are predetermined, and there is not much room for the inhabitants' own initiatives; moreover, it has been made up of set pieces of already existing cities, including a central park, a convention center modeled on Sydney's opera house, and vast shopping malls resembling Dubai's sumptuous duty-free areas. People move *in* Songdo, but are they really moved *by* their city? A retort city like Songdo has the problem that it lacks a history, a sense of place based on a long interaction with the environment; its structures have not grown in laborious processes, but were imposed on a blank slate. This is the primary reason why critics of eco-city projects predict a somber future for these cities as their top-down modeled design misses the cultural aspects of urban ecology and communal identity.[48] In an environmental sense, they also dodge the difficult question of what to do with already existing cities which may not be as sustainable or self-sufficient. What is needed is not so much a look ahead to shiny urban projects, but rather a look back to the environmental history of our cities: where do they come from and how have they negotiated the shifting landscapes of their surroundings? How can they work with their natural environments to reach an ecological equilibrium and a sound base for biophilic well-being? That South Korea's leading politicians are not blind to this issue can be seen in Seoul, where government-led green initiatives have seen the implementation of natural reserves along the shores of the once polluted Cheonggyecheon River. The river and its biosphere have recovered and have turned into a popular recreational area with beneficial effects on Seoul's urban climate and its (non)human inhabitants.[49] When it comes to urban planning, the simplest measures sometimes have the greatest effects. This can, in fact, be seen as an integral part of what constitutes ecological knowledge within our cities: a sensitivity to natural landscape features and an experience in dealing with them based on a long history of nature-culture interaction. There is an inherent danger of romanticizing cities with histories, of course. And it is clear that a place-based urban history can have negative effects as well, for instance by spurring closed community fervor when newcomers (or refugees) are prohibited from settling down in a specific part of the city—as Claudia Mantovan relates in her essay in this collection—or by presenting a hindrance to innovation when laws prohibit the renovation or reuse of abandoned or rundown factory buildings. Nevertheless, it is my belief that history can make a difference, especially where the communities and people come to the fore as the protagonists of city life and as integral parts of the multilayered levels of everyday complexity (and sometimes chaos).[50]

This is also where storytelling comes in as the framework for, and basis of, sustainable urbanism: in eco-cities built from scratch "no pre-existing population and social groups are invited to influence the planning process, advocate, or bargain for their interests."[51] Rather, eco-cities "are designed and marketed as smart cities and intelligent places offering all the related economic and environmental advantages, in terms of competitiveness, environment, quality of life, and safety."[52] In their analysis of the advertisement plans, design graphics, and architectural visions for Dongtan, the Chinese showcase eco-city which was supposed to be built near Shanghai but was stopped due to financial and infrastructural problems, Julie Sze and Yi Zhou uncover the narratives connected to ecological cities.[53] As they make clear, this Chinese eco-city (and, in this respect, it does not differ from other projects around the world) can be "understood as a projection of global fantasies of what an ecological life and experience would look like"[54] while neglecting the specific local conditions and environments on which it is built. The specific "top-down character of the project […] does not put the local community at the center," and its "language of ecology is a very specific index of economic development"; as Sze and Zhou put it, "material prosperity is not only the precondition for building the eco-city; it is also the goal."[55] In the same vein, Mark Jarzombek argues that most eco-city projects use popular notions of ecology as a mere label, "pretending that masterplans are cities." He sees their danger: they can turn into "corporate enclave[s]" with little or no benefit to the majority of the population.[56] What can indeed be witnessed with regard to eco-development on a global basis is that the wealthy classes benefit most from green initiatives. Moreover, the construction of eco-cities is not only a costly enterprise but also a tremendous logistical task, and it remains to be seen whether the transportation of raw materials, the construction of high-technological buildings, and the import of nourishment will really justify the effort, especially in an environmental sense. Rather, it seems as if eco-cities and their visions of urban metabolisms managed by machines reinforce myths of technology as a means of resolving ecological problems when it is, in fact, reproducing processes of capitalist modernization.[57] It is important to stress that these critical voices do not question the necessity of reformulating strategies of urban living and design but challenge contemporary notions of ecological urbanism that use sustainability as a mere label without content.[58]

Storytelling is used to sell and advertise eco-cities, but how can it be used to create what I would like to call "sustainable thinking" with regard to contemporary urbanism? I suggest that there are two possibilities: one is storytelling as a mode of critique, and the other is storytelling as a mode of showing alternatives and integrating already existing urban models. Both have been used in contemporary debates and books that deal with sustainability discourses in cities and that try to confront planners' intentions with the voices of local residents. Andrew Ross, in his 2011 monograph *Bird on Fire: Lessons from the World's Least Sustainable City*, looks at the example of Phoenix, Arizona, and how this desert city navigates the tightrope of being a crowded city with only a small number of natural resources and climatic conditions demanding innovative design methods. He traces how ideas of technological fixes and urban development plans often clash with dwellers' own experiences in dealing with the land. While the city administration and politicians formulated bold plans for their city, the many citizens whom Ross interviewed presented him with another image entirely, namely one in which environmental justice and ecological balance were unraveled by misguided top-down initiatives and economic interests. The stories presented by Ross function as a counter-discourse against "the vested power of the growth machine"[59] that uses ecology as a label for obscuring strategies of gentrification, as well as show that "the greening of cities is a grand act of improvisation,"[60] in which community action and bottom-up processes of urban gardening, recycling strategies, and resource protection can make a big difference. By interviewing community leaders as well as marginalized groups, Ross uncovers the social roots of urban sustainability, along with "the conviction that a community's resilience depends on its capacity to adopt the conditions of its most vulnerable populations as a baseline for green policymaking."[61] Instead of the discursive greenwashing involved in ecological (re)design, he thus argues for involvement of local residents in discussions concerning urban environmental conditions, since a city's sustainability relies on the contribution of every urban dweller and does not solely depend on technological management. In a time when advertising and images have a strong impact and are dispersed easily via social networks, it becomes crucial to question the underlying messages of techno-materialist modernization initiatives and to ask for the (economic and social) costs of these enterprises. Listening to and telling the stories of people immediately affected by these issues can be one way of challenging any great narratives that appease the conscience of consumers with the promise of their actions having minimal to zero environmental impact. As Hagan puts it, "people power must supplement if not replace new-fangled technologies: reduce, recycle, re-use. Citizens must *choose* not to drive a car, must *choose* to insulate their roof, must *choose* to have solar panels."[62] In this context, narratives and communal interaction have to make an appearance, not in the form of didactic measures, but as modes of democratic involvement and debate. "Sustainability," as Ross points out, is then "a social challenge as much as it is a biophysical goal."[63] Alone, "a set of metrics"—data that measures urban metabolisms—"reflects a purely physical understanding of how societies strive to be ecologically resilient. By contrast, there are no indexes for measuring environmental justice, no indicators for judging equity of access to the green life, and no technical quantum for assessing the social sustainability of a population."[64] I would claim that cultural stories told by residents can function in exactly this way, and it is time to incorporate them into our frameworks of what constitutes sustainability in a city.

This may be harder to do with regard to eco-cities built on a blank slate or parcel of land that has no long history of human settlement. But, as Julie Sze has shown in her book *Fantasy Islands: Chinese Dreams and Ecological Fears in an Age of Climate Crisis*, this is still possible when one looks at the specific sociocultural contexts in which eco-cities are built. To Sze, these "ecological fantasy islands" are part of a "utopian branding of the future" and "reflect the values and ideologies of their creators."[65] Sze takes their pursuit of eco-development and change as a metaphor that encapsulates the contradictions of our ecological crisis. Like Ross, she warns "against simple designs or [a] technological fix,"[66] while also claiming that the design dreams of eco-cities fail both at taking "their own ecotopianism seriously"[67] and at imagining true alternatives to our current predicaments. In her view, it is necessary to get rid of static concepts of the "environment" and "to see 'environments' through multiple lenses," thus creating "multiple environmentalisms."[68] This includes a historical, deep perspective, an analysis of present conditions, and an imaginative framework that enables us to seek alternatives.[69] I propose using "eco-cities" in exactly this way: as a mode of cultural ecological narrative that helps us to think about urban sustainability from various angles and in new ways. Our cities will not benefit from big design projects built on a blank slate, but they could profit from applying and adapting some of the goals and measures of eco-cities to their own environments, including renewable energy strategies, a sensitivity to material metabolisms, and issues of public transportation. Some of the urban forerunners of sustainability, including Copenhagen, Curitiba, Freiburg, and others, have long integrated these ideas into the management of their infrastructures and surroundings. They prove that "strong grassroots involvement and participation" are vital to sustainability and "are manifestations of a *culture* of sustainability."[70] This cultural aspect includes "an understanding of site and context" on the one hand,[71] but also the question of how "the tremendous creativity of urban people can be applied to the imperative of sustainable development" on the other.[72] This makes clear why eco-cities drawn up on masterplans are doomed to fail, because they lack a "cultural context"[73]—or rather, they do not take their own cultural place and background seriously enough. As long as eco-cities continue to be sold as pragmatic and manageable forms of terraforming, they will fall short of their initial aims of creating truly livable and sustainable cities. Built-from-scratch cities are often laboratories of technological innovation and economic models of boosting growth, when they could be so much more: laboratories of thought where new modes of urban community, (bio)diversity, and renewable design could be tested. Sustainability is not only about maintaining balance, but, in cultural terms, it is also a dynamic process of the activation of creative energies for imagining alternative lifestyles. In consequence, the stories that we tell about our cities could make all the difference. Are they places beyond redemption, ugly stains on the beautiful carpet of our planet? Or are they places of hope, where pathways to more environmentally just and friendly ways of living can be sought? In the history of humankind, cities have always functioned as nodal points in a network of mobility and trade, but they have also been hubs of creative energy and cultural imagination. To see creativity as a resource in its own right might be key to unlocking the great ecological challenges of the future. The "environmental humanities" should join the debate. The time is now.

NOTES

1. United Nations, "Secretary-General Calls for Practical Achievable Programme to Make Globalization a Positive Force for All World's People," news release, July 5, 2000, http://www.un.org/press/en/2000/20000705.sgsm7479.doc.html.
2. United Nations Department of Economic and Social Affairs, Population Division, *World Urbanization Prospects: The 2014 Revision* (New York: United Nations, 2015), http://esa.un.org/unpd/wup/Publications/Files/WUP2014-Report.pdf.
3. There is now a vast variety of literature on urbanization and how to interpret these staggering figures; see Herbert Girardet, *Cities, People, Planet: Urban Development and Climate Change* (Chichester, England: Wiley, 2008), 254–56.
4. United Nations, "Secretary-General Calls for Practical Achievable Programme."
5. Ibid.
6. Girardet, *Cities, People, Planet*, 1.
7. See Christopher Schliephake, *Urban Ecologies: City Space, Material Agency, and Environmental Politics in Contemporary Culture*, Ecocritical Theory and Practice Series (Lanham, MD: Lexington Books, 2015).
8. C 40 Cities, "The Copenhagen Climate Communiqué," *C40Cities.org*, accessed November 30, 2015, http://www.c40cities.org/news/news/-20091215.jsp.
9. Andrew Ross, *Bird on Fire: Lessons from the World's Least Sustainable City* (New York: Oxford University Press, 2011), 7.
10. Susannah Hagan, *Ecological Urbanism: The Nature of the City* (London: Routledge, 2015), 4.
11. Ibid., 4.
12. Ibid., 4.
13. Tai-Chee Wong and Belinda Yuen, "Understanding the Origins and Evolution of Eco-City Development: An Introduction," in *Eco-City Planning: Policies, Practice and Design*, ed. Tai-Chee Wong and Belinda Yuen (Heidelberg: Springer, 2011), 2–3.
14. Ibid., 2.
15. Hagan, *Ecological Urbanism*, 9.
16. Ibid., 11.
17. See Girardet, *Cities, People, Planet*, 108–9.
18. Richard Ingersoll, "The Ecology Question and Architecture," in *The SAGE Handbook of Architectural Theory*, eds. C. Greig Crysler, Stephen Cairns, and Hilde Heynen (Los Angeles: SAGE, 2013), 574.
19. Girardet, *Cities, People, Planet*, 108.
20. For an overview, see Hagan, *Ecological Urbanism*, 20–31.
21. Ibid., 16.
22. See Ingersoll, "Ecology Question and Architecture," 574.

23. Girardet, *Cities, People, Planet*, 123.

24. For a concise discussion of current debates among ecologists and biologists, see Matt Ridley, "Nature Has No 'Balance' for Us to Keep," *Wall Street Journal*, March 23, 2012.

25. Girardet, *Cities, People, Planet*, 123–30.

26. That is why Girardet speaks of "Mobilisation" (10) in order to render the way in which cities have turned into nodal points for the mobility of products and people; see also ibid., 75, 93, and 111; on agriculture and food imports, see ibid., 236–53.

27. Ibid., 113–14.

28. Peter Head and Debra Lam, "How Cities Can Enter the Ecological Age," in *Eco-City Planning: Policies, Practice and Design*, ed. Tai-Chee Wong and Belinda Yuen (Heidelberg: Springer, 2011), 19.

29. See Thomas Elmqvist, Guy Barnett, and Cathy Wilkinson, "Exploring Urban Sustainability and Resilience," in *Resilient Sustainable Cities: A Future*, eds. Leonie J. Pearson, Peter W. Newton, and Peter Roberts (New York: Routledge, 2014), 19–21.

30. Girardet, *Cities, People, Planet*, 6.

31. See Hagan, *Ecological Urbanism*, 94–95.

32. See Steffen Lehmann, *The Principles of Green Urbanism: Transforming the City for Sustainability* (London: Earthscan, 2010), 139–41.

33. See Hagan, *Ecological Urbanism*, 87.

34. Elmqvist, Barnett, and Wilkinson, "Urban Sustainability and Resilience," 19.

35. Because anthropogenic climate change and the different variables included in urban systems are dynamic phenomena, recent approaches replace "sustainability" with "resilience." Resilience, some commentators claim, is better suited for thinking about issues of "recovery, adaptive and transformative capacities of interlinked social and ecological systems and subsystems" as well as "the ability of an urban system either to adjust or adapt in the face of change." Elmqvist, Barnett, and Wilkinson, "Urban Sustainability and Resilience," 20.

36. Ecological (re)design of cities is not merely a present concern, but as a very long history that includes forebears as Ebenezer Howard, Patrick Geddes, or Paolo Soleri. Although I cannot get into details at this stage, it is interesting to note the utopian potential of their respective visions, itself part of a long cultural-historic process of thinking about humanity's relation to its environment. See especially Ruth Eaton, "Architecture and Urbanism: The Faces of Utopia," in *Utopia: The Search for the Ideal Society in the Western World*, eds. Roland Schaer, Gregory Claeys, and Lyman Tower Sargent (New York/Oxford: The New York Public Library/Oxford University Press, 2000), 298–315.

37. See Girardet, *Cities, People, Planet*, 264–67.

38. Wong and Yuen, "Eco-City Development," 3.

39. See Hagan, *Ecological Urbanism*, 95.

40. Ibid., 95.

41. For a concise overview, see Meine Pieter van Dijk, "Three Ecological Cities, Examples of Different Approaches in Asia and Europe," in *Eco-City Planning: Policies, Practice and Design*, eds. Tai-Chee Wong and Belinda Yuen (Heidelberg: Springer, 2011), 31–33. Hambleton summarizes some of the basic principles of

eco-city design. It "supports an economy that works with the environment not against it; questions the primacy of economic interests over other considerations; is alert to the environmental impact of decisions relating to urban infrastructure (particularly transport, energy, water, waste and buildings in general); and is committed to the principles of sustainable planning and urban development." Robin Hambleton, *Leading the Inclusive City: Place-based Innovations for a Bounded Planet* (Bristol: Policy Press, 2015), 208.

42. Scott Dunn and Walter Jamieson. "The Relationship of Sustainable Tourism and the Eco-City Concept," in *Eco-City Planning: Policies, Practice, and Design*, eds. Tai-Chee Wong and Belinda Yuen (Heidelberg: Springer, 2011), 97.

43. Hagan, *Ecological Urbanism*, 95.

44. See Wong and Yuen, "Eco-City Development," 3.

45. See Ibid., 9.

46. New Songdo is featured prominently in Frédéric Castaignède's documentary series *Cities of the Future* (2014), produced by the European TV channel ARTE. For other examples like Dongtan, Masdar City, or King Abdullah Economic City, see Girardet, *Cities, People, Planet*, 289–94; Lehmann, *Principles of Green Urbanism*, 116-18; Hagan, *Ecological Urbanism*, 97–105.

47. As an example, see Gale International, "Songdo Vision," *YouTube*, April 17, 2008, https://youtu.be/cVtvlQ4Jzbw.

48. See Nicolai Ouroussoff, "In Arabian Desert, a Sustainable City Rises," *New York Times*, September 25, 2010.

49. See Dunn and Jamieson, "Sustainable Tourism," 101.

50. At this point, my essay also touches on the topic of place-making and the question of "non-place." While it would be worthwhile to delve deeper into these issues I cannot do this due to the initial topic and the length of the essay. I have extensively discussed these issues and the literature elsewhere: Schliephake, *Urban Ecologies*, xxiii–xxviii.

51. Nikos Komninos, *The Age of Intelligent Cities: Smart Environments and Innovation-for-All Strategies* (London: Routledge, 2015), 124.

52. Ibid., 125.

53. See Julie Sze and Yi Zhou, "Imagining a Chinese Eco-City," in *Environmental Criticism for the Twenty-First Century*, eds. Stephanie LeMenager, Teresa Shewry, and Ken Hiltner (New York: Routledge, 2012), 218–19.

54. Ibid., 220.

55. Ibid., 226.

56. Mark Jarzombek, "Post-Sustainability," in *Smartcities and Eco-Warriors*, eds. C. J. Lim and Ed Liu (London: Routledge, 2010), 248–49.

57. See Shiloh Krupar and Stefan Al, "Notes on the Society of the Brand," in *The SAGE Handbook of Architectural Theory*, eds. C. Greig Crysler, Stephen Cairns, and Hilde Heynen (Los Angeles: SAGE, 2013).

58. On the heterogeneity of contemporary approaches to sustainable architecture, see Simon Guy, "Introduction: Whither 'Earthly' Architectures: Constructing Sustainability," in *The SAGE Handbook of Architectural Theory*, eds. C. Greig Crysler, Stephen Cairns, and Hilde Heynen (Los Angeles: SAGE, 2013).

59. Andrew Ross, *Bird on Fire: Lessons from the World's Least Sustainable City* (New York: Oxford University Press, 2011), 20.
60. Ibid., 20.
61. Ibid., 17.
62. Hagan, *Ecological Urbanism*, 96 (original emphasis).
63. Ross, *Bird on Fire*, 199.
64. Ibid., 245.
65. Julie Sze, *Fantasy Islands: Chinese Dreams and Ecological Fears in an Age of Climate Change* (Oakland: University of California Press, 2015), 14–15.
66. Ibid., 163.
67. Ibid., 101.
68. Ibid., 159.
69. Cf. ibid., 101.
70. Girardet, *Cities, People, Planet*, 18–19 (original emphasis).
71. Lehmann, *Principles of Green Urbanism*, 118.
72. Girardet, *Cities, People, Planet*, 257.
73. Ibid., 273.

BIBLIOGRAPHY

C 40 Cities. "The Copenhagen Climate Communiqué." *C40Cities.org*. Accessed November 30, 2015. http://www.c40cities.org/news/news/-20091215.jsp.

Dunn, Scott, and Walter Jamieson. "The Relationship of Sustainable Tourism and the Eco-City Concept." In *Eco-City Planning: Policies, Practice, and Design*, edited by Tai-Chee Wong and Belinda Yuen, 93–109. Heidelberg: Springer, 2011.

Eaton, Ruth. "Architecture and Urbanism: The Faces of Utopia." In *Utopia: The Search for the Ideal Society in the Western World*, edited by Roland Schaer, Gregory Claeys, and Lyman Tower Sargent, 298–315. New York/Oxford: The New York Public Library/Oxford University Press, 2000.

Elmqvist, Thomas, Guy Barnett, and Cathy Wilkinson. "Exploring Urban Sustainability and Resilience." In *Resilient Sustainable Cities: A Future*, edited by Leonie J. Pearson, Peter W. Newton, and Peter Roberts, 19–28. New York: Routledge, 2014.

Gale International. "Songdo Vision." *YouTube*. April 17, 2008. https://youtu.be/cVtvlQ4Jzbw.

Girardet, Herbert. *Cities, People, Planet: Urban Development and Climate Change*. Chichester, England: Wiley, 2008.

Guy, Simon. "Introduction: Whither 'Earthly' Architectures: Constructing Sustainability." In *The SAGE Handbook of Architectural Theory*, edited by C. Greig Crysler, Stephen Cairns, and Hilde Heynen, 555–72. Los Angeles: SAGE, 2013.

Hagan, Susannah. *Ecological Urbanism: The Nature of the City*. London: Routledge, 2015.

Hambleton, Robin. *Leading the Inclusive City: Place-based Innovations for a Bounded Planet*. Bristol: Policy Press, 2015.

Head, Peter, and Debra Lam. "How Cities Can Enter the Ecological Age." In *Eco-City Planning: Policies, Practice and Design*, edited by Tai-Chee Wong and Belinda Yuen, 17–30. Heidelberg: Springer, 2011.

Ingersoll, Richard. "The Ecology Question and Architecture." In *The SAGE Handbook of Architectural Theory*, edited by C. Greig Crysler, Stephen Cairns, and Hilde Heynen, 573–89. Los Angeles: SAGE, 2013.

Jarzombek, Mark. "Post-Sustainability." In *Smartcities and Eco-Warriors*, edited by C. J. Lim and Ed Liu, 247–49. London: Routledge, 2010.

Komninos, Nikos. *The Age of Intelligent Cities: Smart Environments and Innovation-for-All Strategies*. London: Routledge, 2015.

Krupar, Shiloh, and Stefan Al. "Notes on the Society of the Brand." In *The SAGE Handbook of Architectural Theory*, edited by C. Greig Crysler, Stephen Cairns, and Hilde Heynen, 247–63. Los Angeles: SAGE, 2013.

Lehmann, Steffen. *The Principles of Green Urbanism: Transforming the City for Sustainability*. London: Earthscan, 2010.

Ouroussoff, Nicolai. "In Arabian Desert, a Sustainable City Rises." *New York Times*, September 25, 2010.

Register, Richard. *Ecocity Berkeley: Building Cities for a Healthy Future*. Berkeley: North Atlantic Books, 1987.

Ridley, Matt. "Nature Has No 'Balance' for Us to Keep." *Wall Street Journal*, March 23, 2012.

Ross, Andrew. *Bird on Fire: Lessons From the World's Least Sustainable City*. New York: Oxford University Press, 2011.

Schliephake, Christopher. *Urban Ecologies: City Space, Material Agency, and Environmental Politics in Contemporary Culture*. Ecocritical Theory and Practice Series. Lanham, MD: Lexington Books, 2015.

Sze, Julie. *Fantasy Islands: Chinese Dreams and Ecological Fears in an Age of Climate Change*. Oakland: University of California Press, 2015.

Sze, Julie, and Yi Zhou. "Imagining a Chinese Eco-City." In *Environmental Criticism for the Twenty-First Century*, edited by Stephanie LeMenager, Teresa Shewry, and Ken Hiltner, 216–30. New York: Routledge, 2012.

United Nations, "Secretary-General Calls for Practical Achievable Programme to Make Globalization a Positive Force for All World's People," news release, July 5, 2000. http://www.un.org/press/en/2000/20000705.sgsm7479.doc.html.

United Nations Department of Economic and Social Affairs, Population Division. *World Urbanization Prospects: The 2014 Revision*. New York: United Nations, 2015. http://esa.un.org/unpd/wup/Publications/Files/WUP2014-Report.pdf.

van Dijk, Meine Pieter. "Three Ecological Cities, Examples of Different Approaches in Asia and Europe." In *Eco-City Planning: Policies, Practice and Design*, edited by Tai-Chee Wong and Belinda Yuen, 31–50. Heidelberg: Springer, 2011.

Wong, Tai-Chee, and Belinda Yuen. "Understanding the Origins and Evolution of Eco-City Development: An Introduction." In *Eco-City Planning: Policies, Practice and Design*, edited by Tai-Chee Wong and Belinda Yuen, 1–14. Heidelberg: Springer, 2011.

Chapter 2

Wilderness as City

Conflicting Values in Framing Sustainable Urbanism

Anirban Adhya and Philip D. Plowright

This essay focuses on wilderness—not wilderness as a physical reality but as a proxy to how we socially construct an approach to the environment and the values that are behind urban sustainability. Wilderness—and thus the American ideological construct of "nature"—only seems the antithesis to "civilization." In actuality, wilderness is imbued with the beliefs and values that have informed the United States' development of cities: domestication, privacy, and property ownership. These fundamental values of urban America are simultaneously reflected in the definition of places of wildness as interpretations of the human-nature relationship and provide a distinct opportunity to understand sustainability in cities.

Urban sustainability is further marked by discourses of public health, debates about diversity of place, and discussions on difference. Expanding on the notion of difference in cities as presence of "otherness"[1] and on acknowledgment of difference having multiple claims on the urban environment,[2] we have used holistic ecology to contend that predictions of sustainability in cities cannot ignore "human-nature integration" and "ecological well-being."[3] Ecology of the city and ecology in the city thus become key conceptual and functional drivers of urban sustainability. Within the context of evolving dimensions of human and nature that are increasingly different, diverse, cosmopolitan, and heterogeneous, interpretation of nature in the context of cities, specifically, assumes varied meanings and significance. As noted by Julian Agyeman,[4] people want access to nature for various reasons, including socialization, recreation, relaxation, solitude, reflection, and refuge.[5]

Irrespective of its purpose, interpretation of nature in our cities within a sustainable ecological framework is critical, especially as landscape considered for nonhuman needs. Western conceptualization of this landscape modality uses the expression "wilderness."[6] However, currently, the concept of wilderness as an interpretation of nature is absent in debates of urban sustainability in discourses of architecture, landscape architecture, urban planning, and civil engineering.[7] Wilderness is a culturally constructed philosophical concept typically presented as a contrast to urban development factors like density and congestion. Following historic Western cultural and religious heritage, wilderness is interpreted as a space lacking improvement and civilization, and is often termed beautiful, bountiful, and dangerous.[8] Divergent to imported European attitudes linked with sublimity and Judeo-Christian religious beliefs, wilderness concepts in the United States are rooted in the notion of frontier and violence against perceived "others," such as nonhumans, wild animals, native Indians, unfaithful pagans, lawless persons, and even African-American slaves.[9]

The contended relationship between the American urban environment and wilderness is captured by Henry David Thoreau's "Walking."[10] Thoreau contends, "The West of which I speak is but another name for the Wild; and what I have been preparing to say is, that in Wildness is the preservation of the World."[11] Wilderness, or "wildness" in Thoreau's terms, is a conceptualization of nature without improvement, intervention, and impact of human action or, in other words, without any human "civilization."[12] Calling for sustenance of this wild domain as "preservation of the world," Thoreau underlines attachment of value to wild nature in human terms. Thoreau's assertion also embodies a call for protection of natural areas for human benefit as a typical response to the limited extent of wild nature that is left, when the frontier is closed, and wild places are threatened by the succession of civilization in the new world. Nineteenth-century American frontier narratives, such as those expressed by Thoreau, are often mythologized as being anti-urban with Arcadian notions of development. Bernard Rosenthal[13] contends that the "Great City" was the sense of destiny for American life, and David Graber[14] explains that wilderness has taken on connotations, as well as mythology, that specifically reflect those later twentieth-century values of a distinctive Anglo-American bent expressed by city building. Rosenthal called the commonly held belief in a nineteenth-century American deep antipathy toward cities a "modern myth" and demonstrated, through his analysis of the social and political rejection of "agrarianism," that the century was permeated by urban values of resource collection (such as land, nature, and slaves), intensification, and management in America. Similarly, but in the context of American national parks, Graber contends that the organizing myth behind protection of these natural areas was wilderness. Values of spectacle, object protection, and historic romanticism of the constructed marvels of the old world played an underlying role in forming the ideal perception and virtuous character of a young nation. In this essay, we examine these two parallel constructions—urbanity and wilderness—in terms of their shared common roots and underlying agencies in order to reinforce a need to reevaluate prevalent models of urban sustainability, which leave their relationship unexamined.

Wilderness now functions to provide solitude and a counterpoint to a technological society in a landscape that is managed to reveal as few traces of the passage of other humans as possible. This argument of wilderness as a psychosocial construct can be found in the efforts leading up to The Wilderness Act of 1964, where a human value-based system is at the root of conceptualizing wilderness as an extended issue of urbanity in terms of human construction of built forms and open spaces in cities rather than as a natural element. Wilderness becomes not an issue of resource preservation, framed by a right to public health, but, rather, a matter of urban issues—the right to private property and right to self.[15] As an idea and an ideal, wilderness is a human construction. It is a myth of the urbane and the urban mind. Wilderness is not, in fact, about nature but a projection of urban values centered on private ownership, human benefit, and human control. As such, wilderness can be seen as the extreme manifestation of the American notion of land control, management, and development. Such discussions highlight the human struggle with ethical debates in imagining the future of urban nature and its human design response.

PROPERTY OWNERSHIP AS THE FUNDAMENTAL WAY OF AMERICAN LAND DEVELOPMENT

The right to property is the fundamental way in which Americans understand human occupation of the landscape. Property rights and ownership have had a long history in the development of North America since the European expansion onto the continent. The current attitude toward land development was acquired from its European, and especially British, heritage. For example, according to English philosopher John Locke,

> Though the Earth, and all inferior Creatures be common to all Men, yet every Man has a *Property* in his own *Person*. This no Body has any Right to but himself. The *Labour* of his Body, and the *Work* of his Hands, we may say, are properly his. Whatsoever then he removes out of the State that Nature hath provided, and left it in, he hath mixed his *Labour* with, and joyned to it something that is his own, and thereby makes it his *Property*.[16]

Initial interaction of indigenous populations with European settlers was an exercise of a conflict of worldviews based on a land ethic. The forcible expulsion of Native American tribal populations and settlement by European immigrants was based on the grounds that, if the land was not improved by the standards of European (British) expectations of agriculture and resource extraction, then the land was not occupied or owned. It was only through evidence of activity based on economic return that ownership could be documented; this generally meant clearing the land of forests and making it agrarian for crop production.[17] As far back as the 1500s, improvement-based occupation became the justification for land acquisition from the Native Americans. Land left "unimproved" was defined as "wasted." The economics of ownership was based on both the actuality, as well as the representation, of activity, for "[a]s the white settlers saw it—eventually taking their cue from Locke—anyone entering such a wilderness who then set about making improvements was entitled to ownership over the property."[18] There is also a basic, practical reason why the rise of privacy through the demarcation of property dominated the American approach to development—put simply, the modality of development is linked with commercial prosperity. Private ownership works to encourage labor and provides a balance between altruism and selfishness. During the settling of America by the Puritan expedition in the 1600s, the colony faced starvation until they moved from communal ownership of the colonies' territory to being assigned specific land to be worked by a family.[19] The individual property ownership, defined by the right to one's own labor, led to increases in quantity of labor, production output, and human survival.

The classical liberal tradition has always stressed an intimate connection between a free society and the right to private property, a key driving force for nineteenth-century urbanity. As Ludwig von Mises[20] put it, "the program of liberalism [...] if condensed to a single word, would have to read: *property*, that is, private ownership."[21] Western politics, drawing extensively on Locke, gave new life to this idea; subsequently, a great deal of political philosophy has focused on the justification (or lack thereof) of natural rights to private property. Classical liberals such as Eric Mack have argued that "the signature right of any rights-oriented classical liberalism is the right of self-ownership."[22] The classical notion of individual property is understood as a bundle of rights: complex layers of entitlements, responsibilities, and liabilities. It was on these principles that the United States was formed to be a country owned by the people who claimed the right of citizenship and enforced that right through owning the land of that country.[23] Property was power, or, as John Adams stated, "Power follows property."[24]

The logic of division and private control that comes with ownership when considering land as an economic resource introduces a sense of conflict in America between communal land (public) and land bounded by improvements (private). The key difference between what was considered public and what was private (owned) was the labor that was applied. There was a customary right to use of unimproved land for hunting and grazing—the major signification of ownership being the fence. Any land, whether legally owned or not, was considered usable for grazing as the "fence laws" during the American colonial period put the burden of proof of ownership on the crop farmer rather than on the livestock rearer. This is tied to the customary right to use a commons and to part of an early welfare action allowing the poor to raise livestock for subsistence.[25] The 1890s saw this attitude change with new American fence laws requiring livestock owners to pen their animals and making them legally responsible if their animals escaped and caused damage. This was part of a larger domestication movement around the end of the nineteenth century, a movement that involved both animals and people, open spaces and built forms. Where areas once shared an indistinct division between public and private, domestication forced a reinterpretation of land with boundaries becoming more defined by hard transitions. The American understanding of land as capital ensured a complete redefinition of its physical resources, including its economic factors. If the return was not cash, then it needed to be something else that was valuable.[26] There was a significant effect on those who were not landowners, making the right to communal use of land less accessible. The right to own and defend land as a right to individual property ownership was reinforced through the decreasing ambiguity of boundaries encouraged by domestication.

DOMESTICATION AND PRIVACY

As land and animals became domesticated, so too did the built environment—moving toward greater isolation and controlled situations. Nineteenth-century architecture in American homes, reflecting shifts in social thinking and norms, moved away from collective gregariousness and carnality (medieval society) to the mind and privacy of the individual.[27] With the human body seen now as more about mind and spirit, nineteenth-century English architecture responded with the design of corridors as passages created for privacy, retirement, individual intimacy, and security. American cities embody the same transformation toward separation and control; hence, as Robin Evans illustrates, it was logical for "Corbusier, Hilberseimer, and the constructivists to use the individual private cell as the basic building block for entire new cities in which all other facilities would be collectivized."[28] The Modern movement thus changed the city through extreme scales of separation and distance in the name of hygiene, sanitation, and efficiency. However, it can be argued that the Modern movement was less about a politics of public health, and more about private values of psychological well-being, independence, and comfort of the individual.

Psychological well-being of the individual was the fundamental non-economic value of property connected to the right to land, the right to one's body, and the right to self—the three key individual rights discussed earlier in relation to property rights. While it took time to be institutionalized through politics and law, a critical cultural shift occurred when the understanding of the land became not economic gain through enclosure and possession but the rights to non-economic benefits through the same mechanisms. The legal structures that enforced the ability to own land or livestock were transferred to the human individual, creating a sense of the ability to own oneself. The right to the body and the right to the self-grew through the ethics movements of the eighteenth and nineteenth centuries and developed with the political structures of democracy, although the roots extend further into history. It was the satirist and pamphleteer Richard Overton who, in 1645, publicly transferred the idea of property ownership to personal ownership, just before John Locke's writings on liberty, which were very influential on the development of American identity. In the context of the Cromwellian control of British politics and the English Civil War, Overton, writing from Newgate Prison, fired out to the population of England,

> To every individual in nature is given an individual property by nature, set to be invaded or usurped by any. For every one, as he is himself, so he has a self-propriety [property in himself], else could he not be himself; and of this no second [person] may presume to deprive any of, without manifest violation and affront to the very principles of nature and of the rules of equity and justice between man and man.[29]

It is through the same concept of land ownership that the right to the body developed, using the same language. Locke's *Two Treatises of Government* in 1681 continued to develop the ideas introduced by Overton's pamphlet, and these would become the basis of American identity—that everyone owned his or her person, and that "We hold these truths to be self-evident, that all Men are created equal, that they are endowed by their Creator with rights; among that these life, liberty and the pursuit of happiness."[30]

Questions of life, liberty, happiness, equity, and justice between individuals were further tested by the development of cities. The rise of the urban metropolis, combined with industrial development, increased population densities and environmental pollution, facilitated the need for increased protections for quality of life. This quality was primarily found not in economic profit but in the ability to maintain isolation from the heightened activities and stimulations that came with the perceived advancement of civilization. As written in 1890 in the *Harvard Law Review*, "[t]he intense intellectual and emotional life, and the heightening of sensations which came with the advance of civilization, made it clear to men that only a part of the pain, pleasure, and profit of life lay in physical things. Thoughts, emotions, and sensations demanded legal recognition."[31] The writings of Georg Simmel addressed the psychological effect of the metropolis. Simmel stated his concerns when he wrote,

> The deepest problems of modern life flow from the attempt of the individual to maintain the independence and individuality of his existence against the sovereign powers of society, against the weight of the historical heritage and the external culture and technique of life. This antagonism represents the most modern form of the conflict which primitive man must carry on with nature for his own bodily existence.[32]

This statement contains issues of power, control, individuality, historical awareness, and alienation that are at the core of the modern metropolitan experience. The rapid acceleration of higher-density urban populations and the escalation of the industrial landscape left humans, psychologically, fully alienated from the natural and "Nature."

Non-economic, substantive due process in a legal context was developed in the United States first through the right to privacy and then through the right "to be left alone."[33] The right to be left alone is central to a contemporary understanding of Western human existence in the American context. This means the right to not be intruded upon by noises, physical touch, or viewing and broadcasting (media and publication) of one's person. It means, in fact, the ability to control surroundings so that one has the power to decide what physical and emotional experiences may be allowed to occur. The call for self-affirmation and self-definition of physical and emotional experiences was instrumental in demanding for natural areas in cities and then actualization of the parks movement led by landscape architect Frederick Law Olmsted and others. Nature as interpreted through natural landscapes of parks, parkways, boulevards, and playgrounds became the new typology for refuge and therapy of the human mind against the increasing congestion and pollution of the city.

NATURE IN CITIES: THE CALL FOR URBAN OPEN SPACES

Creating recognizable assets in urban space for urban dwellers, the large urban movement for parks and playgrounds was enmeshed in issues of privacy and property ownership. Olmsted argued for parks as a new urban planning and design strategy for land use change, investment, and development. Social reformers, like Jane Addams and Jacob Riis, propagated the need for playgrounds for recreational activities to improve social conditions and reduce crime. Though this movement was couched as a demand for natural open spaces as a need for public consumption, it, in fact, extended the non-economic value of land in cities in terms of personal health and individual well-being. Underlying this movement were two critical beliefs: (a) nature is restorative and (b) recreation is therapy.[34] Parks were seen as restorative—an idyllic counterpoint to the congestion of the city—as well as considered as therapeutic—a place to relieve the tension of urban living. Both abilities operate to recharge human attention and to provide relief, catering to the physiological and psychological needs of an individual.

Olmsted's call for pastoral landscapes and natural zones (public parks) within urban environments can thus be seen as the individual needs of health and rejuvenation against the unhealthy urban conditions.[35] The paradox is that, although notions of public space were defined and reinforced by Olmsted, the stress on the health of the individual reinforced urban space as an extension of one's need for privacy. However, this individual private need expressed through the call for parks was not presented as directly based on economic value or profit but as the extension of non-economic values of safety, comfort, cleanliness, and health. Similar to zoning regulations, private, non-economic priorities were critical to Olmsted's urge for public health and sanitation. The argument for parks as public property then arose as a distinct domain for individual gratification, control for cleanliness, and well-being separated from the unclean and unhealthy surrounding urban environment. This reflects American society's growing need for a private domain of security, comfort, and independence.

The park movement became a visual expression of the struggle for the individual human to exist in a modern world conceptualized as a machine, necessitating, as defined through Friedrich Nietzsche and socialism, "[...] the resistance of the individual to being levelled, swallowed up in the social-technological mechanism."[36] While parks were one vector of resistance, another counterforce was retreat into private, domestic space—space considered to be an extension of the individual human body. Adolf Loos responded with a clear distinction between exterior public space and interior private space. The modern interior, for Loos, was one of shelter, protection, and nurturing against the city beyond.[37] As Beatriz Colomina notes, "Loos breaks down the condition of the house as an object by radically convoluting the relation between inside and outside,"[38] where the window was used for light and the mirror used to extend the sense of space, allowing the house to be completely internalized and isolated from chaotic urban space.

In response to the pressure created due to crowding and density in urban space, the parks movement and the call for natural open spaces in cities embodied the same instinct for privacy offered by the internalization and domesticity of space—release from human density and the resultant violation of psychological individuality. While publicness is a concept by which Western liberal society organizes areas of social life, including issues of individual bodily access, agency, and interest,[39] ironically, privacy and the search for individual independence, comfort, and safety are at the root of the evolving constructs of public spaces, including the soon-to-be-defined "wilderness." In other words, discussion about urban open space is not about creating public space, as traditionally framed and politically constructed; rather, it concerns control and the same "right to self" embedded in the demarcation of private land and property ownership.

URBAN DESIGN ZONING AND LAND USE CONTROL AS FACILITATORS OF PRIVACY

Rapid urban development in the nineteenth-century United States was fueled by industrialization and migration, acerbating the need for controllable privacy.[40] With tremendous pressure on existing infrastructure, rapid urbanization also posed questions of sustainability and survival in terms of housing, sanitation, water, transportation, congestion, and disease control. Early forms of formal urban design and planning were solutions and tools to solve these issues as responses to public health issues.[41] The city building process, in many ways, was a reactionary process from the very beginning; it was a complex process enacted by different agents: residents, business owners, investors, workers, industrialists, politicians, legislators, speculators, among many others. Irrespective of their varying roles and power, one consistent pattern and theme that connected these agents of city building was their right to own their land. Protecting and increasing property values was the most important goal in nineteenth-century urban America, pursued primarily through different ways of control and regulation: deed restrictions, land use separation, zoning laws, and comprehensive planning.

Though zoning formally started in 1910 and rapidly developed to become standard urban regulation between 1910 and 1930, the seeds of zoning as an urban design and planning tool can be found in the late nineteenth-century city building process.[42] Beyond voluntary covenants and private deed requirements, some cities tried to address urban problems from population growth, migration, and industrialization in the late nineteenth century by formal laws and regulations. Early examples can be found in: subdivision control in Oak Park, Illinois (1882)[43]; the banning of public laundromats in San Francisco[44]; and the Tenement House Act,[45] by which New York City regulations limited the height of tenement houses throughout the city to one and one-half times the street width. Thus, the nineteenth-century movement for maintaining and improving land and property value paved the way for major zoning legislation in the early twentieth century.

However, extending from the notion of improvement of self (body) and land (property), the new century saw a significant shift from understanding property value as economic to include non-economic factors such as the emotional and psychosocial value of personal health. Growing arguments for healthy living in terms of light, ventilation, and sanitation became the underlying premises for comprehensive zoning policies in urban design and planning. Though public good and public health were touted to be at the forefront of the late nineteenth and early twentieth-century urban movement of land use control and zoning regulations throughout the United States, the underlying intention of this control-based urban development process was maximizing individual property values and ownership rights through control of personal environment. With this history, urban zoning has been an effective tool of exclusion and separation, privileging individual rights over public concerns of sustainability like shared use of space, mixed-use property development, and mixed-density built form.

WILDERNESS AS AN URBAN CULTURAL CONSTRUCT

Susan Griffin, in her 1978 ecofeminist classic, at once links and divides urbanity and wilderness:

> Separation. The clean from the unclean. The decaying, the putrid, the polluted, the fetid, the eroded, waste, defecation, from the unchanging. The changing from the sacred [. ...] Death from the city. Wilderness from the city. Wildness from the city.[46]

The 1964 Wilderness Act is considered a major landmark in the formal protection of nonhuman natural territory through "establish[ing] a National Wilderness Preservation System for the permanent good of the whole people, and for other purposes."[47] There is a perception that wilderness in the United States was set aside for all the other (nonhuman) inhabitants of this ecosystem. It was a balm against what was seen as the tendency of excessive development and abuse of the land by humans and was an important barrier to both save Americans from themselves and from a "disease" infecting the world.[48]

Within environmental history of the United States, the idea of wilderness was linked with a positive identity of the nation, an inherent part of the emotional composition of citizenship and nationhood. However, the language of the Act, as well as the earlier writings of many leaders of the wilderness movement, stress that the purpose of wilderness preservation was the permanent good for humans ("the people") rather than ecosystems, never meaning to isolate the land for only nonhuman use. In the letter used to introduce the Wilderness Act, written by Wallace Stegner in 1960, the major argument made to the government was that of human health and the use of the landscape for non-economic purposes—the same factors used to argue for zoning laws and the development of an urban park system. In fact, the Wilderness Act explicitly creates a territory that is purposeful for human use while disallowing that use to be technology-based infrastructure (roads, power lines, permanent human fixtures) or for direct economic benefit. The Wilderness Act explicitly states that the preservation of "nature" areas was for "the use and enjoyment of the American people in such manner as will leave them unimpaired for future use and enjoyment as wilderness, and so as to provide for the protection of these areas, the preservation of their wilderness character [...]."[49] It may be reassuring to consider the conceptualization of wilderness as a strictly nonhuman territory to balance the excessive development pursued in the human-built environment. However, this position is a myth.

The conservation movement could be said to begin with the Boone & Crockett Club, an organization established in January 1888 by Theodore Roosevelt. This forbearer of twentieth-century conservation organizations started not as a way to "save" nature from humans but for the simple fact that Roosevelt liked to hunt on the Badlands, a non-economic factor but one of his central pleasures in life. In 1885, Roosevelt would write,

> I thought I should take a short trip to the prairie, in the beautiful June weather, and get a little sport and a little fresh meat out of the bands of prong-horn bucks, which I was sure to encounter. [...] In June, fair weather can generally be counted on in the dry plains country.[50]

Overpopulation of cattle in this region during the 1880s was causing the reduction, and even elimination, of those prong-horns (antelope) along with buffalo, wild sheep, and migratory birds in the region. The Boone & Crockett Club would not have been formed if those species starting to disappear didn't also happen to be prime recreational hunting stock. The foundation of a managed approach to wilderness areas and its inhabitants began to change the perception of wilderness from a desolate area untouched by civilization needing improvement (i.e., negative) to a productive landscape (non-economic) needing protection.[51] The conversation that started around big game pursued conservation and the definition of wilderness as a form of domestication. The same attitude that was developing for home, factory, and industry was turned toward nature using scientific management of the former group to define the latter.[52] The ultimate benefit was not ecological health but personal mental health of the (male) hunter.

WILDERNESS AS A NON-PUBLIC PLACE FOR RECUPERATION

There is a strong sense of spiritualism through the American intellectual tradition of Thoreau, Ralph Waldo Emerson, and the Hudson River School. However, we have to be careful with the transcendental aspects of wilderness worship, instead, understanding it as a pursuit of mental health through reflection and sense of being rather than of spirituality and philosophy without immediate physical benefits. Wilderness, in this view, is about emotional health. More than the big game hunting of Roosevelt and worship of the landscape as a place for God found in the writing of Thoreau and John Muir,[53] wildness and natural locations were defined as places for recuperation from the social/industrial life of the city. While being conceived as a nonhuman space, wilderness is more correctly considered as a place for *a single isolated* human. Wilderness has been constructed this way through American mythology—both religious and cultural—because it was useful to American needs as a society to have a place of isolation or retreat from the stresses and psychological pressures of the density of a human urban space.

There were three aspects of wilderness that became important through the rhetoric of the formative cultural debate and conservation movement in the United States. These all revolve around the concept of individual health and sense of wholeness. First, consistently found in the canonical wilderness writing from the late nineteenth century to the mid-twentieth century are expressions of silence, contemplation, isolation, aloneness, mental health, and lack of human presence (besides those experiencing the landscape), including non-presence of human technology or transportation infrastructure. They are the only concepts that can be found in writings about wilderness spanning the decades from the mid-1800s to the late 1960s. Contemplation is always accompanied by solitude and silence, such as Muir's "countless hosts of the forests hushed and tranquil, towering above one another on the slopes of the hills like a devout audience."[54] Current research in the social-psychological theories of self-affirmation has provided evidence of the human benefits of wilderness.[55] Specific research has inferred that through leisure, people actively seek out and create opportunities for self-definition by engaging with domesticated exterior spaces, from urban landscapes, parks, and playgrounds to the extreme example of wilderness.[56] D. R. Williams further elaborates that the self-affirmation process can be thought of as composed of three facets: an affective dimension (self-esteem); cognitive dimensions or self-images (a set of beliefs about ourselves); and a motivational dimension for self-definition (striving to understand ourselves and how others understand us).[57] In the context of wilderness, these three facets of the self-affirmation process construct identity on another three levels: (1) an individual identity that forges a connection with nature, affirming important beliefs about personal self, such as being adventurous, independent, self-reliant, and well-travelled; (2) a collective identity that defines a cultural identity and national ego of preserving and conquering wilderness; and (3) an ecological identity that constructs a biological-self dependent on systemic connection and networks in the biome. In the 1800s, escape from the city into the mountains and other natural settings was culturally codified.[58] In the Romantic tradition, wilderness became a posterchild for a return to certain values seen to be lost by current development and needed to make an individual whole. Wilderness allowed a return to the natural human condition—defined not only as an animal but also as a civilized, American male person; as Stegner notes, "[f]or an American, insofar as he is new and different at all, is a civilized man who has renewed himself in the wild."[59]

Second, wilderness and nature were seen as the foil to urban space, tied deeply into the aesthetics of the landscape. This pulled from the parks movement and the rise of recognition of emotional stress due to extra-body simulations by the industrialized urban condition. It also drew from the European aesthetic tradition. Through the lens of separation, wilderness as a human value is understood as an image and an asset, something to be considered at a distance in its wildness; or, as Olmsted said, "enjoyment of scenery employs the mind without fatigue and yet exercises it; tranquilizes it and yet enlivens it; and thus, through the influence of the mind over the body, gives the effect of refreshing rest and reinvigoration to the whole system."[60] As mentioned above, the idea of separation and exclusion of the individual from industrialized society was central to Olmsted's argument for parks as important typologies of urban development. The issue, in urban and social terms, is the use of the natural environment for health and recuperation while managing the risk of the truly wild. This very use required certain safety measures—the smoothed landscape, the manicured and well-defined path that allowed "viewing" of, but not direct contact with, the "other." Here, wilderness was meant as scenery rather than immersive experience.

Third, the American experience of wilderness has been deeply embedded in the frontier mythology. Continuing the attitude expressed by Roosevelt, early wilderness advocacy was against the consumption of landscape and its commodification by mass tourism.[61] But this didn't mean that the territory was to be unoccupied. It meant that it should be a resort available for personal and private consumption by the few who had the means and resilience to occupy the land. It was not to be a resort with support infrastructure of comfort for the "masses" (guides, porters, bathtubs). The separation was between two types of people: those who were aligned with the rigors of non-urban travel and those who would treat immersion in the landscape as a theme park. The second type of people considered themselves to be "doing

America." These people were dismissed as outsiders as they were seen as destroying natural beauty and heritage through their actions, which were portrayed in terms of "shooting our buffaloes" and "causing themselves honorably dragged up and down our Sierras, with perennial yellow gaiter, and ostentation of bathtub."[62] The value of the land deemed wilderness was to give people a private, domesticated exterior for personal use. It was only when too many people showed up, when the landscape was perceived as public rather than private that the conservation movement pushed for government protection through the Wilderness Act. It was the commercialized attitude by American industrial society and tourism toward natural contexts that generated a response as the preservation movement.[63] Often the experience of wilderness by the individual was cathartic and deeply personal, stressing a sense of isolation: "[n]owhere, not even at sea, does a man feel more lonely than when riding over the far-reaching, seemingly never-ending plains; and after a man has lived a little while on or near them, their very vastness and loneliness and their melancholy monotony have a strong fascination for him."[64] The idea of the frontier and wilderness became integrated into the American psyche as a site for "the spiritual refreshment that being natural can produce."[65] While wilderness was something Western-European civilization had lost, in American development, it was constantly part of the continued industrialization and urbanization of the country. As an idea, it represented an escape from civilization in both personal and material terms and a recovery from modern ailments inflicted by contemporary living. These ailments included not only health effects of pollution and foul air of the cities but also gender issues of domestication. The wilderness became a site for rugged (male) independence while civilization through domestication was seen as becoming more and more feminine (the domestication of the human male; wilderness as escapist fantasy). It was a private sphere, very much parallel to the rising consciousness of the domestic interior as seen through shelter and protection (Loos) and identity and independence (Simmel), but combined with the American wilderness mythology. In this, private space became projected outwards as an urban concept but shaped understanding of the natural environment.

WILDERNESS AS AN EXTENSION OF CITIES

Along with a conceptual isolation of wilderness come constructions of fear and distrust. This is particularly evident in the manner in which Americans treat nature and natural elements in their domestic spaces and private properties. Considering the regimen of maintenance in cities—lawn-mowing, pest removal, and weather-proofing—it is the reverse that emerges as actuality.

Americans constantly strive to maintain, or save, their built environment from the forces of nature. Ironically, an overgrown lawn or shrub is seen as undesirable; evidence of animal presence is seen as repulsive. Americans aren't saving wilderness through preservation; they are protecting a sense of progress that is ultimately an issue of visual refinement, order, and containment as a metaphor for civilization. From early in the 1930s, the argument about wilderness—which preceded the political action—was defined in terms of human benefit (resources, activity, and enjoyment) and not in terms of ethics. This was an argument of aesthetics: wilderness as beauty and experience, or as a human experiential resource.

The Western concept of wilderness—as a philosophical idea, as a policy, and as a law—is a human construct that has reflected cultural values more than any inherent truth of the world. To say it bluntly, if considering the Gaian ecosystem, there is no such thing as wilderness. As a position, Western philosophical and cultural theory has found wilderness to be a necessary concept in order to balance dualistic metaphysics that dominate foundational attitudes, as well as to create the mythology of human-not-animal (i.e., not "natural"). In this way, wilderness becomes the counterpoint to industrialization and human infrastructure. Such a state is not something humans have ever really known; 'wilderness', so imagined, is a foil for Western culture, a romanticized Garden of Eden. The term is constructed when there is very little of wild nature left, as in Europe when explorers leave for exotic places, or in the United States when the frontier is closed and wild places are threatened by the success of civilization. Therein hangs much of its fascination, for wilderness enthusiasts have a kind of archetypal longing for, or archaic vision of, a world with no people in it—or with just one person in it: themselves. As David Lowenthal writes, "wilderness is not, in fact, a type of landscape at all, but a congeries of feelings about man and nature of varying import to different epochs, cultures, and individuals."[66]

As a sociocultural construct, the American experience of wilderness is one that defines the concept as an extension of the individual, and as the most private, rather than the most public, of spaces. Wilderness and wilderness areas contain the underlying philosophy of privacy while being presented as a regulated public space. More than this, in the American philosophy of spatial development, wilderness as a construct is defined through the same mechanisms used to develop American cities. Essentially, the division of space into non-economic spheres of isolation (the grid system) is the same approach that allowed the division of a territory into the "built" and the "untouched." "Wilderness area" is another one of those filter-words with which Americans color the nature they see. The truth is, say the reconstructionists, people cannot know any such people-less world; that is only pretense. Wilderness is not a contrast *to* cities but rather an extension *of* cities.

THE ROLE OF WILDERNESS IN DEBATES ON URBAN SUSTAINABILITY

Our observation is that if Americans continue to interpret nature based on factors of human profit, human improvement, and human self-affirmation, then mainstream conceptions of sustainability need to be challenged through questioning the dynamic, yet skewed, relationships between what is an exclusive human sphere of existence and what is perceived and constructed as nature, as the Other.

The consideration of wilderness as a specific and limited human construct underscores important pragmatic implications for design practices. If privacy and individual private needs are dominant underlying drivers of North American urban development, then we need to clarify and address the evolving definitions and roles of the public realm in cities. Growth management policies, infrastructure planning, and zoning regulations have traditionally been used as tools for exclusion and separation (between types of humans, as well as between humans and nonhumans) as demonstrated by the privileged construction of wilderness. If cities are fundamental sites of proximity and heterogeneity, then analyzing, understanding, and even undermining this falsely constructed human-nature dichotomy are imperative for imagining and speculating about the future of our urban environment.

We need to reexamine our current discussions and reevaluate our policies and proposals for value judgments in sustainable urbanism regarding practices and policies of housing (medium- and high-density housing), land use (mixed-use projects), open space (green infrastructure), and accessibility (public transit, non-motorized transportation). Larger questions remain related to the idealistic notion of urbanity and the social-cultural construct of nature. Do philosophical frameworks, societal value structures, and disciplinary boundaries need to be redefined? Can we leverage our acceptance (and recognition) of human space as essentially the association of spheres of privacy? Are there specific strategies to allow an ecologically based sustainable urbanism to be developed from existing conditions, and what is the role of the natural? Answers to such questions, debated and tested in our cities, will shape the future of sustainability as a practice of our way of life and how we, as humans, inhabit, see, and perceive the world.

NOTES

1. Julian Agyeman, "Entering Cosmopolis: Crossingover, Hybridity, Conciliation and the Intercultural City Ecosystem," *Minding Nature* 7, no. 1 (2014): 20; see also Ruth Fincher and Jane M. Jacobs, eds., *Cities of Difference* (New York: The Guilford

Press, 1998) and Richard Sennett, *The Conscience of the Eye: The Design and Social Life of Cities* (London: Faber and Faber, 1991).

2. Leonie Sandercock, *Cosmopolis II: Mongrel Cities in the 21st Century* (London: Continuum, 2003).

3. Anirban Adhya and Philip Plowright, "Setting Priorities: Sustainability, Environmental Health, and Embedded Value Judgments for the Urban Design Process," in *Cities in Transformation Research & Design: EAAE/ ARCC Milano, International Conference on Architectural Research*, eds. Marco Bovati et al. (Milano, Italy: Il Poligrafo, 2012), 819–27.

4. Agyeman, "Entering Cosmopolis," 24.

5. Agyeman discusses difference in cities, illustrating the "relational, messy, and contested concept of urban nature." The example elaborates how development of wildflower meadows in city parks by the parks department of the City of Bristol (UK) led to different opinions on the new landscape. For example, Asian and African-Caribbean populations refused to use the park due to a residual fear of snakes associated with long grass, though the initiative was considered successful by the majority of the public.

6. William Cronon, "The Trouble with Wilderness; Or, Getting Back to the Wrong Nature," in *Uncommon Ground: Rethinking the Human Place in Nature*, ed. William Cronon (New York: W. W. Norton & Co., 1995), 69–90; Holmes Rolston III, "The Wilderness Idea Reaffirmed," *The Environmental Professional* 13 (1991): 370–77.

7. Our analysis of representative texts of various design disciplines concluded that, within the latent content of current sustainable urbanism projects, a persistent conceptualization of the human-built environment is, at present, completely separate from that of natural systems. In particular, the concept of wilderness, as a core representation of nature, is completely absent from any text in all of the capacities studied. While plants, animals, and ecosystems are discussed in terms of human value, wilderness is a concept outside of the dialogue. It seems, currently, to have no place in human landscapes. See Anirban Adhya and Philip Plowright, "Sustainable Architecture as Ecology: Defining Shared Systemic Priorities of Public Health and Social Ethics in Places," in *Architecture and Sustainability: Critical Perspectives for Integrated Design*, eds. Ahmed Z. Khan and Karen Allacker (Leuven, Belgium and Den Haag, Netherlands: ACCO, 2015), 85–94.

8. David M. Graber, "Resolute Biocentrism: Dilemma of Wilderness in National Parks," in *Reinventing Nature? Response to Post-Modern Deconstruction*, eds. Michael E. Soulé and Gary Lease (Washington, D.C.: Island Press, 1995), 123–35; and Bernard Rosenthal, "The Urban Garden," *Texas Studies in Literature and Language* 20, no. 1 (1978): 119–38.

9. Lynda Schneekloth, "The Frontier Is Our Home," *Journal of Architecture Education* 49, no. 4 (1996): 210–25; Thomas Roderick Dew, "An Address on the Influence of the Federative Republican System of Government upon Literature and the Development of Character," *The Southern Literary Messenger* 2 (March 1836): 266.

10. Henry David Thoreau, "Walking," in *The Writings of Henry David Thoreau*, eds. F. B. Sanborn and Bradford Torrey (Boston and New York: Houghton Mifflin, 1906), 205–48.

11. Ibid., 224.

12. Cronon, "Trouble with Wilderness," 1.

13. Rosenthal, "Urban Garden."

14. Graber, "Resolute Biocentrism."

15. *Fletcher v Peck*, 10 U.S. (6 Cranch) 87, United States Supreme Court, February 1, 1810; Alan W. Ewert, Denise S. Mitten, and Jillisa R. Overholt, *Natural Environments and Human Health* (Boston: CABI, 2014), 48.

16. John Locke, *Two Treatises of Government* (London: Thomas Tegg; W. Sharpe and Son; G. Offor; G. and J. Robinson; J. Evans and Co., 1823), 116.

17. Andro Linklater, *Owning the Earth: The Transforming History of Land Ownership* (New York: Bloomsbury USA, 2013).

18. Ted Steinburg, *Down to Earth: Nature's Role in American History* (Oxford: Oxford University Press, 2002), 14.

19. Linklater, *Owning the Earth*, 25.

20. Ludwig von Mises, *Human Action: A Treatise on Economics* (New Haven, CT: Yale University Press, 1949).

21. Ludwig von Mises, *Liberalism: In the Classical Tradition*, trans. Ralph Raico (New York: The Foundation for Economic Education, Inc./Cobden Press, 2002), 19.

22. Gerald Gaus, "Property," in *The Oxford Handbook of Political Philosophy*, ed. David Estlund (Oxford: Oxford University Press, 2012), 93.

23. Linklater, *Owning the Earth*, 50.

24. John Adams, quoted in Linklater, *Owning the Earth*, 50.

25. Steinburg, *Down to Earth*, 106.

26. Ibid., 109.

27. Using analysis of paintings (by Raphael and William Morris), house plans (by William Morris and Philip Webb, and by Robert Kerr), and diagrams (by Alexander Klein), Robin Evans demonstrates this transformation toward a more individualized private society, where the priority was the right to one's own body leading to the right to self and privacy. Robin Evans, "Figures, Doors and Passages," in *Translations from Drawing to Building and Other Essays* (London: Architectural Association, 1997), 77, 81, 85, 88.

28. Evans, "Figures, Doors and Passages," 84.

29. Richard Overton, quoted in Linklater, *Owning the Earth*, 45.

30. *A Declaration by the Representatives of the United States of America*, in General Congress Assembled (rough draft), 1776, accessed September 15, 2015, http://www.ushistory.org/declaration/document/rough.htm/.

31. Samuel D. Warren and Louis D. Brandeis, "The Right to Privacy," *Harvard Law Review* 4, no. 5 (1890): 195.

32. Georg Simmel, "The Metropolis and Mental Life," in *Modernism: An Anthology of Sources and Documents,* eds. Vassiliki Kolocotroni, Jane Goldman, and Olga Taxidou (Chicago: University of Chicago Press, 1998), 51.

33. The growth of personal privacy through legal measures was summarized first as protection from "battery in its various forms; liberty meant freedom from actual restraint; and the right to property secured to the individual his lands and his cattle" (Warren and Brandeis, 193). It was only later, with the rise of concern for

psychological health through the nineteenth century, that "the scope of these legal rights broadened; and now the right to life has come to mean the right to enjoy life; the right to be let alone; the right to liberty secures the exercise of extensive civil privileges; and the term 'property' has grown to comprise every form of possession-intangible, as well as tangible." Ibid., 193.

34. Alexander Garvin, *The American City: What Works, What Doesn't* (New York: McGraw-Hill, 2002).

35. Frederick Law Olmsted, *Civilizing American Cities* (Boston: Da Capo Press, 1997).

36. Simmel, "Metropolis and Mental Life," 52.

37. Adolf Loos, "Architecture," in *The Architecture of Adolf Loos: An Arts Council Exhibition*, eds. Yehuda Safran, Wilfried Wang, and Mildred Budny, trans. Wilfried Wang, 2nd ed. (London: The Council, 1987), 104–9.

38. Beatriz Colomina, "The Split Wall: Domestic Voyeurism," in *Sexuality & Space*, ed. Beatriz Colomina (New York: Princeton Architectural Press, 1992), 85.

39. Stanley I. Benn and Gerald F. Gaus, "The Public and the Private: Concepts and Action," in *Public and Private in Social Life*, eds. S. I. Benn and G. F. Gaus (New York: St. Martin's Press & Croom Helm, 1983), 25.

40. Both internal migration from the South and external migration from Europe, China, and Japan led to large population increases in American cities. The value of city properties rose for multiple reasons, and city land was in high demand.

41. Thomas Fisher, "Viral Cities," *Places Journal*, October 2010, accessed June 14, 2015, https://placesjournal.org/article/viral-cities/.

42. Deed restrictions and nuisance laws, along with different forms of private covenants, mutual agreements, informal arrangements, and ad hoc resorts were common measures of land use separation and restriction for maintaining property values. They were also useful for keeping a separation between property owners and urban events such as congestion, pollution, undesired activities, and unwanted people. See William A. Fischel, "An Economic History of Zoning and a Cure for Its Exclusionary Effects," *Urban Studies* 41, no. 2 (2004): 317–40.

43. Stuart Meck, "Early Planning and Zoning Law," in *U.S. Rocky Mountain Land Use Institute Conference*, University of Denver, Colorado, March 6, 2013.

44. The banning of laundromats was related to racial tensions between white Americans and Chinese immigrants and was enacted as a municipal ordinance. The ordinance was overturned as an issue of individual civic rights by the U.S. Supreme Court, Yick Wo v. Hopkins, 118 U.S. 356 (1886). See Paul Spickard, *Almost All Aliens: Immigration, Race, and Colonialism in American History and Identity* (New York: Routledge, 2007), 162.

45. The New York State Tenement House Act of 1901 addressed quality of life issues for tenement buildings, requiring occupants to have access to fresh air and sunlight. This was the third Tenement Act addressing this issue and was enacted to close loopholes created by previous housing reform acts, the First Tenement Housing Act (1867) and the Second Tenement Housing Act (1879).

46. Susan Griffin, *Woman and Nature: The Roaring Inside Her* (New York: Harper and Row, 1978), 95; on the concept of wilderness in Susan Griffin's work, see

Michael Zimmerman, "The Blessing of Otherness: Wilderness and the Human Condition," in *The Wilderness Condition: Essays on Environment and Civilization*, ed. Max Oelschlaegar (Washington, D.C.: Island Press, 1992), 245–70.

47. In 1964, the Second Session of the 88th Congress of the United States passed the Wilderness Act, signed into law by President Lyndon Johnson on September 3, 1964. This is the opening statement of the Act, stressing the purpose of the new law. Wilderness Act of 1964, Pub. L. No. 88–577, 16 U.S.C. § 1131–36 (1964).

48. Cronon, "Trouble with Wilderness," 69.

49. Wilderness Act of 1964.

50. Theodore Roosevelt, *Hunting Trips of a Ranchman* (New York and London: G. P. Putnam's Sons, 1885), 231.

51. Steinburg, *Down to Earth*, 145–6.

52. Ibid., 137.

53. Cronon, "Trouble with Wilderness," 75.

54. John Muir, *The Mountains of California* (New York: Century, 1894), 264.

55. D. R. Williams, L. Haggard, and R. Schreyer, "The Role of Wilderness in Human Development," in *Wilderness Benchmark 1988: Proceedings of the National Wilderness Colloquium*, ed. H. Freilich (Asheville, NC: U. S. Department of Agriculture, Forest Service, Southeastern Forest Experiment Station, 1989), 169–80.

56. D. R. Williams, "Personal and Social Meanings of Wilderness: Constructing and Contesting Places in a Global Village," in *Personal, Societal, and Ecological Values of Wilderness: Sixth World Wilderness Congress Proceedings on Research, Management, and Allocation*, comp. A. E. Watson, G. H. Aplet, and J. C. Hendee, vol. 2, October 24–29, 1998 (Bangalore, India. Proc. RMRS-P-14. Ft. Collins, CO: United States Department of Agriculture, 2000).

57. Ibid., 79.

58. Ewert, Mitten, and Overholt, *Natural Environments*, 48.

59. Wallace Stegner, "Wilderness Letter," in *The Wilderness Reader,* ed. Frank Bergon (Reno, NV: University of Nevada Press, 1994), 330.

60. Olmstead, *Civilizing American Cities*, 504.

61. Steinburg, *Down to Earth*, 243.

62. Clarence King, "Mountaineering in the Sierra Nevada," in *The Wilderness Reader,* ed. Frank Bergon (Reno, NV: University of Nevada Press, 1994), 140.

63. Steinburg, *Down to Earth*, 242.

64. Roosevelt, *Hunting Trips*, 233.

65. Stegner, "Wilderness Letter," 330.

66. David Lowenthal, "Is Wilderness 'Paradise Now?': Images of Nature in America," *Columbian University Forum* 7 (1964): 34.

BIBLIOGRAPHY

Adhya, Anirban, and Philip Plowright. "Setting Priorities: Sustainability, Environmental Health, and Embedded Value Judgments for the Urban Design Process." In *Cities in Transformation Research & Design: EAAE/ ARCC Milano, International*

Conference on Architectural Research, edited by Marco Bovati, Michele Caja, Giancarlo Floridi, and Martina Landsberger, 819–27. Milano, Italy: Il Poligrafo, 2012.

———. "Sustainable Architecture as Ecology: Defining Shared Systemic Priorities of Public Health and Social Ethics in Places." In *Architecture and Sustainability: Critical Perspectives for Integrated Design*, edited by Ahmed Z. Khan and Karen Allacker, 85–94. Leuven, Belgium and Den Haag, Netherlands: ACCO, 2015.

Agyeman, Julian. "Entering Cosmopolis: Crossingover, Hybridity, Conciliation and the Intercultural City Ecosystem." *Minding Nature* 7, no. 1 (2014): 20–25.

Benn, Stanley I., and Gaus, Gerald F. "The Public and the Private: Concepts and Action." In *Public and Private in Social Life*, edited by S. I. Benn and G. F. Gaus, 3–27. New York: St. Martin's Press & Croom Helm, 1983.

Colomina, Beatriz. "The Split Wall: Domestic Voyeurism." In *Sexuality & Space*, edited by Beatriz Colomina, 73–128. New York: Princeton Architectural Press, 1992.

Cronon, William. "The Trouble with Wilderness; Or, Getting Back to the Wrong Nature." In *Uncommon Ground: Rethinking the Human Place in Nature*, edited by William Cronon, 69–90. New York: W. W. Norton & Co., 1995.

A Declaration by the Representatives of the United States of America. In General Congress Assembled (rough draft), 1776. Accessed September 15, 2015. http://www.ushistory.org/declaration/document/rough.htm/.

Dew, Thomas Roderick. "An Address on the Influence of the Federative Republican System of Government upon Literature and the Development of Character." *The Southern Literary Messenger* 2 (March 1836): 266.

Evans, Robin. "Figures, Doors and Passages." In *Translations from Drawing to Building and Other Essays*, 55–91. London: Architectural Association, 1997.

Ewert, Alan W., Denise S. Mitten, and Jillisa R. Overholt. *Natural Environments and Human Health.* Boston: CABI, 2014.

Fincher, Ruth, and Jane M. Jacobs, eds. *Cities of Difference.* New York: The Guilford Press, 1998.

Fischel, William A. "An Economic History of Zoning and a Cure for Its Exclusionary Effects." *Urban Studies* 41, no. 2 (2004): 317–40.

Fisher, Thomas. "Viral Cities." *Places Journal*, October 2010. Accessed June 14, 2015. https://placesjournal.org/article/viral-cities/.

Fletcher v Peck. 10 U.S. (6 Cranch) 87. United States Supreme Court. February 1, 1810.

Garvin, Alexander. *The American City: What Works, What Doesn't.* New York: McGraw-Hill, 2002.

Gaus, Gerald. "Property." In *The Oxford Handbook of Political Philosophy,* edited by David Estlund, 93–112. Oxford: Oxford University Press, 2012.

Graber, David M. "Resolute Biocentrism: Dilemma of Wilderness in National Parks." In *Reinventing Nature?: Responses to Post-Modern Deconstruction*, edited by Michael E. Soulé and Gary Lease, 123–35. Washington, D.C.: Island Press, 1995.

Griffin, Susan. *Woman and Nature: The Roaring Inside Her.* New York: Harper and Row, 1978.

King, Clarence. “Mountaineering in the Sierra Nevada.” In *The Wilderness Reader,* edited by Frank Bergon, 136–50. Reno, NV: University of Nevada Press, 1994.

Linklater, Andro. *Owning the Earth: The Transforming History of Land Ownership.* New York: Bloomsbury USA, 2013.

Locke, John. *Two Treatises of Government.* London: Thomas Tegg; W. Sharpe and Son; G. Offor; G. and J. Robinson; J. Evans and Co., 1823.

Loos, Adolf. “Architecture.” In *The Architecture of Adolf Loos: An Arts Council Exhibition*, edited by Yehuda Safran, Wilfried Wang, and Mildred Budny, 104–9. Translated by Wilfried Wang. 2nd ed. London: The Council, 1987.

Lowenthal, David. “Is Wilderness ‘Paradise Now?’: Images of Nature in America.” *Columbian University Forum* 7 (1964): 34–40.

Meck, Stuart. “Early Planning and Zoning Law.” In *U.S. Rocky Mountain Land Use Institute Conference*, University of Denver, Colorado, March 6, 2013.

Muir, John. *The Mountains of California.* New York: Century, 1894.

Olmsted, Frederick Law. *Civilizing American Cities.* Boston: Da Capo Press, 1997.

Rolston III, Holmes. “The Wilderness Idea Reaffirmed.” *The Environmental Professional* 13 (1991): 370–77.

Roosevelt, Theodore. *Hunting Trips of a Ranchman.* New York and London: G. P. Putnam’s Sons, 1885.

Rosenthal, Bernard. “The Urban Garden.” *Texas Studies in Literature and Language* 20, no. 1 (1978): 119–38.

Sandercock, Leonie. *Cosmopolis II: Mongrel Cities in the 21st Century.* London: Continuum, 2003.

Schneekloth, Lynda. “The Frontier Is Our Home.” *Journal of Architecture Education* 49, no. 4 (1996): 210–25.

Sennett, Richard. *The Conscience of the Eye: The Design and Social Life of Cities.* London: Faber and Faber, 1991.

Simmel, Georg. “The Metropolis and Mental Life.” In *Modernism: An Anthology of Sources and Documents,* edited by Vassiliki Kolocotroni, Jane Goldman, and Olga Taxidou, 51–59. Chicago: University of Chicago Press, 1998.

Spickard, Paul. *Almost All Aliens: Immigration, Race, and Colonialism in American History and Identity.* New York: Routledge, 2007.

Stegner, Wallace. “Wilderness Letter.” In *The Wilderness Reader,* edited by Frank Bergon, 327–33. Reno, NV: University of Nevada Press, 1994.

Steinberg, Ted. *Down to Earth: Nature’s Role in American History.* Oxford: Oxford University Press, 2002.

Thoreau, Henry David. “Walking.” In *The Writings of Henry David Thoreau*, edited by F. B. Sanborn and Bradford Torrey, 205–48. Boston and New York: Houghton Mifflin, 1906.

von Mises, Ludwig. *Human Action: A Treatise on Economics.* New Haven, CT: Yale University Press, 1949.

———. *Liberalism: In the Classical Tradition.* Translated by Ralph Raico. New York: The Foundation for Economic Education, Inc./Cobden Press, 2002.

Warren, Samuel D., and Louis D. Brandeis. "The Right to Privacy." *Harvard Law Review* 4, no. 5 (1890): 193–220.

Wilderness Act of 1964. Pub. L. No. 88–577. 16 U.S.C. § 1131–36. 1964.

Williams, D. R. "Personal and Social Meanings of Wilderness: Constructing and Contesting Places in a Global Village." In *Personal, Societal, and Ecological Values of Wilderness: Sixth World Wilderness Congress Proceedings on Research, Management, and Allocation*, compiled by A. E. Watson, G. H. Aplet, and J. C. Hendee. Vol. 2. October 24–29, 1998. Bangalore, India. Proc. RMRS-P-14. Ft. Collins, CO: United States Department of Agriculture, 2000.

Williams, D. R., L. Haggard, and R. Schreyer. "The Role of Wilderness in Human Development." In *Wilderness Benchmark 1988: Proceedings of the National Wilderness Colloquium*, edited by H. Freilich, 169–80. Asheville, NC: U. S. Department of Agriculture, Forest Service, Southeastern Forest Experiment Station, 1989.

Zimmerman, Michael. "The Blessing of Otherness: Wilderness and the Human Condition." In *The Wilderness Condition: Essays on Environment and Civilization*, edited by Max Oelschlaegar, 245–70. Washington, D.C.: Island Press, 1992.

Chapter 3

Unraveling the Poles of Suburb and City

Karim Wagih Fawzi Youssef

In a globalized network, with increasing urbanization of the world population, cities are taking upon themselves new challenges of urban growth and economic development.[1] Within such a scenario, the issue of sustainability comes to the forefront of the agenda of many cities. Cities are becoming more and more under the purview of questioning and assessing their degree of comprehensive sustainability—environmentally, economically, socially, and politically.[2] Many municipalities have lately renewed their urban policies to include sustainability as their overarching goal and their direction moving forward. Solutions to the current state of cities revolve around such projects as the densification of the suburbs, conserving the natural landscape by establishing natural reserves, watershed management, enhancing the walkability of neighborhoods, designing complete streets, improving the energy efficiency of buildings, using renewable energy resources and green infrastructure, and encouraging participatory governance. These lofty goals and the visible changes in the streets of many cities—including, for example, bike lanes, new traffic signals for cyclists, the reintroduction of street cars, and car-sharing services—send the message that the problem of urban sustainability is a well-defined one, whose solution appears merely to be a technical, logistic, or programmatic challenge. Within such a positivistic paradigm, the problem of the city, cast as a presently unsustainable entity, seems well-defined, and the evident solution becomes the physical reorganization of urban form and the rezoning of land uses within the city in order to achieve a social ideal and the most energy-efficient and environmentally friendly form. What is manifestly ignored under such a positivistic paradigm is that sustainability decisions are contextual and value-laden.[3] Aidan Davison expresses this point well: "interest in urban sustainability promises not techniques by which to prove the best way to live, but skills of informed debate and wise judgement."[4] Along these lines, this essay questions the ontological nature of the city and argues that only by doing so can we can begin to understand the relation between the city and sustainability.

Justified as a reaction to modern urban planning, which entailed the separation of residential, commercial, and light-industrial land uses, "mixed-use development" is one catchword in a narrow and positivistic interpretation of sustainability. The rationale is that mixed-use developments use the land more efficiently, avoid monotony in the built environment, and contribute to the liveliness of the city by encouraging walkability, à la Jane Jacobs. Not only has land use planning come under scrutiny to "fix" the city, but also redesigning street patterns of neighborhoods is now part of the approach adopted by municipal planners in many North American cities, including Calgary. Now they advocate a "return-to-the-grid" as a panacea for the ills of enclave urbanism of conventional post-1970s suburban development and as restitution, in their view, of sensible planning that prioritizes connectivity, and thus sustainability, in the face of decades of experimenting with curvilinear streets, loops, and cul-de-sacs.[5]

Universally promoted by the New Urbanism movement as a reaction against urban sprawl, the grid is heralded for its advantages of connectivity, ease of extension, and ease of densification, as well as for promoting walkability and flexibility in transport routes. Such advantages are found to be lacking in the enclave urbanism model, where streets have a hierarchical pattern. The image that enclave urbanism seems to promote is the hierarchical structure of a tree while New Urbanism favors the image of the network.[6] The hierarchical flow of the street system in enclave urbanism is set in opposition to the more flexible flow of the two-dimensional direction of the grid street system. In other words, the structure of the city is transformed from a nested hierarchical structure that implies depth and concatenation to a two-dimensional, shallow structure that implies connectivity.

Casting the problematic of sustainability as largely related to the street pattern of cities may lead us into the fallacy of determinism, where one solution is discarded in favor of another that only seems to promise a better ideal for cities. Using Disney's neotraditional "Celebration" residential development in Orlando, Florida, as his example, Alex Marshall warns about shifting from one form to another without looking at underlying processes. He cautions that opposing the New Urbanist grid-type street pattern to the cul-de-sac street pattern of conventional suburban development overlooks that their dynamic and raison d'être—namely, the transport system—are actually the same.[7] Thus, the problem is not mixed-use development or the grid per se. Rather, the issue is the narrowing of solutions to single organizing principles.

This essay aims to redirect emphasis in the study of urban sustainability from a positivistic, problem-solving paradigm toward a constructivist paradigm, a paradigm that is more suited to the shift from the narrow understanding of environmental sustainability to the wider understanding of comprehensive sustainability. Under such a constructivist paradigm, the question should not be how to reorganize a city in order to foster sustainability, but how to *represent* a city in a way that is conducive to sustainability. Rather than adopting a simple, positivistic approach, one needs to ask the ontological question: can the city be a frame for sustainability? Posed differently, is it valid to speak of the sustainability of a city? The issue at stake is not how to design or *make* cities sustainable, but how to represent the city in the first place. Urban sustainability is an issue of representation rather than an issue of an integrated or upgraded design of a city and its neighborhoods.

The main argument of this essay is that a city can be a frame for sustainability, but that this is contingent upon how the city is represented. Representations of a city are tools not merely for expressing particular aspects of the city but also for framing the relation between sustainability and the city. This means that failure to represent a city means failure to plan for a sustainable city. The possibilities of viewing the city as a frame for sustainability open up when representations capture different aspects of a city. The question of sustainability, then, is not whether representations of a city should be made but how to reconcile multiple, and sometimes competing or contradictory, representations. In other words, no one type of city organization necessarily promotes sustainability better than the other, especially when sustainability is contingent upon the particular, complex, and varied formations of cities.

Combining the city and its multiple representations emancipates the study of urban sustainability from a perspective of a city as merely a physical entity to a perspective of a city as a sociocultural and political entity. This distinction is important insofar as it widens the focus on sustainability from attributes of the physical dimension of a city to include attributes of the social and cultural dimensions. Such an enlarged focus prevents the issue of sustainability of the city from becoming a positivistic and narrow debate to a constructivist paradigm that corresponds with present, postmodern culture. One positivistic view, for instance, is that the more compact the physical aspect of a city is, the more sustainable the city. However, such a view is contested by Michael Neuman, who observes that the city should be conceived in terms of processes rather than espousing a narrow focus on the physical form as derived from the compact city hypothesis.[8]

The purpose of this edited collection is to bring in the humanities on the same plane as that of the sciences and social sciences in the study of the relation between the city and sustainability. Including the humanities is not meant to just add another entry to the debate alongside other fields. Rather, it makes room for an ontological questioning of the city-sustainability nexus. It allows for movement from a problem-solving approach to a problem-defining approach, that is, what, in Kuhnian terms, would qualify as a paradigm shift. Toward this end, this essay proposes that one of the pathways for the entry of the humanities within the city-sustainability debate is through the critical reintroduction of the concept of representation in our conceptualization of the city. I argue that it is only by shifting the mode of thought from a static physical model of the city to multiple dynamic representational models that a paradigm shift, from a positivistic to a constructivist approach to the city-sustainability problematic, is possible.

WHAT IS A CITY?

It may be counterintuitive to begin by questioning the definition of a city, especially since "city" is a very common term and many people live in cities of different sizes and functions. A city may simply be defined by the number of its inhabitants (for instance, having at least 300,000 residents), its geographic size, the density of its population, its population profile, and attributes of its built environment. In effect, understanding what a city is may provide the key to profiling the relation between sustainability and the city. When Steve Pile posed the question "What is a city?," he referred to Lewis Mumford's definition of a city as a social institution or an organism and to Brian Robson's definition of the social significance of a city.[9] Pile teased out an image of a city as an arena or theater where a dynamic unfolds between the association and dissociation of people and the drama and opportunities that accompany this dynamic. This perspective on a city is essentially social, and one may categorize it as belonging to a social network paradigm.

If one were also to look at the physical structure of a city that accommodates its network of social interactions, then a city would appear essentially as a relatively dense urban environment with a widely differentiated land use pattern. An important aspect of this urban environment is the central-decentralist tension that makes up the core dynamics of that environment and is responsible for the center-periphery structure. The centrality aspect manifests physically in the downtown core, which gives identity to a city as high-rise commercial and residential buildings located there create a characteristic skyline for a city. The decentralism aspect manifests physically in a less dense, peripheral, suburban environment, which may also consist of commercial and residential uses but is not as dense as the urban core. The centrality-decentrality aspects are usually represented by a monocentric model characteristic of the structure of many cities and metropolitan areas.[10]

POSITIVISTIC SUSTAINABILITY

The current paradigm of positivistic approaches to sustainability implies that cities, as they presently are, are not sustainable. Thus, there is a need to make cities sustainable, especially in relation to environmental sustainability. This implies that cities need to become energy-conscious, waste-conscious, and pollution-conscious. If we look at the larger equation, so to speak, and the modernization of cities, this consciousness has been the case given the practicalities of particular situations. Cities were constructed using available technology and available resources in a cost-conscious and impact-conscious mindset as far as cultural sensitivity and industry regulations allowed. This is not a new perspective on sustainability, as some authors have posited that sustainability should be "interpreted as a socio-historical process rather than an end state."[11]

As different technologies are becoming available and renewable resources are being harnessed, the same mindset is at work to build cities of tomorrow: those cities will be built based on a wise use of energy and resources. In a post-sustainable era, new technologies and new resources will be available in such a manner that we cannot isolate the current drive toward positivistic sustainability as a new trend. The pursuit of positivistic sustainability seems to be a bandwagon for many cities to jump on. Dedication of whole budgets are then justifiable for the accelerated push toward such a sustainability that do not necessarily make economic sense to the present population or even in the long term relative to returns on investment due to the fluctuating and dynamic nature of markets. Perhaps we should question this accelerated push, as Joseph Desjardins has.[12]

The teleological outcome of this rationale results in cities pushing toward a sustainable future, which is inevitable, even though they are not sustainable as they currently stand. Residualization of the majority of the population is at stake in such a premature pursuit. It is an imbalanced drive toward sustainability, the pursuit of an image rather than the substance, a pursuit characterized as a type of situated rationality.[13] Seeking such a future state becomes the basis of present expedited human action. In order to free us from the narrow-mindedness of the positivistic paradigm of sustainability, we have to understand the city, not merely as a physical entity, but also as a social, cultural, and political entity. If we are cognizant of these other dimensions in addition to the physical, it readily becomes evident that there is no singular representation of the city.

THREE MAIN REPRESENTATIONS OF THE CITY

From the plethora of ways of representing the built environment of a city, there are at least three main representations, after Henri Lefebvre, which are of significance to inhabitants of a city. These are the representations of perceived, conceived, and lived space. What interests us here is how these representations may relate to sustainability.

For the first type, perceived space, inhabitants' perception of a particular built environment becomes a process contributing to their identity and expression of their values. One can thus speak of the *cultural sustainability* of a city. For instance, the physical layout of Tianjin, an eco-city in China, while embracing stringent energy efficiency standards and grand sustainability principles (from a positivistic stance), lacks a human scale due to giant residential blocks that would have otherwise resulted in a pedestrian friendly environment. To echo Christopher Schliephake's argument about eco-cities in his essay in this collection, Tianjin calls into question top-down approaches to sustainability that overly rely on the use of technology rather than paying attention to the cultural needs of inhabitants and the local community. The problem of following the "new city model" for eco-cities is building the whole city as a finished project rather than letting it develop organically.[14]

Meanwhile, the perceptual qualities of the physical environment may also be an instrument of class identity and class separation.[15] Aesthetization of the landscape and built environment may be used as a strategy to express prestige.[16] For Iris Young, it is also a form of cultural imperialism and social oppression.[17] The aesthetization of the built environment includes what Young and Sharon Zukin refer to as a form of symbolic violence through the use of dominant landscapes. Matthew Rofe shows that symbolization of prestige could be so profuse to the point of being signified as a cognitive barrier preventing outsiders from entry; that is, symbolization acts as a virtual wall and initiates a process of self-othering for outsiders.[18]

For the second type, conceived space, the city is likened to a physical map, and one can thus speak of the *structural sustainability* of a city insofar as one can orient oneself and commute from one part of the city to the other. At the scale of the city, the representation of conceived space is described by Kevin Lynch as an *environmental image*, a *structural understanding*, which is the product of immediate perception, as well as of memory.[19] Perceived and conceived spaces are not separate but have interconnections between them. For Lynch, it is a question of not only familiarity with a physical setting but also of its distinctiveness and legibility. Lynch raises the importance of the *imageability* of the city through the use of urban elements such as nodes that act as focal points and edges that act as boundaries. The imageability advocated by Lynch follows a gestalt visual logic for the ordering of the environment. It relies on ease of identification of points of activity within the environment through the use of landmarks as well as ease of identification of different areas within the city through defining their boundaries by edges. Lynch argues that ease of identification of the physical form of the city facilitates reading of, and orientation within, the city (the city becomes legible) and memorability of the city (the city becomes more imageable). However, the postmodern era has undermined the role of nodes and edges (as representative of centrality) in favor of liminal and third spaces (as spaces for innovation and decentralization).[20] The dilemma is set by a tension between formal and informal spaces, between a modern rational structure and a postmodern labyrinthine structure.

The third representation, the behavioral or lived space, involves the day-to-day dynamics and routines taking place within a city. Lived space is one of interaction between people, space, and activities taking place within the space. Although the term may seem neutral and straightforward, the city as a lived space is actually a contested place. The city becomes a space of appropriation and territorialization.[21] This is empirically evident at the large scale of the city when contrasting the urban core, the inner city, and peripheral suburban environments. It is evident as well at smaller scales within a city. For example, Dominika Polanska explains the emergence of gated communities in post-communist Poland as a reaction, in terms of the form of development and housing types, to Poland's communist past.[22] This reminds us of Pile's definition, mentioned above, by which the city is seen as a pattern resulting from tensions of associations and disassociations. One can thus speak of the *social sustainability* of a city in terms of the social interaction of inhabitants and their social networks.

STRUCTURAL SUSTAINABILITY

Of these three main representations of the city, the remainder of this essay elaborates on the second as an example of how a different *representation* of a city, from a structural perspective, impacts the city-sustainability nexus. As stated above, the structure of the city is usually represented as a center-periphery one, where the center has the highest concentration of employment and high-rise office buildings and skyscrapers while the periphery has the highest concentration of residential single-family houses and ground-oriented development interspersed with suburban shopping malls and commercial strip malls. This monocentric representation is characteristic of many North American cities, especially medium-sized ones, such as Calgary, as well as many other cities worldwide.

However, with the decentralization of industry from the city center and the suburbanization of commercial activity, such as the proliferation of large commercial centers in the suburbs, the city is taking on a more polynuclear structure albeit one in its infancy. Examples of polynuclear cities include Los Angeles and Toronto. Why a polynuclear structure has not materialized as a dominant structure for cities may be explained by the power of capitalism at work to consolidate such a center-periphery structure.[23] Sustaining a center-periphery structure has not been autonomous but, rather, enforced through a capitalist regime and the increasing role of finance capital in influencing the built environment. Uneven geographical development at the regional, as well as the metropolitan, scale is not merely an outcome of physical differentiation of space but is maintained and actively constructed by capitalism in order to ensure its survival.[24] As Edward W. Soja reminds us, spatial structure plays a role in the reproduction of capitalism and class relations.

In order to liberate city structure from this narrative of the hegemony of neoliberal capitalism and its influence on the continual reproduction of such a structure,[25] and in order to provide an arena for a real consideration of the city-sustainable nexus, another representation of the city is direly needed. Rather than being conceptualized merely as a static structure of the modern city, the core-periphery dynamic needs to be viewed as a *process* in the postmodern city.[26] As Alan Walks explains, referring to Lefebvre, urban space results from a first- and second-order synthesis that could manifest similarly in urban and suburban geographies. The first-order synthesis is a dialectic tension between centrality and dispersion whereas the second-order synthesis is a dialectic tension between difference and compartmentalization. Walks seeks to identify the underlying conceptual processes that work along an urbanism-suburbanism axis that need not lead to a singular association between suburbanism and the suburbs or between urbanism and the city core. Rather, the dialectic tension between such processes may manifest elements of suburbanism in the urban or, alternatively, may manifest elements of urbanism in the suburban. Understanding this multiple dynamic of association is key to having a new meta-representation of the city that allows for a reconceptualization of the city-sustainability nexus. It is only by understanding the interrelation and multiple dependencies between the suburban and the urban vis-à-vis suburbanism and urbanism that one can begin to delineate broad outlines for a sustainable city.

Walks's second-order synthesis is manifested in the compartmentalization of residential developments and the emergence of gated communities, for instance. Such compartmentalized developments may take the form of an exclusive suburb, a totalizing suburb, or a hyper-segregated ghetto.[27] According to Peter Marcuse, such developments, in turn, also have an intrinsic center-periphery structure at their smaller scale, with the nature of the center, rather than the periphery, as a differentiating factor.[28] At the center of such developments are the social, economic, and political relationships to power and wealth. As Marcuse observes, the stronger the divisions in society, the clearer the demarcations between the residential developments even without using manifest physical boundaries. Put differently, the gestalt logic of the city with the demarcation of clear boundaries, as advocated by Lynch, becomes unnecessary.

THE POSTMODERN CITY: POLES, NOT NODES

Nodes and the connections between them that form a pattern or network may produce a legible city for ease of navigation within it. Such an approach may enhance the imageability of a modern city by means of clarity of function and visual extent of boundaries of elements in a city. However, in a postmodern era where hierarchical organization in many aspects of modern life is giving way to a rhizomatic organization characterized by a horizontal ontology and an instantaneous network paradigm,[29] the sustainability of the city needs to be based on a different aesthetic than that proposed by Lynch, which is in many ways a gestalt and modern aesthetic.

The postmodern condition, or what Zygmunt Bauman calls liquid modernity, of our times articulates a sense of loss of place and identity, as well as class separation. A metaphor for such a condition is *the labyrinth*,[30] an artificial complexity which stratifies people according to their degree of mobility. Those who know the laws of the labyrinth are at the top of the social pyramid while the majority are left to navigate the obscurity of the labyrinth. Bauman's use of the labyrinth metaphor may seem incongruent with the condition of *liquefaction* that he advances: that of constant precariousness and instability.[31]

Posing this otherwise, how is an individual's freedom to choose and to act embedded within the rigid complex structure of the labyrinth? Perhaps a more congruent metaphor is that of *the reticulum*,[32] impermanent constructions or assemblages of heterogeneous elements, or the metaphor of *the rhizome*. Bauman clearly emphasizes that complex societies have become very rigid as a result of intertwining the freedoms of subsystems and agents through deregulation, flexibility, and liberalization in such a way that, paradoxically, the system and agents are disengaged and bypass each other. This bypassed system is captured in Erik Swyngedouw's notion of the "post-political condition," a condition that forecloses the political, and in which neoliberal capitalism is accepted consensually as an inevitable economic system.[33] Urban problems are not cast as problems of the system or of uneven power relations but, instead, are dealt with in a managerial mode of governance. A post-political populism and a post-democratic condition that employ a universal language (e.g., the creative city, the sustainable city, the green city, the eco-city, etc.), rather than address particular claims of social groups or classes, emerge. How, then, should a comprehensive urban environment be structured if it is to emancipate itself from a centralized structure and, at the same time, not be lost in a labyrinth? The dilemma is set by a tension between formal and informal spaces, between well-defined identities and rhizomatic identities. Which structure is socially just and which is socially unjust?

Perhaps the solution to the dilemma resides in a postmodern understanding of nodes as poles and an understanding of defined edges as permeable fringes. Moving from a paradigm of nodes to a paradigm of poles emancipates the structure of a city from a Euclidean and Cartesian representation to a non-Euclidean and Hegelian dialectic one. The center of the center-periphery structure is neither represented as a static node nor defined by an edge or boundary but, instead, becomes a pole with a sphere of influence extending beyond the periphery. Likewise, peripheral suburban developments are not static districts or areas, nor are they defined by rings concatenated within each other, but, instead, they are poles with a sphere of influence reaching and converging toward the center. That is, one returns to Walks's paradigm of interpenetration of the urban and the suburban, that is, elements of suburbanism in the urban, and elements of urbanism in the suburban. Such a dynamic is what gives a city the quality of a totality, a quality without which defining the sustainability of a city becomes a struggle or, at best, a diffuse feat.

The use of the term *pole*, rather than the simple use of the term *node*, is deliberate. A node is usually associated with a network and functions as a locus of concentration and an intersection of flows within the network. In urban geography terms, a node functions as a place that comes into being at the intersection of social and infrastructural networks. In contrast to the view of a node as a mere outcome of concentrated density of interchange and activity in a network is the term pole, which conveys a more active role, as well as implies a wider area of influence than a physically delimited node. In this sense, the city, with its urban and suburban areas, is not merely a physical agglomeration, but a constitution of poles that perform a complex function of attracting flows. That is, a pole is not an outcome of a network; rather, it is a network's generator. In addition, the pole has a property of an extensive field of influence as opposed to the node, which has a property of an intensive field of activity. Such a view of extensity of poles allows for overlaps and interpenetrations between them, as well as a relational view of identity, rather than an absolute one, at the city scale.

Rather than defining the suburban as a diminished version of the urban, the suburban is identified as an entity in itself. Stephen Marshall suggests the term "semi-urban" instead of "suburban" to describe suburban environments.[34] L. Vaughan et al. advocate the same approach by viewing suburban space as a distinctive domain.[35] Such a view moves away from a centripetal-centrifugal paradigm and the binary debate on whether a compact city model versus a diffuse one is more sustainable. Rather than seen as a structure with one center and its periphery, the city is represented with two poles: an urban pole and a suburban pole, which interpenetrate. Such a conceptualization of the city will determine the mode of the future development of the monocentric model toward a truly polycentric one, as opposed to a pseudo-polycentric one. Rather than circumventing the real problem of the city by reducing it to concerns of physical expansion and the location of polycentric nodes (which may only multiply the monocentric model and its problems), the drive should be toward allowing for a multiplicity of representations that are, nevertheless, coherent along physical, social, and cultural dimensions. This will only be possible when we realize that the issue is not a physical and logistic one, but has more to do with the meta-representation of the city in the first place as one involving the relation between the urban and the suburban. Without such a conceptualization, however, the continually expanding city will exist as an amorphous entity with negative impacts on the social and cultural dimensions of the sought-after sustainable city model.

NOTES

1. Nancy B. Grimm et al., "Global Change and the Ecology of Cities," *Science* 319, no. 5864 (2008): 756–60, doi:10.1126/science.1150195.
2. Kent E. Portney, *Taking Sustainable Cities Seriously: Economic Development, the Environment, and Quality of Life in American Cities*, 2nd ed. (Cambridge, MA: Massachusetts Institute of Technology, 2013).
3. Lawrence Martin, "Incorporating Values into Sustainability Decision-Making," *Journal of Cleaner Production* 105 (2015): 146–56, doi:10.1016/j.jclepro.2015.04.014.
4. Aidan Davison, "Stuck in a Cul-de-Sac?: Suburban History and Urban Sustainability in Australia," *Urban Policy and Research* 24, no. 2 (June 2006): 213, doi:10.1080/08111140600704137.
5. Karim Wagih Fawzi Youssef, "Gated-Ness, Income Segregation and Neighbourhood Cohesion in Two Western Canadian Metropolises" (PhD diss., University of Calgary, 2015), https://www.researchgate.net/publication/272680490_Gated-ness_Income_Segregation_and_Neighbourhood_Cohesion_in_Two_Western_Canadian_Metropolises.
6. Christopher Alexander, "The City Is Not a Tree," *Architectural Forum* 122, no. 1 (1965): 30–32.
7. Stephen Marshall, "The Emerging 'Silicon Savanna': From Old Urbanism to New Suburbanism," *Built Environment (1978-)* 32, no. 3 (January 2006): 267–80, doi:10.2307/23290105.
8. Michael Neuman, "The Compact City Fallacy," *Journal of Planning Education and Research* 25, no. 1 (September 1, 2005): 11–26, doi:10.1177/0739456X04270466.
9. Steve Pile, "What Is a City?," in *City Worlds*, eds. John Allen, Doreen Massey, and Steve Pile (London and New York: Routledge, 1999), 3–52.
10. Daniel Arribas-Bel and Fernando Sanz-Gracia, "The Validity of the Monocentric City Model in a Polycentric Age: US Metropolitan Areas in 1990, 2000 and 2010," *Urban Geography* 35, no. 7 (October 3, 2014): 980–97, doi:10.1080/02723638.2014.940693.

11. Andrea Colantonio and Tim Dixon, *Urban Regeneration and Social Sustainability: Best Practice from European Cities* (Hoboken, NJ: Wiley-Blackwell, 2010), 21.

12. Joseph DesJardins, "Is It Time to Jump off the Sustainability Bandwagon?," *Business Ethics Quarterly* 26, no. 1 (January 15, 2016): 117–35, doi:10.1017/beq.2016.12.

13. Jens Beckert, "Imagined Futures: Fictional Expectations in the Economy," *Theory and Society* 42, no. 3 (March 17, 2013): 219–40, doi:10.1007/s11186–013–9191–2.

14. Jonathan Kaiman, "China's 'Eco-Cities': Empty of Hospitals, Shopping Centres and People," *The Guardian*, April 14, 2014, https://www.theguardian.com/cities/2014/apr/14/china-tianjin-eco-city-empty-hospitals-people.

15. Şerife Geniş, "Producing Elite Localities: The Rise of Gated Communities in Istanbul," *Urban Studies* 44, no. 4 (April 2007): 771–98, doi:10.1080/00420980601185684; Sabine Meier and Lia Karsten, "Living in Commodified History: Constructing Class Identities in Neotraditional Neighbourhoods," *Social & Cultural Geography*, 2012, 1–19, doi:10.1080/14649365.2012.704641.

16. Sharon Zukin, *Landscapes of Power: From Detroit to Disney World* (Berkeley, CA: University of California Press, 1991).

17. Iris Marion Young, *Justice and the Politics of Difference* (Princeton, NJ: Princeton University Press, 1990).

18. Matthew W. Rofe, "New Landscapes of Gated Communities: Australia's Sovereign Islands," *Lanscape Research* 31, no. 3 (2006): 309–17.

19. Kevin Lynch, *The Image of the City* (Cambridge, MA: MIT Press, 1960).

20. Edward W. Soja, "Margin/alia: Social Justice and the New Cultural Politics." In *The Urbanization of Injustice*, eds. Andy Merrifield and Erik Swyngedouw (New York: New York University Press, 1997), 180–99.

21. Michel de Certeau, *The Practice of Everyday Life*, trans. Steven Rendall (Berkeley, CA: University of California Press, 1984).

22. Dominika V. Polanska, "Gated Communities and the Construction of Social Class Markers in Postsocialist Societies: The Case of Poland," *Space and Culture* 13, no. 4 (November 1, 2010): 421–35, doi:10.1177/1206331210374140.

23. Edward J. Soja, "The Socio-Spatial Dialectic," *Annals of the Association of American Geographers* 70, no. 2 (1980): 207–25, doi:10.1111/j.1467–8306.1980.tb01308.x.

24. Edward W. Soja, *Seeking Spatial Justice* (Minneapolis: University of Minnesota Press, 2010).

25. Peter Marcuse, "The Forms of Power and the Forms of Cities: Building on Charles Tilly," *Theory & Society* 39, no. 3/4 (2010): 471–85, doi:10.1007/s11186–010–9117–1.

26. Alan Walks, "Suburbanism as a Way of Life, Slight Return," *Urban Studies* 50, no. 8 (June 2013): 1471–88, doi:10.1177/0042098012462610.

27. Peter Marcuse, "The Ghetto of Exclusion and the Fortified Enclave: New Patterns in the United States," *American Behavioral Scientist* 41, no. 3 (November 1997): 311–26, doi:10.1177/0002764297041003004.

28. Ibid.

29. Gilles Deleuze and Félix Guattari, *A Thousand Plateaus: Capitalism and Schizophrenia* (Minneapolis: University of Minnesota Press, 1987).

30. Zygmunt Bauman, *Liquid Modernity* (Cambridge, UK: Polity Press, 2000), 138.

31. Youssef, "Gated-Ness, Income Segregation and Neighbourhood Cohesion."

32. Leslie Jaye Kavanaugh, *The Architectonic of Philosophy: Plato, Aristotle, Leibniz* (Amsterdam: Amsterdam University Press, 2007).

33. Erik Swyngedouw, "The Antinomies of the Postpolitical City: In Search of a Democratic Politics of Environmental Production," *International Journal of Urban and Regional Research* 33, no. 3 (2009): 601–20, doi:10.1111/j.1468–2427.2009.00859.x.

34. Marshall, "The Emerging 'Silicon Savanna.'"

35. Laura Vaughan et al., "Do the Suburbs Exist? Discovering Complexity and Specificity in Suburban Built Form," *Transactions of the Institute of British Geographers* 34, no. 4 (October 2009): 475–88, http://www.jstor.org/stable/40270732.

BIBLIOGRAPHY

Alexander, Christopher. "The City Is Not a Tree." *Architectural Forum* 122, no. 1 (1965): 30–32.

Arribas-Bel, Daniel, and Fernando Sanz-Gracia. "The Validity of the Monocentric City Model in a Polycentric Age: US Metropolitan Areas in 1990, 2000 and 2010." *Urban Geography* 35, no. 7 (October 3, 2014): 980–97. doi:10.1080/02723638.2014.940693.

Bauman, Zygmunt. *Liquid Modernity*. Cambridge, UK: Polity Press, 2000.

Beckert, Jens. "Imagined Futures: Fictional Expectations in the Economy." *Theory and Society* 42, no. 3 (March 17, 2013): 219–40. doi:10.1007/s11186–013–9191–2.

Colantonio, Andrea, and Tim Dixon. *Urban Regeneration and Social Sustainability: Best Practice from European Cities*. Hoboken, NJ: Wiley-Blackwell, 2010.

Davison, Aidan. "Stuck in a Cul-de-Sac?: Suburban History and Urban Sustainability in Australia." *Urban Policy and Research* 24, no. 2 (June 2006): 201–16. doi:10.1080/08111140600704137.

de Certeau, Michel. *The Practice of Everyday Life*. Translated by Steven Rendall. Berkeley, CA: University of California Press, 1984.

Deleuze, Gilles, and Félix Guattari. *A Thousand Plateaus: Capitalism and Schizophrenia*. Minneapolis: University of Minnesota Press, 1987.

DesJardins, Joseph. "Is It Time to Jump off the Sustainability Bandwagon?" *Business Ethics Quarterly* 26, no. 1 (January 15, 2016): 117–35. doi:10.1017/beq.2016.12.

Geniş, Şerife. "Producing Elite Localities: The Rise of Gated Communities in Istanbul." *Urban Studies* 44, no. 4 (April 2007): 771–98. doi:10.1080/00420980601185684.

Grimm, Nancy B., Stanley H. Faeth, Nancy E. Golubiewski, Charles L. Redman, Jianguo Wu, Xuemei Bai, and John M. Briggs. "Global Change and the Ecology of Cities." *Science* 319, no. 5864 (2008): 756–60. doi:10.1126/science.1150195.

Kaiman, Jonathan. "China's 'Eco-Cities': Empty of Hospitals, Shopping Centres and People." *The Guardian*, April 14, 2014. https://www.theguardian.com/cities/2014/apr/14/china-tianjin-eco-city-empty-hospitals-people.

Kavanaugh, Leslie Jaye. *The Architectonic of Philosophy: Plato, Aristotle, Leibniz.* Amsterdam: Amsterdam University Press, 2007.

Lynch, Kevin. *The Image of the City*. Cambridge, MA: MIT Press, 1960.

Marcuse, Peter. "The Forms of Power and the Forms of Cities: Building on Charles Tilly." *Theory & Society* 39, no. 3/4 (2010): 471–85. doi:10.1007/s11186–010–9117–1.

———. "The Ghetto of Exclusion and the Fortified Enclave: New Patterns in the United States." *American Behavioral Scientist* 41, no. 3 (November 1997): 311–26. doi:10.1177/0002764297041003004.

Marshall, Alex. "A Tale of Two Towns: Kissimmee versus Celebration and the New Urbanism." In *How Cities Work: Suburbs, Sprawl, and the Roads Not Taken*, 270. University of Texas Press, 2001.

Marshall, Stephen. "The Emerging 'Silicon Savanna': From Old Urbanism to New Suburbanism." *Built Environment (1978-)* 32, no. 3 (January 2006): 267–80. doi:10.2307/23290105.

Martin, Lawrence. "Incorporating Values into Sustainability Decision-Making." *Journal of Cleaner Production* 105 (2015): 146–56. doi:10.1016/j.jclepro.2015.04.014.

Meier, Sabine, and Lia Karsten. "Living in Commodified History: Constructing Class Identities in Neotraditional Neighbourhoods." *Social & Cultural Geography*, 2012, 1–19. doi:10.1080/14649365.2012.704641.

Neuman, Michael. "The Compact City Fallacy." *Journal of Planning Education and Research* 25, no. 1 (September 1, 2005): 11–26. doi:10.1177/0739456X04270466.

Pile, Steve. "What Is a City?" In *City Worlds*, edited by John Allen, Doreen Massey, and Steve Pile, 3–52. London and New York: Routledge, 1999.

Polanska, Dominika V. "Gated Communities and the Construction of Social Class Markers in Postsocialist Societies: The Case of Poland." *Space and Culture* 13, no. 4 (November 1, 2010): 421–35. doi:10.1177/1206331210374140.

Portney, Kent E. *Taking Sustainable Cities Seriously: Economic Development, the Environment, and Quality of Life in American Cities*. 2nd ed. Cambridge, MA: Massachusetts Institute of Technology, 2013.

Rofe, Matthew W. "New Landscapes of Gated Communities: Australia's Sovereign Islands." *Lanscape Research* 31, no. 3 (2006): 309–17.

Soja, Edward W. *Seeking Spatial Justice*. Minneapolis: University of Minnesota Press, 2010.

———. "The Socio-Spatial Dialectic." *Annals of the Association of American Geographers* 70, no. 2 (1980): 207–25. doi:10.1111/j.1467–8306.1980.tb01308.x.

———. "Margin/alia: Social Justice and the New Cultural Politics." In *The Urbanization of Injustice*, edited by Andy Merrifield and Erik Swyngedouw, 180–99. New York: New York University Press, 1997.

Swyngedouw, Erik. "The Antinomies of the Postpolitical City: In Search of a Democratic Politics of Environmental Production." *International Journal of Urban and Regional Research* 33, no. 3 (2009): 601–20. doi:10.1111/j.1468–2427.2009.00859.x.

Vaughan, Laura, Sam Griffiths, Mordechai Haklay, and Catherine Emma Jones. "Do the Suburbs Exist? Discovering Complexity and Specificity in Suburban Built Form." *Transactions of the Institute of British Geographers* 34, no. 4 (October 2009): 475–88. http://www.jstor.org/stable/40270732.

Walks, Alan. "Suburbanism as a Way of Life, Slight Return." *Urban Studies* 50, no. 8 (June 2013): 1471–88. doi:10.1177/0042098012462610.

Young, Iris Marion. *Justice and the Politics of Difference*. Princeton, NJ: Princeton University Press, 1990.

Youssef, Karim Wagih Fawzi. "Gated-Ness, Income Segregation and Neighbourhood Cohesion in Two Western Canadian Metropolises." PhD diss., University of Calgary, 2015. https://www.researchgate.net/publication/272680490_Gated-ness_Income_Segregation_and_Neighbourhood_Cohesion_in_Two_Western_Canadian_Metropolises.

Zukin, Sharon. *Landscapes of Power: From Detroit to Disney World*. Berkeley, CA: University of California Press, 1991.

Chapter 4

The Rhetoric of Sustainability

Teaching Writing to (Mostly) Engineering Students in an Urban Setting

Iuliu Ratiu

In the Fall of 2012, as a teaching fellow at the Georgia Institute of Technology (Georgia Tech) in a Writing and Communication Program (WPC) that emphasizes process, multimodality, digital literacy, and humanistic perspectives in a technological world, I developed a first-year composition (FYC) course titled "The Rhetoric of Waste and Sustainability." Both the setup of the WCP and the profile of the institute created for me the optimal conditions to teach sustainability-based courses. Because I was willing to take risks with the material that I presented, there was a clear pay-off in student engagement. I know I was able to create classroom situations where students were both engaged and enthusiastic about discussing concepts such as waste, sustainability, and the notion of living green as they pertained to their own personal and professional goals by using blogs, class presentations, discussion boards, and multimedia artifacts.

In this essay, I reflect on how I used the rhetoric of waste and sustainability to teach writing and multimodal communication to engineering students in an urban setting. While, on the surface, this essay may seem to only tangentially refer to the topic of sustainability and the city, it does in fact directly address the role that pedagogy must play in promoting comprehensive sustainability in a global world in which the urban writ large has taken precedence over the rural to the detriment of the natural. In doing so, I first plan to show the discursive gap between responsible institutional policies and failing political, economic, and technological practices as they relate to sustainability. I then argue that students become better communicators and professionals in their fields when working on projects that they find important for better understanding both the positive and the negative impacts that their majors have on controversial topics such as climate change and resource use. More ambitiously, my contribution to this edited collection is meant to intervene where environmental education, policy, and practice have failed: winning the hearts and minds of (engineering) students and giving them the tools to address sustainability from a multifaceted perspective.

Irrespective of whether they come from rural, suburban, or urban backgrounds, from the United States or from other countries, first-year students at Georgia Tech soon learn to adapt to living in a large city: they realize this not only when they leave the campus, situated in Atlanta, and venture off into the outer world, but also when they are on campus working on writing projects that, as I show below, propose to make the campus more sustainable. Investigating the sustainable aspect of campus operations, such as grounds and waste management, food supply, grid maintenance, or transportation, students realize that their campus is the perfect model on which to try out ideas for more eco-friendly urban environments. An arts, tech, and innovation hub, Atlanta also has serious infrastructure and social problems, and efforts to address them often come from Georgia Tech graduates, one of the most widely known examples of such initiatives being the Atlanta BeltLine (a sustainable redevelopment project along 22 miles of disaffected railroad tracks). Next to vocational training, a rhetoric and communication approach to sustainability benefits students in that it simulates the microcosm needed to understand some of the technical and societal problems created by unsustainable practices. In addition to the Calculus basics that they obtain in a large lecture hall or to the hands-on experience of problem-solving that they acquire in a small invention lab, students learn to master critical thinking skills in a composition classroom. One that focuses on sustainability allows future engineers to find the proper balance, for instance, between technical innovation and the equal distribution of resources. When they leave the campus for good, students help shape the conversation with regard to comprehensive sustainability not only because they have expert technical skills but also because they know how to make sound arguments in order to promote meaningful change.

Long considered an offshoot of environmental awareness and geopolitical economic growth, sustainability has recently been adopted and appropriated as a policy vehicle and marketing tool. It has also made its way into education simply because the study of sustainability lends itself well to an interdisciplinary approach and includes instruction across various disciplines, ranging from scientific and technical fields, such as sustainable development, ecology, environmental policies, landscape architecture, city and regional planning, and natural resources, to the liberal arts, including geography, economics, sociology, anthropology, and ethics. However, until recently, sustainability has not really been the focus of composition and writing classes. Therefore, my essay, about the course that I designed in order to remedy this absence at Georgia Tech, offers suggestions on how to teach multimodal composition courses with a focus on sustainability. In a sense, during the three years that

I spent teaching at Georgia Tech, I was fortunate enough to interact with two worlds that only occasionally intersect: the rigorous promotion of critical thinking skills coming from the writing program and the solution-based approach of a technological institute.

For the purposes of this essay, I plan to draw on both my theoretical expertise and my pedagogy in teaching a sustainability studies writing course to engineering students in order to answer two (simple) questions: (1) how sustainable is a tech college situated in an urban area when, after more than twenty years of research on photovoltaic cells, there were only three buildings with solar panels on their roofs on the whole campus? And (2) can technology solve some of the sustainability problems that it created?

According to various rankings, including "The Princeton Review" and Sierra Club, Georgia Tech is constantly mentioned as having one of the most sustainable campuses in the nation. Yet, until recently, an institute-wide focus on sustainability was almost non-existent despite long-standing (and mostly independent) research clustered around and within excellence programs and centers such as the Center of Excellence for Photovoltaics, the Sustainable Design and Manufacturing Program, and the Center for Quality Growth and Regional Development. This situation changed with the 2015 adoption of a five-year Quality Enhancement Plan (QEP) with the programmatic theme of "Serve, Learn, Sustain." The adoption and future implementation of the QEP clearly show how sustainability practices have recently been embraced and promoted through an interesting collaboration between popular demand (alumni in this case) and greenwashing strategies. Part of the ten-year review process of Georgia Tech's reaffirmation for the Southern Association of Colleges and Schools Commission on Colleges, the QEP also offers valuable insight into how sustainability becomes a key factor to account for state and regional educational practices, corporate responsibility, and political, if not ethical, standards.[1]

Take, for example, the prestige, clout, and practical influence that a person such as G. Wayne Clough can bring to the way in which sustainability becomes one of the core issues of the institute. Secretary Emeritus of the Smithsonian Institution and President Emeritus of Georgia Tech, Clough served as the twelfth secretary of the Smithsonian from 2008 to 2014 and as the tenth president of Georgia Tech between 1994 and 2008. In fact, Clough is the only Georgia Tech graduate to become president of the institution. As recognition for his accomplishments, the Undergraduate Learning Commons, one of the few Platinum LEED Certified buildings on campus, was named after him in 2011. Most importantly, Clough returned to campus just in time: when the push for a transition to a cleaner, greener, low-emission, energy-based economy has arguably reached a point of no return. Interestingly enough, as if to mark his return to Georgia Tech, in March 2015, Clough delivered a distinguished lecture, "From the Eocene to the Anthropocene: An Engineer's View of Climate Change," the first installment of the Kenneth Hyatt Distinguished Alumni Leadership Speaker Series organized by the School of Civil and Environmental Engineering, the very department that graduated him in 1964 with a BS degree in Civil Engineering. The fact that a reputable scholar such as Clough is doing the legwork to promote an engineering approach to solving climate change speaks volumes about the current push toward a more sustainable way of living.[2]

Contrast this approach with my provisional answers to my two questions above. My first question problematizes the sustainability of a technological school that lags behind when it comes to implementing the solutions advanced by its research centers, while the second question refers to technology as part of the solution, not the problem, when dealing with sustainability-related issues. In this respect, my aim is to interrogate the importance of research and development attitudes and to address the possible divide between noble goals and less-than-noble practices. As a case in point, I question—rhetorically, of course—the scarcity of solar panels on campus after more than twenty years of research on photovoltaic cells. Similarly, of equal rhetorical relevance is analyzing the institution's impacts on not only academic research but also policy making and the development of corporations headquartered or doing business in the Atlanta area.

For instance, in 2003, Georgia Tech dedicated the Ford Environmental Science and Technology Building following a donation of US$10 million from the Ford Motor Company. As of March 20, 2016, according to information provided by the website *Philanthropy at Tech*, a student-led initiative of the Georgia Tech Student Foundation, the Ford ES&T Building is part of Georgia Tech's interdisciplinary Life Science and Technology Complex, and it used to be the largest academic building on campus. Housing classrooms, labs, and offices for the Environmental Engineering program, Earth and Atmospheric Sciences, Environmental Biology, Environmental Chemistry, Biomedical Engineering, and Chemical and Biomolecular Engineering, the building was the first university building in the University System of Georgia to be funded by a corporation. And herein lies the (discursive and rhetorical) problem. Even though this and many other corporations (operating in or not in Atlanta) hire Georgia Tech graduates and fund programs at the institution, this is a good example of a different kind of greenwashing: efforts made by corporations to fund eco-friendly research in order to atone for their negative impacts on the environment.

Another factor to take into account when analyzing corporate accountability and greenwashing is to assess their transparency when it comes to their communication strategies. For instance, given the fact that the Coca-Cola Company's world headquarters are right next to Georgia Tech's campus (or the other way around), one of my students working on an assignment about plastic waste—in particular, on the reliance on single-use plastic water bottles—took it upon himself to contact the Coca-Cola Company in order to investigate the claim that the plant-based glycol used in manufacturing the PET for a certain brand of water bottles is a step in the right direction toward reducing the negative impact of plastic waste. While the student didn't get any official response, he found out that, though the science behind this change is accurate, its morality is not: in a world where a billion people don't have access to safe drinking water, the manufacturer wastes two gallons of water for every single gallon that it produces. I was impressed with the student's research skills, and I think he made a compelling case when he questioned the efficiency of technological advances.

While these are mere anecdotal and circumstantial examples of the complex relation between academic research and corporate interests, they also showcase the practical steps that students can take to change things for the better. Even when doing research sponsored by corporations whose role in promoting unsustainable practices is undeniable, students can become part of a larger effort to improve the situation by acknowledging the harm done and by using their expertise to design more sustainable products. As my students were quick to realize, rhetoric helped them to see through the smokescreen of greenwashing, while research promoted higher standards and technical innovations to solve some of the problems caused by industries employing unsustainable practices.

In what follows next, I will discuss sources relevant to teaching a FYC course based on sustainability using rhetoric, process, and multimodality; I will present the argument for a FYC course based on sustainability; and I will share my reflections on proposing, teaching, and revising such a course.

SELECT SOURCES RELEVANT TO TEACHING FYC COURSES BASED ON SUSTAINABILITY

While there is consensus and overall agreement with regard to how sustainability may be defined, researched, and promoted, there have been only a handful of investigations into how sustainability should or may be approached in a composition class. The term ecocomposition was first introduced by Randall Roorda, Lee Smith, and Michael McDowell on a panel at the Conference on College Composition and Communication (CCCC) Annual Convention in 1998 in Chicago. To emphasize how ecology, environmental concern, and sustainability have become major parts of composition studies, Sidney Dobrin and Christian Weisser edited, in 2001, *Ecocomposition: Theoretical and Pedagogical Approaches*, a collection of essays which posited ecocomposition as the study of the relationships between environments and discourse, with an emphasis on the importance of developing the critical thinking skills needed in the reception, as well as the production, of discourse. Following in their footsteps, I advocate for a practical approach to ecocomposition, a method that pertains well to engineering students who are particularly savvy at solving problems.

This solution-based approach bypasses some of the challenges facing environmental communication writ large in that it leaves behind more controversial aspects of environmentalism and sustainability studies by focusing on problems at hand. Influential environmental scholars, such as Robert Cox, the former president of the Sierra Club, consider both the theoretical and the practical aspects of this apparent divide. On the one hand, in "Nature's 'Crisis Disciplines': Does Environmental Communication Have an Ethical Duty?" (2007), Cox considers that scholars need to play an ethical role in raising awareness about the ultimate issues of environmental destruction, such as limited resources, pollution, or endangered species; on the other hand, in his textbook *Environmental Communication and the Public Sphere* (2009), Cox transforms advocacy into pedagogy and invites instructors and students not only to engage in discursive practices but also to actively contribute to green activities on campuses and in their communities.[3]

This solution-based approach is also favored by Bryan Norton, who argues in "Pragmatism, Adaptive Management, and Sustainability" that "sustainable outcomes are not definable in advance but must emerge from a program of active social experimentation and learning."[4] Interestingly enough, Norton is a philosophy professor in the School of Public Policy at Georgia Tech. He has written extensively about the "conceptual disorder" that makes communication difficult between ecologists, scientists, and policy makers. For instance, in "Improving Ecological Communication," Norton first critiques the failed implementation of the wetlands policy and the ecologists' disregard for social values in their choice of research topics; he then argues for ecologists to engage in a search for solutions to the problem of communication "by the introduction and general use of one or more indicator terms that both represent measurable characteristics of ecological systems and also reflect impacts on social values."[5] While Norton's work deals specifically with analyzing policy discourse, his proposal for social learning and measurable characteristics of ecological systems is applicable not only to ecologists and scientists but also to engineers. Therefore, bringing his recommendations to a composition class focused on sustainability helps to better prepare future engineers, scientists, and policy makers to avoid communication errors, to propose solutions that align well with communal values, and, by doing so, to create sustainable outcomes.

Two other important sources that I find useful in showcasing the approach of sustainability to composition studies and to other disciplines are a special issue titled "Teaching Earth" of *Transformations: The Journal of Inclusive Scholarship and Pedagogy*, guest edited by Anthony Lioi in 2010, and a collection of essays titled *Teaching Sustainability, Teaching Sustainably*, edited by Kristen Allen Bartels and Kelly A. Parker in 2012. Even so, it took large publishing houses some time until they commissioned single-themed readers and handbooks dedicated to the study of sustainability through composition and communication lenses. The authors whose works have been excerpted in these textbooks have been actively thinking, writing, and advocating for an interdisciplinary approach to sustainability. For instance, Oxford University Press released Carl Herndl's *Sustainability, A Reader for Writers* in 2013, while Macmillan Education launched Christian Weisser's *Sustainability, A Bedford Spotlight Reader* in 2015. Both Herndl and Weisser, together with their collaborators, have been promoting environmental rhetoric and ecocomposition for the last two decades.

In 1996, Carl Herndl and Stuart Brown edited *Green Culture: Environmental Rhetoric in Contemporary America*, a collection of essays by various authors that analyzes environmentalism as a triad: nature as resource, nature as spirit, and nature as object. More focused in scope and quite comprehensive in its programmatic outline, the *Sustainability Reader* that Herndl edited seventeen years later shows in fact the momentum behind sustainability studies. Structured around broad topics, the reader has the following sections with excerpts ranging from the work of now classic environmental writers, such as Rachel Carson, to the valuable blog contributions of concerned undergraduate students: (1) Us: How We Live With Each Other and With the World; (2) Trash: The Costs of Throwing "Stuff" Away; (3) Food: A Different View of the Food Chain; (4) Climate Change: What It Is, How It Affects Us, and Why We Argue About It So Much; (5) Energy: Supply, Demand and Invisible Consequences; and (6) Soil and Water: Resources We Take for Granted.

This theme-based structure is also followed in Weisser's textbook, which connects sustainability to various topics, such as politics, urban environments, and international relations. Whereas Herndl's reader has most section titles in the form of topic statements, with only two tentative questions, Weisser's reader uses only questions: What Are the Foundations of Sustainability?; How Is Sustainability a Political Issue?; How Do Crises and Disasters Create Challenges for Sustainability?; How is Sustainability Connected to Local and Urban Environments?; How is Sustainability a Transnational Issue?; and How Is Tourism and Recreation Connected to Sustainability? Starting with a foundational overview of sustainability and ending with the ways in which sustainability affects the hospitality industry, Bedford St. Martin's reader pertains better to the expectations and lifestyles of young students. While many of them come to college having taken Advanced Placement Sustainability classes and show real interest in sustainability studies, not all of them are keen to engage in heavy theoretical and sometimes divisive political and ethical debates, especially given the fact that sustainability issues remain mainly invisible to them. However, after they learn from the section dedicated to tourism and recreation that unsustainable practices, such as overfishing or extreme logging, negatively impact the natural places that they visit on vacation, students feel encouraged to get involved and make a difference.

From the two readers presented above, I have only had a chance to use the one edited by Carl Herndl, which was published a year after I began teaching my sustainability-focused classes at Georgia Tech. When and if given the opportunity to teach sustainability-focused composition classes again, I will also assign Weisser's textbook because it seems less cluttered and user-friendly by virtue of being a more concise source book. However, beyond the general pros and cons of using a reader in college courses, one of the primary reasons for assigning a sustainability-themed reader to a composition class is that it lends particularly well to showcasing the various aspects of sustainability.

In addition, using a sustainability reader in a composition class taught at an institute of technology makes it easier for students with different majors both to target sustainability problems in their own fields and to get an overall picture of the grim situation caused by unsustainable practices. In this respect, my contribution to ecocomposition has mainly to do with shifting the focus from the importance of acquiring critical thinking skills in a composition class to integrating them and infusing responsibility into a practical, problem-solving process. I believe firmly that it is imperative to offer more composition and communication classes focused on sustainability to engineering students if we truly want to address the insidious equivalency between greening the economy and greenwashing marketing tools. Ideally, engineering students working toward landing a well-paying job with a resource-wasting corporation will be better equipped to resist depletion at all costs. Early exposure to sustainability problems within their chosen fields of study will instill in them the need to make sustainability a priority. Tentatively, an engineer with a good understanding of sustainability problems will be a more competitive professional in a green economy whose main driving force will no longer be profit-making but sustainable development. Similarly, when found in a leadership position, professionals with a solid understanding of sustainability, as well as strong communication skills, will be better at convincing their partners while, at the same time, motivating those working for them with regard to the beneficial adoption of green practices.

THE ARGUMENT FOR A FYC COURSE BASED ON SUSTAINABILITY

Given the magnitude of the problems facing our global community, my intention, from the time that I arrived at Georgia Tech, was to use the composition class as a fertile, discussion-based ground to imagine some possible technical solutions to those problems. I, therefore, continued my research in sustainability with a focus on pedagogy, and my experience with teaching writing and communication classes included inviting engineering and other students to identify and propose solutions to problems of sustainability and resource waste. Thus, my students interviewed representatives from the land and management office to address the irrigation system on campus and the ruts caused by students taking shortcuts on campus lawns; they interviewed their chemistry professor to find out how biohazard waste is properly disposed; and they drafted white papers proposing the installation of roof-top gardens or solar panels on some of the buildings on campus. Students realized that analyzing sustainability in their personal and professional lives served a purpose that went beyond the requirements of a writing and composition class: they learned critical thinking skills in order to serve the public good by providing solutions to technical, economic, and social problems.

From the start, this is also what I had hoped to achieve by proposing the topic of sustainability for a composition class. Even though many students had prior knowledge of environmentalism and climate change, they were still interested to examine how discourse and communication frame debates about sustainability, resource waste, and the green economy in academia, in their personal and professional lives, and in social media. I knew that, coming from diverse national, political, and socioeconomic backgrounds, Georgia Tech students were in a perfect situation to assess, both critically and practically, underlining issues dealing with unsustainable practices.

Furthermore, it was my hope that my course would show other instructors both how to provide practical and meaningful contexts for the students to engage with sustainability early on, as the 2015 QEP recommends, and how to integrate the topic of sustainability into teaching multimodal communication by helping students to become better communicators and collaborators, which is what English composition teachers are trained to do. Although I should take pride in teaching a course on a topic shortly before it was adopted institute-wide, I also feel I should have done better than merely anticipating the implications of teaching a given subject matter. Last but not least, the benefits of teaching a sustainability-focused composition course to engineering students include not only that they will become better professionals in their field but also that they will develop the communication skills needed to argue for, and propose new solutions to, sustainability problems in a convincing way. All major assignments had built into them a process that motivated students to collaborate (online and in person) with each other and to keep track of each other's work. For instance, each student had to write, in blog form, a project proposal detailing their composition process and a reflection assessing their own work. In short, with the exception of the final drafts of their projects, which were turned in separately, the students were exposed to each other's thoughts and ideas.

This was significant because it allowed students to converse using various platforms and, in doing so, to test publically their own ideas regarding sustainability. Not only could they keep track of their own work, but they also could witness their colleagues' investment in, and involvement with, similar topics. For instance, most of them learned to appreciate the practical aspects of the task at hand and started to pay more attention to how they could make their lives more sustainable. They analyzed their personal waste and looked for alternatives to create a more efficient campus recycling system. They looked at the campus transportation system, mapped out the bus routes and bike lanes, and made the case for the most sustainable means of transportation, whether by bus, by bike, or on foot. And they tested the campus meal plan, together with its alternatives (ordering in, cooking, or eating off campus) to see what was more sustainable. Because they could read each other's proposals, first drafts, and reflections, they also could reflect on similarities and differences between their work and that of their peers. Both individually and collectively, the students were in this way experimenting with composing argumentative pieces about sustainability. They took the time to research the topic and to role-play solutions to some of the sustainability issues that they had discovered in their personal and would-be professional lives.

Here then is an overview of my efforts to propose, develop, and teach a course that introduced the approach of sustainability to the first-year writing program at Georgia Tech. Doing so made me start to understand the needs and challenges of humanistic undergraduate instruction at Georgia Tech, and I became familiar with the institution's mission and processes.

MY COURSE DESCRIPTION

Following the requirements of the guidelines of Georgia Tech's Writing and Communication Program, my course description reads as follows:

The Rhetoric of Sustainability addresses the nature of multimodality by analyzing the elements of the so-called WOVEN approach to communication: Written, Oral, Visual, Electronic, and Nonverbal in relation to sustainability. This course develops analytical reading and writing skills through the investigation of methods used in cultural and literary studies and the application of those methods to specific artifacts. In a sequence of three major assignments, including a rhetorical analysis, a visual artifact, and a group project, students analyze their personal sustainable practices; they create visual artifacts documenting the sustainability of their chosen field of study; and, finally, they work in groups to develop multimedia projects where they propose improvements that would make Georgia Tech's campus and community more sustainable.

The required readings in the course included chapters from Georgia Tech's custom-made composition manual *WOVENText* (multimodal synergy, argumentation, research, plagiarism, working with visuals, etc.), together with select chapters from the followings three books: (1) Colin Beavan's yearlong project *No Impact Man*—which is, in turn, a living experiment, a blog, a book, a documentary film, and a publicity stunt about a family trying to live off the grid in New York City (no waste, no electricity or public transportation, eating only local food and buying only used things); (2) in Fall 2013, Evgeny Morozov's critique of technology and IT, *To Save Everything, Click Here: The Folly of Technological Solutionism*; and (3) in Fall 2014 and Spring 2015, Carl G. Herndl's *Sustainability, A Reader for Writers*, which takes on key issues, including climate change, food, water, soil, energy and resource management, as well as trash.

Since I already presented Herndl's reader above, here are my brief considerations on how I used the two books by Beavan and Morozov, together with provisional thoughts on how rhetoric, process, and multimodality pertain to the teaching of sustainability. During my first two semesters of teaching English 1101 courses at Georgia Tech, I assigned Colin Beavan's book *No Impact Man: The Adventures of a Guilty Liberal Who Attempts to Save the Planet and the Discoveries He Makes about Himself and Our Way of Life in the Process*. As the title suggests, the book narrates the yearlong experiment that the author and his family (his wife, their two-year-old daughter, and the family's dog) conducted in order to lead a sustainable life in New York City. They ate only food that they cooked themselves; they stopped using any means of public transportation; and they got rid of their TV set and started to spend more time with each other and with their friends. The book pertains very well to a discussion of sustainability in that it offers students a gradual and personal approach to small changes that they can make in order for their lives to become more sustainable. Easy to read, the experiment is also easy to follow. While it does not necessarily offer technical and scientific solutions to larger and more stringent sustainability problems, the book does a good job of showcasing the problems.

The book thus exposes engineering students to ideas that they might find beneficial later on in their professional careers. For instance, when Beavan compares the pluses and minuses of urban public transportation, students become interested in finding practical solutions to issues such as congested traffic, unreliable and overcrowded buses and trains, ride-sharing, or bike-sharing programs. More importantly, Beavan's experiment has a built-in composition and multimodal approach in that its record started out as a blog, which then became a book, which then became a video documentary. As they read the book, consult the blog, and/or watch the documentary, students learn about the importance of developing ideas and how the same subject matter is presented using different modes and media. They also discern Beavan's shifting urban identity as he finds ways to combat climate change, interacts with urban farmers, and tries to raise awareness with regard to the benefits of a waste-free, low-tech, and slow-paced life in Manhattan.

Another book that I adopted a few semesters later was Evgeny Morozov's *To Save Everything, Click Here: The Folly of Technological Solutionism.* I felt Morozov's skepticism regarding our overreliance on technology to be a perfect counterpart to Beavan's enthusiastic DIY approach to sustainability. Without specifically addressing Beavan's tentative withdrawal from a consumerist society in his book, Morozov challenges the solutionist mantra of IT gurus and venture capitalists. To Beavan's critique of industry's push to manufacture demand, Morozov adds Silicon Valley's tendency to manufacture solutions to (mostly) trivial problems. Confronted with these two extremes, the students were then able to judge for themselves the extent of their own involvement with sustainability, approaching the issues from a communication perspective but with personal and professional applications in mind. Ultimately, students learned that we should not rely exclusively on technology to solve the waste problem and that we should do our best to live a less wasteful and more sustainable life.

REFLECTING ON THE VALUE OF THE COURSE

In the English 1101 courses (Multimodal Engineering, Sustainable Engineering, The Rhetoric of Waste and Sustainability, Solutionism and the Rhetoric of Sustainability) that I taught for five semesters, I invited my students to explore waste and sustainability as reflections of who they are, both as individuals and as members of larger communities. Students used multimodal communication as a tool to identify and propose solutions to problems of sustainability and resource waste. Students worked individually and in groups to document and, as much as possible, reduce waste from their personal lives while at the same time to imagine and work hard toward achieving a more sustainable profession in the future. They learned that technology is both the problem and the solution: while it can solve many sustainability problems and help to reduce waste, it also comes at a price that we might not be able or willing to pay. Most of them were impressed by the prospects of a more sustainable personal and professional life. They became more determined to make sustainability one of their priorities.

More importantly, they were intrigued by the fact that oftentimes, as engineers, they would be tasked to design faster, rather than more sustainable, cars, for instance. In this respect, tentatively bridging the discursive gap between technological advancements and sustainable practices became one of the goals of my classes. Rather than being pressured, shamed, or guilted into adapting to an economically efficient but resource wasteful industry by a status quo that disregarded sustainability, my students were challenged and rewarded by at least imagining and championing sustainable technologies. In the end, the rhetoric of waste and sustainability became a validation of engineering, technology, and problem-solving. Doing more with less, designing more efficient engines, creating more sustainable systems, cutting down costs, and saving money, time, and resources were no longer empty signifiers but crucial tools meant to help students become both better communicators and better professionals in their fields of study. For example, students improved their writing skills, as well as their presentation skills, when they were asked to visually lay out their ideas on a poster. They had to consider how the placement of facts, claims, and details best challenged the audience's choice of conserving resources by showcasing the ability of technology to reduce waste. As a case in point, they analyzed the pros and cons of designing and posting notes next to light switches versus the installation of motion sensors in order to assess the more efficient and sustainable way to reduce electricity consumption. In doing so, they also discovered the bigger picture: while personal habits (such as turning off the lights when leaving a room) and technology (using motion sensors to reduce electricity use) contribute significantly to cutting down resource waste, they pale in comparison to the efforts needed to drastically create a more sustainable energy grid. Thus, they learned that creating such a grid requires a collective effort, which, in turn, implies policy change, technology advances, and political will to stop relying on fossil fuels in generating electricity and to switch to renewables.

In particular, students majoring in industrial and systems engineering are quick to realize the relationship between waste and inefficient systems. Many of the products that industrial engineers strive to optimize include people, money, knowledge, information, equipment, energy, and materials. Some of the problems that industrial engineers deal with are efficiency, profitability, and improving the quality of products. For instance, industrial engineers find ways to minimize wasted material when creating layout patterns for clothing; they determine the best locations for new fire stations, and they design the admissions procedure at hospitals; they also regulate systems for shipments, evaluate the effects of adding robots in a manufacturing line, and design the flow of materials to the assembly line in a car factory. In short, industrial engineers collaborate with other engineers and specialists in various fields to get the work "done" most efficiently. In a sense, then, students learn from one another that they need each other's future expertise in order to get the work "done", and, in doing so, they not only make instruction more efficient but also are more prepared to tackle complex sustainability problems.

Mechanical engineering is often called the jack-of-all trades among the engineering professions. It requires education and skills that span a broad range of technical, social, environmental, and economic problems. Mechanical engineers focus on product design, research and development, manufacturing, systems management, energy, and marketing. Mechanical engineers deal with the generation, distribution, and use of energy. They are also involved with the processing of materials and the control and automation of manufacturing systems. Last but not least, they work to maximize efficiency with minimal waste, thus relating to the topic of sustainability.

Biomedical engineers focus on sustainable medicine and the idea that humans are a part of a whole that includes the entire global ecosystem. As students often realize, if we do not work to be sustainable, then there will be fewer people left for whom to provide healthcare. Therefore, when disease occurs, sustainable medicine gathers information from the whole ecosystem, looking for patterns and cycles, not just individually occurring symptoms. Similarly, chemical engineers are equipped and determined to solve sustainability problems. They try to eliminate all products that can be detrimental to the population, like any type of pollutant, or any substance that can cause or release pollution. Not only do chemical engineers find newer, sustainable practices, but they also find a better, more efficient, and more environmentally friendly way to perform old tasks. For instance, it is a chemical engineer's job to minimize fuel consumption.

These examples demonstrate my passion for developing new sustainability initiatives that involve students and for instilling in them excellent communication and teamwork skills. In short, it is my hope that, after taking a composition class focused on sustainability, students are better communicators and more informed professionals. A great example of their improved communication skills is given by their use of creative titles, such as "Debugging the Problem," "Coding the Environment," and "Will Plastic's Problem Degrade Away?" More importantly, though, with a focus on sustainability, they have a better grasp of their major, finding out, for instance, that when we buy toilet paper bleached with chlorine, we are unknowingly supporting pulp and paper mills that pollute our environment with dioxin.

These examples and countless more allow me to write with confidence about how my teaching and understanding of sustainability have been challenged and significantly improved for the past few semesters when overseeing and managing my students' group projects. That is why I think that a composition course focused on sustainability will benefit both the students taking it and the instructor teaching it. For instance, the students will be better prepared to tackle technical efficiency problems in the real world if they start looking for solutions early in their university career. As future engineers, administrators, or concerned citizens, they will be in a better position to promote and implement more sustainable standards, policies, and practices.

While I am grateful for the opportunity given to me to teach a topic of my choosing, I couldn't have done it without the help of the students whom I had the good fortune to teach and learn from. Here, then, is some information about who they were and where they were coming from. Rather than having a statistical breakdown of the students, I would prefer to present here a composite image. I am confident that my impressions are consistent with the impressions that my students began to have about one another as they started to become more comfortable both with the topic of the class and with their new surroundings as first-year students.

First and foremost, the students entered class being apprehensive with regard to the expectations of an English course. Needless to say, for most of them, English as a subject was not one of their priorities, nor was it necessarily one of their strong suits. This much can be said about sustainability, as well. They more or less agreed that they needed to have optimal critical thinking skills and be good communicators in order to have a successful academic career. They also agreed we should all live more sustainable lives, though they couldn't see how they, as seemingly powerless students, could do anything to change that. Slowly but surely, with the help of prompts such as the rhetorical analysis of the last item that they had thrown away in the trash bin, the students started to interact with one another and to realize the importance of sustainability both as a topic in a composition class and as a factor in their future profession. A seemingly trivial piece of waste like a banana peel often times launched us into fruitful conversations. It showed us the variants of English (American, British, international) present in the classroom since what is commonly called a "peel" in the United States is called a "skin" in the United Kingdom. It revealed who among us was familiar with composting (some students were coming from households in which a banana peel didn't end up in the trash but in the compost bin). It made us investigate how many industries were involved in the production, transportation, marketing, and selling of bananas. And, last but not least, it showcased diversity, as well as commonality of purpose because the students in my classes were a good sample of future professionals charged to make our lives more sustainable.

Fortunately, my last semester of teaching a FYC course based on sustainability at Georgia Tech in Spring 2015 coincided with a momentum wherein the clean-energy economy almost hit its stride: for the first time, the price of alternative fuels was comparable to that of fossil fuels. Although there are still many hurdles to pass, such as creating new supply grids and storing facilities for alternative energy, the change seems to be irreversible. To its merit, Georgia Tech decided to confront this tooth and nail with the recent adoption of the QEP and the promoting of so-called Serve-Learn-Sustain (SLS) courses. As a matter of fact, the inaugural SLS courses were offered starting in Spring 2016. Among the many offerings, two courses offered within the honors program stand out for being team-taught and interdisciplinary, according to the SLS webpage as of March 20, 2016. The first course, Sustainable Community & Systems (GT 2803 HP1), will explore sustainability from a system dynamics perspective, looking at issues like climate change, energy, and forestry, and the role of economic incentives, regulation, and public participation as they relate to communities in Atlanta and around the world. The second one, Sustainable Community Principles (GT 2803 HP2), will explore community connections to sustainability, focusing on methods for working with community partners—like asset mapping, participatory design, and ethnography—around issues including water, food, and education.

Even though these two courses count toward free electives, the program coordinators have high expectations and hope to be able to offer a certificate and/or a minor in this area of concentration. What is interesting, from a programmatic point of view, typical for the top-down corporate approach of dealing with sustainability, is not only the fact that these courses are offered under the guise of the Honors Program but also the division of labor that the coordinators assign to various majors. Thus, the first course has a top-down, bring-sustainability-to-community approach and is taught by instructors coming from business, public policy, earth and atmospheric sciences, and chemical engineering. The second course investigates the sustainable demands of community and focuses on bottom-up change, and instructors come from architecture, computer science, interactive computing, mechanical engineering/industrial design, and public policy. Although public policy instructors are the only ones present in both camps, unintentionally perhaps, it seems that rather than being the sole responsibility of one discipline or another, addressing sustainability in an academic setting should be the province of a top-down and bottom-up, interdisciplinary, collective effort. However, while there is no guarantee that students with a grasp of sustainability will make a change for the better when in leadership positions, there is still hope that they will be, at least, better informed and will, therefore, use information to the best of their knowledge.

NOTES

1. "History," *Serve-Learn-Sustain*, Georgia Tech, accessed March 20, 2016, http://serve-learn-sustain.gatech.edu/history/.

2. "Faculty Profile: G. Wayne Clough," *School of Civil and Environmental Engineering*, Georgia Tech, accessed March 20, 2016, http://ce.gatech.edu/people/Faculty/6425/overview/.

3. For more on sustainable communication, see Tema Milstein, "Green Communication," in *Greening the Academy: Ecopedagogy Through the Liberal Arts*, eds. Samuel Day Fassbinder, Anthony J. Nocella II, and Richard Kahn (Rotterdam, The Netherlands: Sense Publishers, 2012), 161–74.

4. Bryan G. Norton, "Pragmatism, Adaptive Management, and Sustainability," *Environmental Values* 8, no. 4 (1999): 461.

5. Bryan G. Norton, "Improving Ecological Communication: The Role of Ecologists in Environmental Policy Formation," *Ecological Applications* 8, no. 2 (1998): 351.

BIBLIOGRAPHY

Bartels, Kirsten Allen, and Kelly A. Parker, eds. *Teaching Sustainability, Teaching Sustainably*. Sterling, VA: Stylus Pub., 2012.

Beavan, Colin. *No Impact Man: The Adventures of a Guilty Liberal Who Attempts to Save the Planet and the Discoveries He Makes About Himself and Our Way of Life in the Process*. New York: Farrar, Straus and Giroux, 2009.

Cox, Robert. *Environmental Communication and the Public Sphere*. Thousand Oaks, CA: Sage, 2009.

———."Nature's 'Crisis Disciplines': Does Environmental Communication Have an Ethical Duty?" *Environmental Communication: A Journal of Culture and Nature* 1, no. 1 (2007): 5–20.

"Faculty Profile: G. Wayne Clough." *School of Civil and Environmental Engineering*. Georgia Tech. Accessed March 20, 2016. http://ce.gatech.edu/people/Faculty/6425/overview/.

Herndl, Carl George, and Stuart C. Brown, eds. *Green Culture: Environmental Rhetoric in Contemporary America*. Madison: University of Wisconsin Press, 1996.

Herndl, Carl. *Sustainability, A Reader for Writers*. Oxford: Oxford University Press, 2013.

"History." *Serve-Learn-Sustain*. Georgia Tech. Accessed March 20, 2016. http://serve-learn-sustain.gatech.edu/history/.

Lioi, Anthony. *transFORMATIONS: The Journal of Inclusive Scholarship and Pedagogy* 21, no. 1 (2010): 14–22.

Milstein, Tema. "Green Communication." In *Greening the Academy: Ecopedagogy Through the Liberal Arts*, edited by Samuel Day Fassbinder, Anthony J. Nocella II, and Richard Kahn, 161–74. Rotterdam, The Netherlands: Sense Publishers, 2012.

Morozov, Evgeny. *To Save Everything, Click Here: The Folly of Technological Solutionism.* New York: PublicAffairs, 2013.

Norton, Bryan G. "Improving Ecological Communication: The Role of Ecologists in Environmental Policy Formation." *Ecological Applications* 8, no. 2 (1998): 350–64.

———. "Pragmatism, Adaptive Management, and Sustainability." *Environmental Values* 8, no. 4 (1999): 451–66.

Weisser, Christian R. *Sustainability: A Bedford Spotlight Reader.* New York: Bedford/St. Martin's, 2015.

Weisser, Christian R., and Sidney I. Dobrin, eds. *Ecocomposition: Theoretical and Pedagogical Approaches.* Albany: State University of New York Press, 2001.

Part II

IN SITU SUSTAINABLE URBANISM

Chapter 5

In a Fragile Context

Exploring the Challenges of Informal Sector Urban Gardening in Rio de Janeiro, Brazil

Lea Rekow

A lack of imaginative and creative policy involvement with the informal populations of the world's cities and the urban spaces in which they live represents one of the most complex crises of the twenty-first century. In the city of Rio de Janeiro, the residents of the densely populated inner-city favelas, as shown in figure 5.1, have literally cemented over entire mountainsides. Yet, though they blanket large swaths of the city, they continue to receive little institutional infrastructure investment or support.

Despite historic state abandonment and neglect, favela communities[1] have developed their own networks and governance structures for the hundreds of informal "cities" that they have independently built throughout the larger metropolitan region. Their infrastructures include residential and commercial districts with transport systems, residents' associations, banking systems, mail delivery services, day care centers, telecommunications companies, utilities services, and third sector organizations. They exist in labyrinths of invisible networks, shadow economies, and social exchanges that play out in

Figure 5.1 Favela housing, Rio de Janeiro. *Source*: Photo by the author, 2011.

a multitude of incidental arrangements of people and things that cross over between the regulated and the illicit, the private and the public, the loud, the lively, and the chaotic. Their residents have the adaptable capacity to navigate through extreme difficulties to build material configurations that meet real-world problems with the use of very few resources. This tenacity offers flexible opportunities to envisage how ecologically penurious urban spaces can be impacted through creative configurations of networked action. Therefore, it could be assumed that this type of DIY city-making would provide fertile ground for establishing micro-gardens and other sustainable "green" initiatives. However, an absence of capital resources, deficient health and sanitation services, a lack of available land (and land tenure security), and the many complexities associated with living in poverty, sociopolitical instability, and armed conflict that the people in Rio's irregular settlements encounter day-to-day preclude the concept of green and cared-for public space from being perceived as viable—that is, from holding any long-term (or even medium-term) option value (the concept that space will continue to remain available for such use in the future).

Despite the difficulties in creating sustainable green spaces inside favelas, these endeavors do, however, provide a practical opportunity for exploring how better practices could be developed under such difficult circumstances and what the obstacles are which currently prohibit them. Thus, this essay seeks to articulate the gap between the theoretical potential and the harsh reality of cultivating green public or semi-public space in favelas by articulating recent on-the-ground actions involving two different approaches to establishing gardens. It illustrates how these projects were able to emerge in resource-scarce, urban contexts where people are living under extreme environmental, social, and economic stress. It also delineates how various forms of violence, intervention, conflict, and abandonment have impacted the viability of such projects and the communities in which they are implemented. Lastly, this essay illustrates two very different approaches toward gardening in Rio's favelas. It demonstrates how each project navigates the lower rungs of the "participatory ladder"[2] to provide a better understanding of how any form of social action in Rio's favelas is deeply political and determined by access (or lack thereof) to a range of rights, including the right to the city and the rights associated with multiple pillars of human security.

Interpreting the value of "participation" within each project can be determined by how groups of actors hold more or less power. According to Sarah White,[3] actors at the "top" and "bottom," are organized by different perceptions, interests, and goals that can be characterized as belonging to one or more of four systems of participation: (1) nominal participation, whereby the more powerful actors at the "top" use the language of participation to legitimize development, and whereby less powerful people are involved in projects through a desire for inclusion. White insists this is little more than theater and does not result in social change; (2) instrumental participation, whereby community participation is used as a means to achieve a stated end, often through the exchange of skills and knowledge at the community level, brought about during a project's implementation; (3) representative participation, whereby community members have a voice in the decision-making and implementation process of projects or policies that affect them, which increases the likelihood of intervention being sustainable for more powerful actors, and which may increase leverage for less powerful actors; and (4) transformative participation, whereby the empowerment of those involved results in the transformation of power structures, relations, and institutions and ultimately leads to closing marginalization and exclusion gaps.[4]

This essay both summarizes the conditions in favelas, and analyzes the community gardening efforts of these case studies in relation to the real-world challenges and limitations of participatory engagement. What it offers, by way of social significance, is a lens through which to understand in what form, and to what degree, productive green spaces can be cultivated to provide food, tranquility, and stress relief to people living in the shadow of conflict and instability, as well as what stands in their way, and what aids in their collapse. This holds value for human-centered city-making practitioners from a range of disciplines who operate in or alongside local government agencies or civic and third sector organizations, especially inside informal and fragile contexts. This essay details some of the socioeconomic, spatial, and political realities that cannot be avoided when engaging with urban agriculture or land use restoration projects within these contexts and intends to give the reader a clear idea of what it means to go into a favela and achieve what would typically be, in the Global North, the relatively straightforward task of establishing a community garden or urban farm.

BACKGROUND

Rio de Janeiro is the second largest city in Brazil, with a municipal population of 6.5 million people (and a metropolitan population of 12 million), of whom approximately 27% live in 763 densely populated, irregular, and informal settlements that grip to the fringe of industry and capital.[5] It has one of the greatest disparities of wealth in the world.[6] Rio's favelas develop like coral reefs, informally and organically clustering throughout the city in self-reliant and frantic constellations, wild and intense in their spatial, social, economic, and governance structures and networks.

In a country with one of the highest wealth disparities in the world, where half of the people survive on only 10% of the income, and more than a quarter live below the federal poverty line, it is hardly surprising that so many people live in substandard housing under severe environmental, economic, and social strain, with little access to public facilities, utilities, infrastructure, or services. Yet anywhere from 17 to 40% of the country's GDP may be generated by Brazil's informal sector through its shadow economies.[7] Clearly, therefore, the people living in favelas constitute a significant part of the socioeconomic landscape of Brazil and are an important sector for the government to recognize.

Still, the State has a historic legacy of operating in Rio's favelas in predictable ways—through military incursion, top-down intervention, and population removal. Physical infrastructure projects are designed to profit a handful of private construction conglomerates rather than the sectors they aim to serve. Social services are usually short lived—falling prey to underfunding, institutionalized corruption, and the empty promises that accompany each electoral cycle.[8]

The State's latest floundering attempt at public security policy is its police "pacification" campaign (UPP), a military intervention tactic aimed at taking control of strategically located favela territory under the banner of the "war on drugs." However, only a fraction of Rio's favelas has been pacified under this scheme (thirty-eight in total). These are the favelas that encroach on the affluent neighborhoods of Rio's North Zone, that have been co-opted to satisfy the ongoing flurry of land grabs, or that were bulldozed, gentrified, and/or redeveloped in order for Rio to re-craft itself as a mega-events host city, which ultimately culminated with the 2016 Olympics.

Under pacification, fluctuating levels of gun violence between rival drug trafficking gangs, between factions within gangs, between gangs and militias,[9] and between gangs or militias and police have created a complex and shifting territorial battlefield where favela residents pay the price.[10] Not only is this logic of armed intervention unsustainable, but also the drug trade continues to flourish as issues of public health and safety, sanitation, education, and the provision of decent socioeconomic opportunity continue to be ignored. Neither have these informal communities been emancipated through the city's sustainable development agenda,[11] which has blazed a neoliberal trail of accumulation by dispossession through the favelas of Rio de Janeiro.[12]

RESEARCH PROPOSITION

Given the difficult, multidimensional constraints of working in a favela, creating gardens with and for residents—and how people's lives are impacted as a result—generates many questions that deserve to be investigated. Do these endeavors lead to improving social space, or are they prone to generate conflict? What does it mean to create productive green space inside communities that resemble internal armed conflicts? Can land use reform take place in such a fragile context? How is space accessed without land tenure security to protect it? How are these spaces governed? Are the processes inclusive? In what ways should their values be measured? And can they be sustained?

SCOPE AND METHODOLOGY

To examine these concepts, fieldwork was undertaken with Green My Favela (GMF), an environmental restoration project and action research network that collaborates with favela residents and a range of public, private, and third sector organizations to establish gardens in favelas in Rio. GMF was launched in 2010 by the author, and has coproduced a dozen gardens in five favelas throughout the city, and advised the state on how to implement sustainability initiatives within this context. As such, GMF has a unique window from which to observe how socio-environmental intervention can unfold, and how it currently flounders.

The fieldwork for this study was undertaken between November 2010 and May 2016 and has yielded a wealth of qualitative information that has been correlated against a wide range of publically available statistical and scholarly reports. The two projects discussed share a variety of common markers yet differ substantially from one another. They are constituted by very different partnerships and levels of support and are defined by very different working methods, results, and impacts. One project is located in the Rocinha favela, the other in Manguinhos. Each reveals a unique set of problems and challenges that arise from distinct circumstances relating to different types of intervention, and each exposes the difficulties of enacting projects in large favela communities. The Rocinha and Manuinhos favelas were chosen for this case report because they are sizeable and prominent low-income communities, because they illustrate important similarities and differences in Rio's favela landscape, and because they are where key projects have been created.

INTRODUCTION TO THE CASE REPORTS

Rocinha, shown in figure 5.2, is generally considered one of the largest favelas in Brazil. It is located in Rio's South Zone, between two of the city's wealthiest suburbs. Its exact population is unknown, and estimates vary wildly, putting numbers anywhere between 69,000[13] and 250,000, athough the reality is likely midway in-between. Rocinha exists on less than a square kilometer of land, as a cascade of concrete and brick that spills down a steep

Figure 5.2 Detail of Rocinha favela. *Source*: Photo by the author, 2011.

mountainside to slam up against one of Rio's major arterial highways below. For decades preceding pacification, the favela was under the armed control of drug traffickers—first the Cammando Vermelhos (CVs), then later the Amigos Dos Amigos (ADA) after a bloody takeover in the mid-2000s. It was pacified by the state in late 2011 and is currently occupied by 700 UPP troops, although the ADA still has substantial control over the socioeconomic landscape as trafficking continues largely unabated.

The Manguinhos Complex, shown in figure 5.3, in geographical contrast to Rocinha, is built on flat terrain. It is a large and expansive cluster of ten to fifteen irregular neighborhoods, occupations, and government relocation projects located in Rio's poor North Zone, in what used to be a sprawling industrial area. Its population has been estimated to be between 35,000 and 70,000, although most reports put the population at around 50,000. It was previously a stronghold (and later headquarters) of the CV drug trafficking gang and was pacified in late 2012. It is currently occupied by 588 UPP troops, although the CVs remain in command of much of the socioeconomic landscape in this favela, as well.

The task of building gardens in these densely populated, urban favelas is difficult to accomplish and even more difficult to sustain, especially under conditions that are exasperated by armed conflict. The repressive

Figure 5.3 Detail of the Manguinhos favela complex. *Source*: Photo by the author, 2012.

sociopolitical climate that generates a culture of fear and terror perpetuates the structural violence of poverty and social exclusion, making any form of sustainable social action, especially one that involves the development of public space, a risky and difficult endeavor.

The urban, social, cultural, economic, historical, spatial, and political complexities that define how these projects unfold are multifaceted. At their most superlative, these projects bind social function to environmental restoration in novel ways to link disconnected social and economic threads to benefit favela communities. Yet their physical space cannot be separated from the ongoing production of volatile sociopolitical relations. These projects must directly confront issues of self-determination, identity, violence, and colonial oppression—all of which brutally impact those who live within favela boundaries.

CONTEXT

These gardens are a world away from the organic food networks and community gardens that develop in the Global North.[14] The chronic buildup of garbage, chemical contamination, and the density of the built environment have reduced the functional quality of the soil to the point that it has been lost in many parts of the favelas. Erosion and flash flooding, frustrated by the overuse of cement, have led to housing collapses and the abandonment of land, and these, in turn, have led to more instability, a larger development footprint, and an increase in urban density in places with already tightly packed substandard housing.

Enacting equitable and inclusive operational strategies to sustain remediated favela environments through social mobilization is not an easy task. The complexities of economic exclusion and armed conflict present pressing priorities that drive day-to-day choices and concerns of residents. Therefore, those who work to establish or advocate for gardens in these territories need to be able to not only articulate their social value but also evaluate their limitations and assess how conflict and risk may be not only reduced but also generated by them. Thus the premise for this work is conceptually based on reimagining better ways of being located in places subject to conflict without the ideology of sustainable social transformation attached.

There are many studies concerning the social, economic, and therapeutic benefits of gardening, as well as the practical knowledge exchange and collective action in informal settlements, yet few examine the challenges of operating in internal armed conflicts.[15] In order to stimulate collaborations that open up new opportunities for people, materials, ideas, and organizational structures to form and function, at least temporarily, it is imperative that space (and place) be read in a multiplicity of ways. What the production of these gardens offers most is the chance: to illustrate how it is possible to create useful space from land that has long been prone to abuse and neglect; to examine under what circumstances these projects flourish and collapse; and to convey the reality of producing green space within these contexts. Though not without failures and risks, these projects aim to convey what it means to try to connect integrated levels of stakeholder action to the reclamation of public space in order to cultivate productivity from derelict land inside some of the most crowded and fragile informal settlements in all of Brazil.

ORGANIZATIONAL MODELS

There is, at most, a handful of organizational models that produce gardens in Rio's favelas. GMF and Hortas Cariocas (HC) are the two models that are examined in this essay. Both aim to improve the physical quality of underutilized and degraded lands, mitigate erosion, provide better drainage, and cultivate productive and appealing public space with, and for, the benefit of the informal sector in practical ways. Both attempt: to strengthen informal frameworks; to build skills-sharing exchanges; to work with municipal policy through collaborative action; to foster alternative socioeconomic models; and to create land use reclamation templates that can be used to promote civic mobilization in resource-scarce environments.

GMF works with various individuals, NGOs, residents' associations, and state and municipal entities. It also acts as a mediator between on-the-ground local needs and top-down policy initiatives, and has advised NGOs and government on how to meet some of the basic critical needs of communities, such as how to orchestrate trash removal and consolidation, how to implement recycling programs, how to develop food and water security initiatives, how to reclaim degraded space, and how to build gardens in different contexts. GMF projects begin either by invitation or by identifying potential spaces for gardens and talking with residents to ascertain interest and find working partners. GMF then brings in the soil, organic seeds, tools, compost, worm bins, water tanks, outside volunteer labor, and so forth, to work with residents to clean spaces of trash, build infrastructure, and prepare spaces for planting. The organization helps residents to plant and grow food, make worm bins or compost, and build retaining walls, diversion channels, biodigesters, bioremediation systems, and compost. It continues to provide labor and resources (tools and equipment, soil, compost, seeds, fencing, shade material, workshops, etc.) to residents once a space has been established if the residents want or need GMF's continued involvement.

CASE REPORT 1: ROCINHA

GMF's work in Rocinha began at the same time as military pacification was taking place. Despite having less than 5% of the city's total population, this favela accounts for around a third of Rio's overall population growth over the last ten years, so space is at a premium. Shown in Figure 5.4, its compact and congested warrens of cement labyrinths are sandwiched between two of the city's most affluent neighborhoods, São Conrado and Gávea. It falls toward the bottom of the Municipal Human Development Index (IDHM)—the sixth lowest out of 126 regions in the city.[16] While Gávea presents the same IDHM as Canada, Rocinha has the same IDH as the Republic of Ghana. Rocinha's three municipal public schools, where 17.1% of children between the ages of six and fourteen have already dropped out,[17] illustrate the need for alternative forms and places for children to learn and feel safe. During times of gunfire exchange, even the favela's one emergency health clinic has come under fire. Its façade is riddled with bullets, signifying how volatile the security landscape here can be.

Figure 5.4 Breaking a hole in the wall to access the space to begin work on the Rocinha Mais Verde garden in Rocinha. *Source*: Photo by the author, 2011.

Though Rocinha residents maintain their lively resilience even under these conditions, the state military takeover of the favela added to the already stressed conditions to make creating a garden under these circumstances even more difficult. Shown in figure 5.5, the Rocinha Mais Verde project took shape in September 2011, shortly before the UPP occupation, when the ADA gang was still in visible armed control of Rocinha. The garden was developed as a collaboration between GMF and the late Lino dos Santos Filho, affectionately known as Tio (Uncle) Lino. Tio Lino was a local community leader who worked for more than twenty-five years with children to discourage them from joining gangs by making art out of recycled waste materials. Tio's efforts changed the lives of hundreds of young people and children in Rocinha.

After a series of conversations with GMF, Tio Lino gained access to a steep, vacant lot of land belonging to the Alegria das Crianças crèche, located next to his art school. The fifty-square-meter area had been used for years as a garbage dump by local residents. At least two meters of trash had built up on top of the site, and another two meters had been compressed into the soil. The space is located in an area known as the Valão (literally meaning ditch and referring to the open sewerage channels), which is one of the poorest

Figure 5.5 Groundbreaking at the Rocinha garden. *Source*: Photo by the author, January 2012.

neighborhoods, situated at the bottom of Rocinha, where the bulk of the open sewers flow. Tio sought approvals from the local drug boss and other community stakeholders to move forward with the project, but in November of 2011, shortly before groundbreaking began, state security troops descended upon the favela in a surprise tactical invasion. All entry and exit points were cut off, and Rocinha was isolated for seventy-two hours. Approximately 3,000 soldiers from different Special Ops units and the armed forces rolled in on foot and in tanks, with police horses, dogs, and helicopters. After decades of gang rule, this sudden change was very difficult and dangerous for residents to negotiate.

In January of 2012—amidst the chaos and violence of military pacification—fourteen GMF volunteers, the children from Tio Lino's art school, and some local residents began remediating the space that was to become Rocinha's first community garden. First, the trash was removed, literally tons of it, manually by taking it out through the steep winding alleyways in wheel barrows. Materials that could be repurposed were used to build retaining walls to create a series of terraces, steps, pathways, and garden beds. The remaining trash was sifted from the top meter of soil and tested for heavy metals and nutrient value and to determine its pH level. Fresh topsoil, mulch, and organic fertilizer were then added to the beds. An old water supply that had been filled with trash was cleaned and sealed, and pipes were laid to a safe water supply and connected to a simple homemade drip irrigation system crafted by one of the local residents. Over a period of months, flowers, shrubs, fruit trees, and organic edibles were planted, as shown in figure 5.6. Hands-on educational workshops were held on how to cultivate and maintain the garden. These introduced children to a range of concepts to connect them to seed-to-table ideas and plant lifecycles. These sessions included painting "still lifes" of fruits and vegetables, studying seeds, plants, and worms under microscopes and through magnifying glasses, planting seeds, transplanting starter plants into the ground, creating compost and worm bins, making take-home herb terrariums out of recycled plastic soda bottles, tasting and smelling fruits, herbs, and vegetables, and maintaining the garden in general. GMF continued to work in the space for several months until, eventually, governance of the space transitioned entirely into the hands of Tio Lino and a local resident who managed the garden with the children.

The space, shown in figure 5.7, grew more stunningly beautiful with each passing month, flourishing with medicinal herbs and functioning as

Figure 5.6 Rocinha Mais Verde planting day. *Source*: Photo by the author, April 2012.

Figure 5.7 Rocinha Mais Verde. *Source*: Photo by the author, May 2012.

a therapeutic learning space for the children. In June of 2012, the United Nations Rio+20 Summit on Sustainable Development showcased the garden as an exceptional project, and it was awarded United Nations Development Programme funding for the work to continue. In May of 2014, however, Tio Lino, aged 63, passed away due to health complications related to diabetes. With Tio's death, a great social void opened up. Conflict between community stakeholders began arising over control of the space, and the project has since deteriorated and remains in peril.

Analysis of Case Report 1

The situation with the garden in Rocinha demonstrates many difficulties that relate to the longevity of such projects. The disarray that the project now finds itself in reflects a level of intra-community turmoil, or at least a lack of ability to resolve internal conflict in the absence of strong community leadership. Working in conditions of poverty can elevate levels of conflict and diminish individual and group decision-making capabilities and disrupt the mechanisms of building intra-community trust.[18] Trust is a complex interpersonal and inter/intraorganizational construct that can suffer at the hands of any social, economic, and political relationship.[19] In the favelas, and elsewhere, much at the heart of any dispute is the struggle for control of territory. The relationship between people's perceptions of control over land (regardless of holding customary or formal titles to property) is symbolic of power and control over other resources and is much more complex than is conventionally assumed.[20]

Planners and other urban actors must comprehend the complex vernacular of public space in favelas if they hope to work in them. At a micro-level, people do not spend money on the exteriors of their homes. For most, their resources simply do not stretch this far; therefore, they are unlikely to extend themselves further by putting effort and resources into developing public space. Some people may not want their homes to be easily identified or likely to draw attention from police. Public space located in proximity to a drug selling point (a *boca*) brings both risk and tension with it (from both police and traffickers). The likelihood of a strategically located site becoming a control point for police or traffickers makes public spaces more dangerous. As there is little to no garbage collection, public spaces are normally used for dumping trash. As well, incidents of theft of even the most low value items in public space are to be expected. On a macro-level, the threat of territorial seizure by the State is high. Since pacification, real estate prices have spiked dramatically, which has also led to gentrification in that people are being pushed out of their homes. Forced removals and other forms of coerced relocation resulted in approximately 110,000 people being dispossessed of their homes in the pre-Olympic period because of the city's mega-events. The overwhelming majority of these people come from favelas and other lower-income sectors inside the municipal boundaries. Within the larger metropolitan region, the numbers reached as high as 800,000.[21] Therefore, developing space that may be taken away at any moment is simply not pragmatic. Even if people are able to remain in their homes or neighborhoods, most do not have the luxury to volunteer their time toward non-essential endeavors. When people are not working, many are either looking for work or caring for their families. Many also suffer from developing major health issues because of inadequate access to healthcare. This may impede their mobility or capacity to work.

Tio Lino's death brings to light this relationship between project sustainability and the health of the individual. A lack of access to satisfactory healthcare plays a major factor in the long-term viability of projects, many of which suffer when ill health befalls project custodians or participants. There are many health inequalities that are suffered by people, especially the elderly, who live in Rio's favelas. Men, on average, live 12.8 fewer years than their counterparts in formal neighborhoods. Even more shocking, for both men and women who are over sixty-five years old, healthy life expectancy in favelas is less than half when compared to Rio's wealthiest sector.[22]

It is important to understand how issues related to individual health impact project sustainability. Poor health, a scenario that the overwhelming majority of favela residents must tackle at all stages of life, is embedded in the key characteristics of living in informal conditions: degraded environmental conditions; a chronic lack of sanitation and related infrastructure; undernutrition (especially in babies)[23]; and critically deficient public health services, which force many to rely on poor-quality, informal, and unregulated health services. For example, it is not uncommon for people to be forced to treat cancer with herbal remedies because treatment and medicines via the public health system are unlikely to be available due to an overstressed and underfunded system which places the poorest at the lower rung of the receivership ladder.

An unstable and potentially dangerous public security landscape also inhibits developing public space. Though visible armed control of Rocinha passed from the ADA to the State in late 2011, the UPP for the most part still only monitors the main thoroughfares and commercial districts that are accessible by vehicle. Fluctuating levels of armed conflict and trafficking continue under the UPP, as pacification focuses on disarmament and territorial control, not on putting an end to drug trafficking per se. As a result, frequent gun violence continues to erupt between gang factions vying for control, and between gangs and the UPP. In addition, incidents of theft, domestic violence, and rape—previously prohibited under gang law—now proliferate, making public space even more volatile. Therefore, the use of public space tends to happen spontaneously according to users' needs and can dissolve again just as quickly. Thus the value of public space is seen as temporary and always in flux; hence maintaining cared-for public space is considered impractical. Consequently, social projects such as community gardens must overcome many challenges brought about by the underlying structural, economic, behavioral, environmental, and public service access barriers that influence sustainable productivity.[24] Recognizing the linkages between these issues allows for a more thorough understanding of how the multiple dimensions of urban poverty are exasperated by a low human security threshold and how projects are affected by suites of interrelated factors.

CASE REPORT 2: MANGUINHOS

In Brazil, intervention in the favelas tends to be top-down and driven to protect State interests. However, in rare cases, government-implemented social infrastructure is being designed for citizen-centered use. HC is an example of one program that delivers social infrastructure to provide assets to favela communities in multiple ways. HC is a municipal-level, organic agroecology project launched by Rio de Janeiro's Department for the Environment in

2006. Since that time, more than thirty gardens have been established in collaboration with residents in multiple favelas and underserved public schools throughout the city. The program is directed under the tenacious lead of its remarkable project director, Julio Ceasar Barros.

HC is a flexible yet structured urban agroecology program with definitive procedural strategies, guidelines, partnerships, and goals. Its purpose is to increase food security for Rio's most vulnerable sectors through establishing organic gardens that produce food without the use of any kind of agrochemicals such as mineral fertilizers or pesticides. The favela gardens are located in public space, on steep, underutilized, mistreated, or abandoned space. Many exist on land located under high-voltage transmission lines. Fifty percent of the produce is donated to school lunch programs, kindergartens, and at-risk families, as determined by local Residents' Associations. The remaining 50% is sold to generate additional income for gardeners or to purchase additional equipment and resources.

HC builds the basic infrastructure and supplies the seeds, fertilizers, soil, compost, tools, and other materials necessary for each project. Each HC beneficiary receives a stipend of R$380 per month. In addition, each project has a coordinator who is responsible for the additional administrative functions (who receives R$480 per month). Beneficiaries are expected to carry a full-time workload, though, for the most part, they have the freedom to schedule their own hours. The ultimate socioeconomic goal of the project is to achieve self-sustaining gardens over the medium-term by reaching production levels that provide consistent incomes for the gardeners through the sale of produce while still donating half to the more vulnerable sectors of the favelas. The most ambitious of all HC projects is the Manguinhos garden.

The Manguinhos Complex of favelas is located in the poorer, sprawling North Zone of Rio, just a few kilometers north of Rio's central business district and the city's famous Maracanã sports stadium. Most of the industries that functioned in Manguinhos were closed or moved to other locations in the 1980s. As a result, the district suffers from chronic unemployment with rates languishing between 30 and 50%. The favela ranks among the lowest on Rio's IDHM: at 122 of 126 regions in the city.[25] Public health issues are acute. Teenagers here are five hundred times more likely to develop cancer and neurological disabilities from exposure to lead than children and teenagers from other neighborhoods[26]; dengue is four times higher than in affluent parts of the city[27]; leptospirosis, a blood disease transferred via rats and sewer water presents a health risk after episodes of flooding; and mild undernutrition has been found to affect one in four children under five years of age.[28] Levels of violence are also extreme.

For decades, Manguinhos was governed by the city's largest and most powerful drug trafficking gang, the CVs. Hundreds of heavily armed drug

Figure 5.8 Building the infrastructure for the Manguinhos garden. *Source*: Photo by the author, May 2013.

traffickers and thousands of addicts occupied a string of shanties built on a garbage dump that functioned as Rio's largest "crackolândia." The area was known as Rio's "Gaza Strip" due to its high levels of violence and presented, for all intents and purposes, as an internal armed conflict. Open warfare, stray bullets, occasional public executions, the open disposal of bodies, public displays of every form of vice including prostitution, and the sale and consumption of crack cocaine and other drugs spilled into the streets to devastate almost every aspect of public life.

This social landscape was markedly altered when Manguinhos was pacified in late 2012. Two thousand armored marines, Special Ops troops, special action dogs, and air-sea support battalions augmented with helicopters and tanks moved in to take control. Addicts were taken away by police and interned by the Municipal Department of Social Welfare in a withdrawal facility for a few days before dispersing back to the streets. The UPP installed 600 officers in the favela. The crackolândia was bulldozed, and the CVs retreated from visible armed dominance, moving their headquarters and trafficking underground.

The infrastructure for a massive organic food garden running a kilometer in length under the high-voltage transmission lines was constructed, as shown in figure 5.8, to replace the crackolândia. As a result, the area has undergone a dramatic transformation, although the ongoing process of pacification remains extremely violent, and the intensity of drug trafficking remains moderately high.

HC received a portion of federal funds allocated to redevelop a larger area in connection to mega-events. Consequently, the program coordinated the project as part of a multi-tiered PPP involving City Hall, the Manguinhos Residents' Association,[29] the Municipal Department for the Environment, and Light, Rio's private electricity provider, who grants the legal right to use the space under the transmission lines. The city began the project by removing 450 tons of garbage and an additional half-meter of contaminated soil from the surface of the space and replacing it with gravel to assist with filtration and drainage. More than three hundred raised garden beds were built, fresh topsoil was brought in to fill them, and eight water tanks were installed and connected to the city water supply to provide an irrigation source. In addition, simple shade structures were built over some areas as temperatures easily reach almost 50°C at times, even in winter months. The Residents' Association advised HC on the hire of two-dozen workers, including ex-addicts, who were trained to cultivate and maintain the immense space.

GMF was invited to develop the northern third of the space as a volunteer community garden with retirees. This section, consisting of forty-eight ten-meter-long garden beds, became self-sustaining within months of getting off the ground and is now fully governed and maintained by volunteer residents who each cultivate their own set of garden beds and collectively maintain the space. Residents living at this northern boundary appreciate the transformation that the garden has brought about. They are no longer scared to leave their homes, and they enjoy having the garden to spend time in. They invest their own time, labor, and economic resources into the project voluntarily, with GMF continuing to augment their efforts with small donations of materials, equipment, and supplementary labor.

As of early 2016, GMF has also established a southern extension to the southeast boundary of the HC garden in Manguinhos. Working with children, residents, and international volunteers, GMF cleared an additional area around 450 square meters in size, and established a fruit orchard and medicinal garden, which is now fully managed and maintained by local residents who live adjacent to the space.

Analysis of Case Report 2

The social and economic impact from the reclamation of these physical spaces in Manguinhos is substantial. The project, shown in figure 5.9, is now generally considered to be the largest organic urban food garden in Latin America in terms of physical size. The public is free to enter at any time and to pick free food for meals, and there has, surprisingly, been very little tragedy of the commons. Residents watch over the space and respect the area for the social and use value it brings. HC employees take home between two and three bags of produce weekly to augment their income. This is worth approximately R$40 of vegetables per week that they would not otherwise be able to afford. In addition, the produce from the GMF volunteer section at the northern end of the garden supplements the tables of around 100 favela residents.

Other notable changes include the difference in the smell of the neighborhood, the reduction of vermin, and improved drainage, which results in less standing water and, therefore, reduces the threat of dengue. According to all of the beneficiaries, the garden's foremost value is the therapeutic relief it brings. Gardeners take pride in the spaces that they cultivate and enjoy giving food to community members that stop by. Gardeners and residents who receive produce all say it has increased their daily intake of vegetables, and thus of nutrients, as a consequence. They also appreciate that the produce is pesticide-free. Some residents whose homes are directly opposite the garden have also started to cultivate small plots in areas outside their front doors.

The promise of the Manguinhos garden indicates the ability of these sorts of endeavors to create local producer-consumer food production and

Figure 5.9 Manguinhos garden. *Source*: Photo by the author, June 2014.

distribution networks, advance education and skills-building, promote care for the vulnerable, and relieve stress and hypertension for informal sectors that live in conflict environments. In addition, it has opened up alternative forms of exchange, such as the currency of food, and points the way to how barter markets, micro-cooperatives, and informal-to-formal market chains can grow. From a human perspective, the garden has fostered positive interpersonal relations, created bridges between societal and cultural tiers, and generated non-monetary wealth. If these results are achievable over the long term, new forms of solidarity financing, micro-financing, and ethical banking opportunities could extend from such practices. Other opportunities to advance economically may also be possible; for example, informal producer to formal market chain barriers could be breached. This could also lead to the cultivation and sale (and thus financial renumeration) of higher-value crops.

However successful it has been, the Manguinhos project still faces substantial challenges. Because the anticipated dredging of the open sewer channel, or river, never took place (state funds were misappropriated three times in under two years), during flash floods, a corner of the HC space the size of around fifty square meters is compromised when the channel overflows, carrying potential risk from waterborne pollutants, including bio-accumulators such as lead that may migrate from the nearby oil refinery, and from leptospirosis transferred via sewerage. The high degree of heavy automobile and rail traffic that slices through this area also generates airborne bio-accumulators that may build up on the leaves of plants. To date, however, there has been no soil monitoring to test for any forms of potential contamination.

The Comlurb waste collection agency—the city department responsible for garbage collection—has also destroyed fencing around the HC garden, which stalled production in a third of the space for a period of several months. The agency's inadequate trash removal does little to mitigate the ongoing and overwhelming trash buildup that is generated by 50,000 people having no satisfactory way to adequately or safely dispose of their garbage. Even though Manguinhos has started to receive some trash removal since pacification, it remains critically inadequate. An acute buildup of garbage continues to be a problem in this area. Though Comlurb has been asked several times to haul growing mountains of debris away, it rarely does so. Therefore, residents continue to dump their trash in whatever space is available. Much of it continues to be disposed of in open burns, or is picked up by intense windstorms, to be redistributed throughout the garden. Therefore, keeping the garden free from trash continues to be a major issue. Comlurb is also charged with periodically supplying recycled soil to replenish the garden beds, but at times their infrequent deliveries are so substandard that they cannot be used. At best, the soil contains high amounts of cut glass.

In addition, the HC project budget is increasingly suffering under a federal, state, and municipal climate of negative growth, economic recession, and a government in fiscal and political crisis. As time goes by, fewer tools are available to gardeners, a situation that gradually worsens with each tool that is broken. Furthermore, due to severe municipal budget cuts in the first quarter of 2016, HC gardeners have had payment of their monthly stipends delayed for months at a time. This has led to decreased productivity and reduced confidence in the government's ability to meet its financial obligations to employees. This, in turn, has reduced the commitment of all but the most loyal of its employees. Even so, Mayor Marcello Crirella supports expansion of this deserving program.

GMF's southern extension to the garden is located in the middle of the favela, and, as such, its social and economic landscape is much more complex than the GMF space at the northern boundary (which is also the boundary of the favela). The residents of this area are much poorer, and there are several *bocas* (drug selling points) located in this area. Because of this complexity, it is also much more dangerous and volatile. Incidents of armed conflict and homicides by police are escalating at an alarming rate. In the past several years, numerous incidents of extrajudicial killings have been documented in Manguinhos, a situation that is accelerating daily in this section of the favela. For example, on the evening of March 12, 2016, eleven-year-old Caio Daniel was murdered at the hands of the UPP. In the same action, the police shot three other teenage youths. Three days later, on the morning of March 15, an hour before GMF began work, two more young men were shot by police next to the garden. One died immediately, while the other was shot in the leg, hunted down, and stabbed to death. The following day, the police drew their guns in the garden while GMF and residents were working in the space. Gardeners had to flee and hide in adjacent homes to wait for the armed offensive to end. Two weeks later, on April 4, in the same block of housing, João Batista Soares de Souza (twenty-nine years of age) was murdered by police just before lunch. His wife, who had been worried about their children climbing on a chair near the window of their apartment, asked her husband to remove it and check on police activity on the street. The moment that de Souza moved to the window he was shot in the head. The shooting was witnessed by his entire family. By all accounts, the murdered party was a good parent, partner, neighbor, and provider for his family. These incidents exemplify the daily reality of life in Manguinhos, especially for black young men. The cumulative negative impact from these harsh realities, and many other forms of structural violence, as well as the trivialization and normalization of gun violence and extrajudicial killings in favelas at the hands of police, have contributed to a culture whereby residents have low expectations of their citizenry rights.

PUBLIC SECURITY OBSTACLES

Public security policy, validated by propaganda that supports the "war on drugs," justifies an extraordinarily high number of deaths at the hands of State military troops in Rio. Of the 1,275 registered cases of killings by on-duty police throughout the city between 2010 and 2013, 99.5% of homicide victims were men, 79% were black, and 75% were aged between fifteen and twenty-nine.[30] In the favelas, the official number of deaths may have declined at the beginning of pacification; however, since 2012, numbers have been on the rise. In addition, many killings go unreported and uninvestigated, and disappearances cover up incidences of murder—the numbers of "disappeared" have risen to the equivalent degree that homicide figures have declined.[31] Furthermore, of the 6,034 disappearances recorded in Rio between November of 2012 and October of 2013, only one gained public visibility resulting in the arrest of officers. This demonstrates the direct relationship between armed violence and the difficulties associated with achieving forms of social stability in such conditions—let alone the demands associated with advancing sustainability projects under such circumstances. This state of affairs also illustrates Rio's historically brutal approach toward community policing.[32]

Conflict-related and lethal violence lowers the levels of human development, inhibits poverty alleviation, and impedes achieving sustainable development.[33] In Rio, with public security funding (and the State's need) for the UPP almost completely drying up, it is expected that the policy will be almost fully dismantled by 2017. Thus residents' level of political trust, as a central indicator of the public's underlying confidence[34] in the State, remains low. Political trust has ebbed with the enactment of every policy decision and at every level of intervention. The negative perceptions held by residents regarding State intervention are not only embedded in the collective experience of residents but also continue to be undermined by a police culture which is known for perpetrating torture, rape, murder, disappearance, assault, corruption, drug dealing, and destruction of property.[35] The majority withdrawal of the UPP from the favelas following the Olympics is also resulting in a power vacuum whereby gang conflict is creating a situation where conflict is arising over the fight for territory, with dramatic consequences for those living in Rio's informal settlements.

ANALYSIS OF PARTICIPATION

It is widely assumed that in order to develop links between sustainability and peace-building in public or semi-public space, placemaking strategies such as the cultivation of gardens must be supported by a wide swath of government policies that help to mitigate social instability. Yet, in the favelas, these key building blocks have been, for the most part, neglected for generations. This reflection of poor governance and weak State capacity, which ultimately limits the scope of sustaining community services such as gardens,[36] in contrast, also leaves favela residents open to use their tenacity to find solutions to their own problems. Only time will tell if residents will have the adaptable capacity to overcome the challenges required to sustain these community assets over the long term. From within these challenges, the language of participation emerges to expose a plethora of possible meanings that are embodied by a range of different actors with different, duplicitous, or contradictory political backgrounds, motivations, agendas, and/or objectives.

In Rio de Janeiro, especially in regard to the favelas, interventions are almost exclusively driven by powerful actors over which the citizens of favelas have little to no control. Any discussion of collaboration or intervention claiming to make use of participatory approaches must be framed by an understanding of limitations on citizens' freedoms, what resources are available to (and denied) them, and when and how they are able to participate in any given venture. It helps to understand citizen participation inside favelas as something deeply political. Favela residents are subject to all forms of structural violence, and their lives are complicated by the challenges associated with conflict and poverty and made even more difficult by low expectations of rights and empowerment.

However, on the one hand, in the case of favela gardening, local actors can affect outcomes, and their participation is crucial to the success or failure of these projects. On the other hand, their voices are not always heard and responded to, nor are their needs always fulfilled. Barriers are hard to shift; real-life situations in regard to health, the demands of work, and other responsibilities change; political parties and third sector programs are vulnerable to withdrawals of support.

Nominal forms of participation have been used by Rio's development actors to legitimize larger-scale redevelopment plans in Manguinhos. At the same time, however, less powerful local actors have become involved from the bottom-up to ultimately drive social change. In the case of the Rocinha garden, community participation was clearly used as a means to achieve a stated vision that could only be realized through the efficient use of skills and the knowledge of community members to execute the project. In both cases, there was some degree of instrumental participation involving community members having a voice in the decision-making of the implementation process of projects.

Conversely, the ability to create a platform to enable representative or transformative forms of participation that result in a substantive degree of empowerment for the local actors involved has not been achieved. In addition, there has been no alteration in any institutional power base, or any meaningful change in the culture of structural violence, police brutality, economic marginalization, or social exclusion that residents confront every day. All the same, the agendas and dynamics between actors within this greater framework of conflicts, differences, and collaborations between bottom-up and top-down flows have been more clearly revealed and may lead to a clearer understanding of the complexity of the politics of participation within urban sustainability initiatives that exist inside, or alongside, low-intensity conflicts.

CONCLUSION

This essay addresses the complex relationships between participation, empowerment, sustainability, violence, and intervention currently being enacted in the favelas of Rio de Janeiro. Using the multi-sectoral and multi-contextual experience of working in two of Rio's largest and most vulnerable informal sector settlements, it identifies some of the major influences that determine how participatory intervention can potentially lead to sustainability, and how it is prone to collapse.

As favelas grow, and as disparities in wealth and resources widen, urban gardening projects potentially offer low-cost, manually implemented, adaptable, and scalable solutions for restoring land. They present themselves as a viable model through which to build integrated partnerships and inclusive management practices; moreover, the benefits they deliver are numerous. They have the capacity to promote food security, increase water filtration, encourage biodiversity, and reduce greenhouse effects. They offer social, leisure, therapeutic, and educational opportunities and have the potential to generate income. However, significant doubt remains as to whether or not these gardens can be actualized over the long term due to their susceptibility to a slew of environmental, social, economic, political, health-related, and security-related circumstances that can fluctuate rapidly.

Nevertheless, this should not be read as a sign that the local community is not functioning as best as it can. These flaws, moreover, can be attributed to failures by the State, which implements inequitable security and development agendas with little regard to how they impact favela residents. As a consequence, residents have, for the most part, resigned themselves to living with low citizenship expectations that are further restricted by some sort of armed control. Given that the city's favelas continue to urbanize eight times faster than the formal city and have grown by 27.65% in a decade, very little action is being undertaken to reduce the risks for the people living in them.[37] At the time of this writing, gardens are one of the only initiatives that represent any form of socio-environmental efforts being enacted inside favelas to improve the lives of residents.

This is important to note because urban gardens are often overlooked in slum upgrading initiatives, especially in regard to their non-use and option values, including how they may play into poverty alleviation. Studies suggest that decreasing the stress related to poverty can lead to an increase in cognitive capacity and more effective decision-making capabilities.[38] At the very least, gardens bring much needed therapeutic respite to those living in favelas, and this makes an important difference for those who interface with them.

The aim of introducing the reader to these types of micro-gardening initiatives has been to provide a window to viable actions that, if they were to be given enough social support, could stand a chance of becoming sustainable and producing long-term value for favelas. Above all, they present small opportunities that, if only in a very limited way, point toward an alternative future for favela residents—one that attaches to policies already adopted by the city and state governments of Rio. Yet, for these sorts of projects to have a broad impact, key changes aimed at institutional reform and transparency must occur. There is no doubt that the multiple forms of violence and conflict present in Rio's favelas act as sustainability disablers. Along with the lack of access to public health and sanitation services, these obstacles present a major threat to human well-being, and consequently to achieving any sustainability agenda. The government must not only draft policies that utilize the language of sustainability, but more importantly, must provide the conditions that enable them. This can only come about if there is authentic political desire and consistent and adequate institutional support in both hard and soft infrastructure, including the reduction of security tensions through accountable policing procedures that protect the informal sector, rather than terrifying it. However unlikely, if that could be achieved, these gardens could flourish as vibrant, participatory, and sustainable enterprises managed by the people living in favelas. At present, perhaps more than anything else, they simply function as practices that strive for the advancement of fundamental justice issues that connect to a range of concerns, stretching from environmental policy and urban restoration to participatory planning and political reform—lofty but worthy goals that are only attainable through a suite of integrated efforts that embody the working processes of coproduction, cooperation, and organization of all those involved.[39]

NOTES

1. Within the context of this essay, the word "community" is used to describe a favela or favela complex collectively or may refer to one or more smaller residential

or mixed-use subcommunities that are located within the immediate vicinity of the project that is being referenced.

2. Sherry Arnstein, "A Ladder of Citizen Participation," *Journal of the American Institute of Planners* 35, no. 4 (1969): 216–24.

3. Sarah White, "Depoliticising Development: The Uses and Abuses of Participation," *Development in Practice* 6, no.1 (1996): 6–15, http://www.jstor.org/stable/4029350.

4. Michel Lyons, "Participation, Empowerment and Sustainability: (How) Do the Links Work?," *Urban Studies* 38, no. 8 (July 2001): 1233–51, doi:10.1080/00420980125039.

5. Instituto Brasileiro de Geografia e Estatistica (IBGE), 2010 Census.

6. In 2012, Rio's Gini coefficient (a comparative inequality index) measured 0.6, on par with the most unequal societies on Earth.

7. Barbosa de Holanda and Fernando Filho, "An Estimation of the Underground Economy in Brazil," IBRE/FGV (The Organisation for Economic Co-operation and Development (OECD), 2012), 1, http://www.oecd.org/ctp/tax-global/An%20Estimation%20of%20the%20Underground%20Economy%20%20in%20Brazil.pdf.

8. Lea Rekow, "Pacification & Megaevents in Rio de Janeiro: Urbanization, Public Security, and Accumulation by Dispossession," *Journal of Human Security* 12, no. 1 (February 2016): 21.

9. Rio's militias are infamous networks of off-duty and former security forces that control favelas or public housing complexes. These gangs or mafia-like organizations are linked to murders and mass murders, extortion, torture, and a range of other criminal activities. The number of militias in control of Rio's favelas has risen dramatically since pacification and presents a growing problem. For more, see also Nicole Froio, "Rio de Janeiro's Militias Grow in Power," *Rio on Watch*, October 10, 2014, http://www.rioonwatch.org/?p=18413.

10. Rekow, "Pacification & Megaevents in Rio de Janeiro."

11. Eduarda de La Rocque and Petras Shelton-Zumpano, "The Sustainable Development Strategy of the Municipal Government of Rio de Janeiro" (Washington, D.C.: Woodrow Wilson International Center for Scholars, 2014), https://www.wilsoncenter.org/sites/default/files/Paper_ La Rocque and Shelton-Zumpano_2014_0.pdf.

12. Luanda Vannuchi and Mathieu Van Criekingen, "Transforming Rio de Janeiro for the Olympics: Another Path to Accumulation by Dispossession?," Special Issue, *Articulo: Journal of Urban Research* 7 (2015), http://articulo.revues.org/2813.

13. IBGE, 2010 Census.

14. Larch Maxey, "From 'Alternative' to 'Sustainable' Food," in *Alternative Food Geographies: Concepts and Debates*, eds. Damian Maye, Lewis Holloway, and Moya Kneafsey (Amsterdam: Elsevier, 2007), 55.

15. For a list of literature reviews associated with this subject, see Lea Rekow, "On Unstable Ground: Issues Involved in Greening Space in the Rocinha Favela of Rio de Janeiro," *Journal of Human Security* 12, no. 1 (2016): 52–73.

16. Índice de Desenvolvimento Humano Municipal (IDH), "Tabela Nº 1172," 2000, http://portalgeo.rio.rj.gov.br/indice/flanali.asp?codpal=54&pal=DESENVOLVIMENTO SOCIAL/HUMANO.

17. Hilka Telles. "Evasão escolar é maior nas favelas do Rio," *O Dia*, August 18, 2014, http://odia.ig.com.br/noticia/riosemfronteiras/2014-08-18/evasao-escolar-e-maior-nas-favelas-do-rio.html; see also United Nations Development Programme (UNDP), *O Índice de Desenvolvimento Humano Municipal Brasileiro* (Brasília: PNUD, Ipea, FJP, 2013), http://www.pnud.org.br/arquivos/idhm-brasileiro-atlas-2013.pdf.

18. Lea Rekow, "Fighting Insecurity: Experiments in Urban Agriculture in the Favelas of Rio de Janeiro," *Field Actions Science Reports* 8 (2015): paragraphs 53–55, https://factsreports.revues.org/4009.

19. Peri K. Blind, *Building Trust in Government in the Twenty-First Century: Review of Literature and Emerging Issues*, November 2006, http://unpan1.un.org/intradoc/groups/public/documents/UN/UNPAN025062.pdf.

20. Flavio de Souza, "The Future of Informal Settlements: Lessons in the Legalization of Disputed Urban Land in Recife, Brazil," *Geoforum* 32, no. 4 (November 2001): 483–92, http://www.sciencedirect.com/science/article/pii/S0016718501000148.

21. Rekow, "Pacification & Megaevents in Rio de Janeiro," 10.

22. Célia Landmann Szwarcwald et al., "Health Inequalities in Rio de Janeiro, Brazil: Lower Healthy Life Expectancy in Socioeconomically Disadvantaged Areas," *American Journal of Public Health* 101, no. 3 (2011): 517–23, http://www.ncbi.nlm.nih.gov/pmc/articles/PMC3036699/.

23. Malaquias Batista Filho, Denise O. Silva, and Helena Sousa, "Desnutrição em crianças de áreas faveladas: Manguinhos, Rio de Janeiro," *Cadernos Saúde Pública* 8, no. 1 (1992): 69–76, http://www.scielo.br/scielo.php?script=sci_arttext&pid=S0102-311X1992000100009.

24. Szwarcwald et al., "Health Inequalities in Rio de Janeiro."

25. IDH, "Tabela Nº 1172."

26. Rita de Cássia Oliveira da Costa Mattos et al., "Avaliação dos fatores de risco relacionados à exposição ao chumbo em crianças e adolescentes do Rio de Janeiro," *Ciência & Saúde Coletiva* 14, no. 6 (2009): 2039–48.

27. Oswaldo Cruz Fundação, "A Dengue e a Desigualdade na Cidade Maravilhosa," FIOCRUZ, 2009, http://www.conhecendomanguinhos.fiocruz.br/files/T&CdenguebibliotecaBETA_0.pdf.

28. Batista Filho, Silva, and Sousa, "Desnutrição em crianças de áreas faveladas."

29. Residents Associations typically act as the mediating body between the State and the gangs to move projects forward. This set of arrangements is too complex to address in this essay.

30. Amnesty International, "Brazil: You Killed My Son: Homicides by Military Police in the City of Rio de Janeiro," August 3, 2015, https://www.amnesty.org/en/documents/amr19/2068/2015/en/.

31. Although these figures may not present a causal link, they may suggest a correlative one.

32. Lea Rekow, "Police, Protests, and Policy in Rio de Janeiro—Mega-Events, Networked Culture, and the Right to the City," in *Citizen's Right to the Digital City: Urban Interfaces, Activism, and Placemaking*, eds. Marcus Foth, Martin Brynskov, and Timo Ojala (Singapore: Springer, 2015), 119–35.

33. A 2013 Geneva Declaration study reports that, worldwide, 1.5 billion people live in fragile areas, conflict zones, or areas dominated by large-scale, organized criminal violence, and that "no low-income fragile or conflict-affected country" was able to achieve "a single United Nations Millennium Development Goal [and will continue to impede] the post-2015 development agenda." See Geneva Declaration on Armed Violence and Development, "More Violence, Less Development: Examining the Relationship Between Armed Violence and MDG Achievement—Preliminary Findings," September 2010, http://www.genevadeclaration.org/fileadmin/docs/MDG_Process/MoreViolenceLessDevelopment.pdf.

34. Distinctions can also be made between the concept of "confidence" (defined as a more passive emotion accorded to the overall sociopolitical system), and "trust" (which is associated with a group of more dynamic beliefs and commitments accorded to people). See, for example, Anthony Giddens, *The Consequences of Modernity* (Cambridge, UK: Polity Press, 1990).

35. Maria Alves and Philip Evanson, *Living in the Crossfire: Favela Residents, Drug Dealers, and Police Violence in Rio de Janeiro*, Voices of Latin American Life (Philadelphia: Temple University Press, 2011).

36. United Nations Department of Economic and Social Affairs Division for Sustainable Development (DESA), "Developing National Sustainable Development Strategies in Post-Conflict Countries," Development Account Project (ROA 105) Guidance Notes, 2011, https://sustainabledevelopment.un.org/content/documents/1039guidancenotes2.pdf.

37. IBGE, 2010 Census.

38. Rekow, "Fighting Insecurity."

39. Paulo Freire, *Pedagogy of the Oppressed*, trans. Myra Bergman Ramos (Reprint, New York: Continuum International, 2000), 87–183.

BIBLIOGRAPHY

Alves, Maria, and Philip Evanson. *Living in the Crossfire: Favela Residents, Drug Dealers, and Police Violence in Rio de Janeiro*. Voices of Latin American Life. Philadelphia: Temple University Press, 2011.

Amnesty International. "Brazil: You Killed My Son: Homicides by Military Police in the City of Rio de Janeiro." August 3, 2015. https://www.amnesty.org/en/documents/amr19/2068/2015/en/.

Arnstein, Sherry. "A Ladder of Citizen Participation." *Journal of the American Institute of Planners* 35, no. 4 (1969): 216–24.

Batista Filho, Malaquias, Denise O. Silva, and Helena Sousa. "Desnutrição em crianças de áreas faveladas: Manguinhos, Rio de Janeiro." *Cadernos Saúde Pública* 8, no. 1 (1992): 69–76. http://www.scielo.br/scielo.php?script=sci_arttext&pid=S0102-311X1992000100009.

Blind, Peri K. *Building Trust in Government in the Twenty-First Century: Review of Literature and Emerging Issues*. November 2006. http://unpan1.un.org/intradoc/groups/public/documents/UN/UNPAN025062.pdf.

de Holanda, Barbosa, and Fernando Filho. "An Estimation of the Underground Economy in Brazil." IBRE / FGV. The Organisation for Economic Co-operation and Development (OECD), 2012. http://www.oecd.org/ctp/tax-global/An%20Estimation%20of%20the%20Underground%20Economy%20%20in%20Brazil.pdf.
de La Rocque, Eduarda, and Petras Shelton-Zumpano. "The Sustainable Development Strategy of the Municipal Government of Rio de Janeiro." Washington, D.C.: Woodrow Wilson International Center for Scholars, 2014. https://www.wilsoncenter.org/sites/default/files/Paper_ La Rocque and Shelton-Zumpano_2014_0.pdf.
de Souza, Flavio. "The Future of Informal Settlements: Lessons in the Legalization of Disputed Urban Land in Recife, Brazil." *Geoforum* 32, no. 4 (November 2001): 483–92. http://www.sciencedirect.com/science/article/pii/S0016718501000148.
Freire, Paulo. *Pedagogy of the Oppressed*. Translated by Myra Bergman Ramos. Reprint, New York: Continuum International, 2000.
Froio, Nicole. "Rio de Janeiro's Militias Grow in Power." *Rio on Watch*, October 10, 2014. http://www.rioonwatch.org/?p=18413.
Geneva Declaration on Armed Violence and Development. "More Violence, Less Development: Examining the Relationship Between Armed Violence and MDG Achievement—Preliminary Findings." September 2010. http://www.genevadeclaration.org/fileadmin/docs/MDG_Process/MoreViolenceLessDevelopment.pdf.
Giddens, Anthony. *The Consequences of Modernity*. Cambridge, UK: Polity Press, 1990.
Índice de Desenvolvimento Humano Municipal (IDH). "Tabela N° 1172." 2000. http://portalgeo.rio.rj.gov.br/indice/flanali.asp?codpal=54&pal=DESENVOLVIM ENTO SOCIAL/HUMANO.
Instituto Brasileiro de Geografia e Estatistica (IBGE). 2010 Census.
———. "Nota técnica: estimativas da população dos municípios Brasileiros com data de referência em 1° de Julho de 2014." http://www.ibge.gov.br/home/presidencia/noticias/pdf/analise_estimativas_2014.pdf.
Lyons, Michel. "Participation, Empowerment and Sustainability: (How) Do the Links Work?" *Urban Studies* 38, no. 8 (July 2001): 1233–51. doi:10.1080/00420980125039.
Mattos, Rita de Cássia Oliveira da Costa, Márcia Aparecida Ribeiro de Carvalho, Helena Ramirez Domingos Mainenti, Ely Caetano Xavier Junior, Paula de Novaes Sarcinelli, Leandro Barreto Vargas de Carvalho, Renato Marçullo Borges, Simone Lorena Quitério, Simone Mitri Nogueira, Isabele Campos Costa, and Maria de Fátima Malizia Alves. "Avaliação dos fatores de risco relacionados à exposição ao chumbo em crianças e adolescentes do Rio de Janeiro." *Ciência & Saúde Coletiva* 14, no. 6 (2009): 2039–48.
Maxey, Larch. "From 'Alternative' to 'Sustainable' Food." In *Alternative Food Geographies: Concepts and Debates*, edited by Damian Maye, Lewis Holloway, and Moya Kneafsey, 55–76. Amsterdam: Elsevier, 2007.
Oswald Cruz Foundation. "A Dengue e a Desigualdade na Cidade Maravilhosa." FIOCRUZ. 2009. http://www.conhecendomanguinhos.fiocruz.br/files/T&C denguebibliotecaBETA_0.pdf.

Rekow, Lea. “Fighting Insecurity: Experiments in Urban Agriculture in the Favelas of Rio de Janeiro.” *Field Actions Science Reports* 8 (2015). https://factsreports.revues.org/4009.

———. “On Unstable Ground: Issues Involved in Greening Space in the Rocinha Favela of Rio de Janeiro.” *Journal of Human Security* 12, no. 1 (2016): 52–73.

———. “Pacification & Megaevents in Rio de Janeiro: Urbanization, Public Security, and Accumulation by Dispossession.” *Journal of Human Security* 12, no. 1 (February 2016): 4–34.

———. “Police, Protests, and Policy in Rio de Janeiro—Mega-Events, Networked Culture, and the Right to the City.” In *Citizen's Right to the Digital City: Urban Interfaces, Activism, and Placemaking*, edited by Marcus Foth, Martin Brynskov, and Timo Ojala, 119–35. Singapore: Springer, 2015.

Szwarcwald, Célia Landmann, Jurema Corrêa da Mota, Giseli Nogueira Damacena, and Tatiana Guimarães Sardinha Pereira. “Health Inequalities in Rio de Janeiro, Brazil: Lower Healthy Life Expectancy in Socioeconomically Disadvantaged Areas.” *American Journal of Public Health* 101, no. 3 (2011): 517–23. http://www.ncbi.nlm.nih.gov/pmc/articles/PMC3036699/.

Telles, Hilka. “Evasão escolar é maior nas favelas do Rio.” *O Dia*, August 18, 2014. http://odia.ig.com.br/noticia/riosemfronteiras/2014-08-18/evasao-escolar-e-maior-nas-favelas-do-rio.html.

United Nations Department of Economic and Social Affairs Division for Sustainable Development (DESA). “Developing National Sustainable Development Strategies in Post-Conflict Countries.” Development Account Project (ROA 105) Guidance Notes. 2011. https://sustainabledevelopment.un.org/content/documents/1039guidancenotes2.pdf.

United Nations Development Programme (UNDP). *O Índice de Desenvolvimento Humano Municipal Brasileiro*. Brasília: PNUD, Ipea, FJP, 2013. http://www.pnud.org.br/arquivos/idhm-brasileiro-atlas-2013.pdf.

Vannuchi, Luanda, and Mathieu Van Criekingen. “Transforming Rio de Janeiro for the Olympics: Another Path to Accumulation by Dispossession?” Special Issue, *Articulo: Journal of Urban Research* 7 (2015). http://articulo.revues.org/2813.

White, Sarah. “Depoliticising Development: The Uses and Abuses of Participation.” *Development in Practice* 6, no.1 (1996): 6–15. http://www.jstor.org/stable/4029350.

Chapter 6

Back to the City

Creativity and Sustainability as a New Approach to Revive the City from the Inside

Heide Imai

In most cities, public bodies are concerned with social, economic, cultural, and political integration of marginal urban areas.[1] One of the main strategies to achieve this purpose is to consider sustainability as one of the engines of great urban transformations to support the realization of different scales of urban renewal projects. As these occur in the form of new urban entertainment and economic and cultural clusters in both central and marginal urban areas, a reevaluation of the relationship of sustainability and the city will be necessary.

In times of globalization, ordinary places are increasingly contested and urban experiences fragmented, resulting in ordinary places that facilitate everyday life becoming marginalized and turned into places that are left behind yet rediscovered, excluded yet revived, forgotten yet remembered. It is important to consider ordinary places' changing roles and functions as this allows us to understand how different places in the city are used over a period of time. Thus, by studying the changing roles, meanings, and functions of ordinary places, we are able to understand how processes of globalization are affecting the formation of different kinds of public places that are used for different purposes, including isolation, resistance, consumption, and creativity.

This essay focuses on the urbanity and everyday practices of in/exclusion of marginal, ordinary places that are especially affected by globalization and rising inequalities. I am interested in a better understanding of sustainability in the city following three issues from an empirical viewpoint and making use of an innovative methodology.

As a first issue, this essay aims to understand everyday practices of in/exclusion and the changing role of the everyday dwellers of Tokyo. Second, this essay aims to reflect critically on the commodification of marginal urban places to understand how a recent dilution/distortion of the term sustainability is affecting everyday practices and temporary uses, as well as different urban identities within the city. This will be done by discussing different ways of occupying the city before following the everyday trail of different urban dwellers, including natives, newcomers, and returnees. I will apply a vignette-style method to introduce the different urban identities, which will help to provide practical examples to answer my research questions and allow the reader to follow the everyday trails of different users in the city of Tokyo.

Based on ethnographic material, this essay will discuss different theoretical and practical perspectives on creativity and sustainability, as different users consider the terms differently in relationship to the fast transforming city. Attempting to explore the potential and activities of marginal ordinary places as drivers or hinderers of creativity and sustainability in the city, I will introduce and discuss specific concepts of reclaiming space (urban commons) and examples of urban identities that take place within the dynamic urban landscape of contemporary Tokyo. In summary, the rising interest and importance of ordinary places will be analyzed, before relating this to their rediscovery as new hot spots and drivers of creative urbanism (with its positive and negative outcomes from a rising number of self-reliant communities/organizations and new forms of isolation, displacement, and gentrification).

Recently, many scholars have begun referring to ordinary, public, and often derelict places as "urban commons." The rising interest in urban commons raises the question, what are urban commons and why they are in the current spotlight? People claim and campaign that everybody can have access to public space and other common facilities in the city, not just stakeholders, developers, investors, and urban elites. Many urban commons, which have emerged in recent years and are a result and a product of collective and social collaborations, often struggle as more and more people participate in them, but mutual frameworks and municipal support appear to be missing from them. Using the dynamics of these social activities and creative movements can be a crucial steppingstone to achieve sustainability and to revive our cities from the inside.[2]

RESEARCH CONTEXT: BETWEEN URBAN POETICS AND URBAN POLITICS

Urban Commons and Ordinary Places

Urban commons are spheres, spaces, and places of human interaction, cooperation, and negotiation. Making urban commons is a process and should be understood as a verb as everything public is becoming a part of the productivity of the city being negotiated between activists, the state, and capital.[3] Many movements, whether the international Occupy Movement or Occupy Gezi in Istanbul, try to express their rights to the city and initiate commons, which can be groups of small pioneers or entrepreneurs, producers of subcultures, or social activities providing networks and services to the poor and marginalized, whether native, new, or temporary urban dwellers. Urban commons are often understood as initiatives focusing mainly on urban gardening, but the versatility of such activities, groups, and movements is much more varied. Governments started to recognize their potential and to encourage, if not support, such movements to upgrade derelict urban spaces.[4] Yet the creative and productive capacity of urban commons is also weakened by the increasing control, institutionalization, and decreasing autonomy of social movements crucial for the formation and productivity of commons. Others go so far as to argue that once local grassroots movements lose their spontaneity and momentum and turn into global streams of protests and justice (often supported by the Internet), the mistrust and rejection of political/economic control coming from the movements' participants hinder the development of the creativity, horizontal character, and togetherness of movements and commons.[5]

Despite urban commons not just being about collective gardening and the resistance of urban hipsters,[6] it remains a question how the city can implement the concept of the commons on a larger scale, making the city and its alternative spaces available to all types of users and uses.[7] Furthermore, urban commons are not just common resources but also need a fundamental set of regulations and strategies for how to be managed to keep a balance between individuality/creativity and direction/supervision.[8] Martina Löw states in her recent work *Managing the Urban Commons: Public Interest and the Representation of Interconnectedness* that the urban commons do not exist as a shared resource but, rather, are negotiated by different (social) media, which constantly offer new images of urban commons without actually addressing (and critically discussing), for example, the gap between rich and poor and rising inequality in the city.[9]

Whether we talk about knowledge commons,[10] cultural commons,[11] or urban commons,[12] it is the political dimension of commons that makes the concept so interesting as commons seem to offer alternatives in times of crisis, stress, and conflict. Furthermore, whether commons can continue to exist will largely depend on the ways in which commons are performed (e.g., in the form of radical activities, peaceful resistance, or harmonious, open-to-all initiatives). The existence, conservation, and preservation of commons also depend on the urgent need for such a resource as the commons, the resource itself, and the presence of a community with a manageable, secured population and dense social networks. Many recent examples, such as the urban commons called Agrocité Project developed by the Atelier d'Architecture Autogérée in Colombes, Paris, show that initiatives can go beyond urban gardening, including urban farming, modern recycling systems, and eco-housing, reaching a complexity and size which ask for the participation of all users.[13]

Urban commons become more common, everyday features with an alternative economy, society, and space. Yet the big question remains: who do they actually include, and are urban commons really open to all? What if they become new urban enclaves, hip places for one group (causing new cycles of gentrification) that exclude other users? Stavros Stavrides writes in his recent book *Common Space: City as Commons* that urban commons should always stay open to newcomers, to new ideas supported by new political initiatives, and to ideas which have yet to be tested as it is crucial to find the right balance between creativity, political support, and continuation.[14]

In comparison to urban commons, this essay will refer to the diversity of public spaces as ordinary places as not everybody takes part in such active collaborations or movements. Many people fight a lonely battle and still have the right to claim their space in the city. Additionally, we can question whether, in times of increasing consumption and commodification of public places, the direct relationship between public space, the formation of civic society, and expression of public voice still exist.[15]

Ordinary places facilitating everyday urban life are increasingly contested and fragmented, affecting the formation of place identities, place attachment, and sense of place, which determine how we can navigate and orientate ourselves in a rapidly transforming urban landscape.[16] To discuss the changing meaning of the creativity and sustainability of the city, I will make use of this concept to help to improve the understanding, recognition, and commitment to ordinary places, which are situated between urban politics and poetics and are the backdrop of the real stories of everyday life. All in all, we are talking about a diversity of concepts, scales, and meanings of urban places, which facilitate different uses and invite different users. The concept of the urban commons is used here to show that ordinary places are situated between different areas of conflict. The cases/actors introduced in this essay are facing these challenges every day.

The urban alleyway was chosen as one form of ordinary place as this urban form is recently experiencing a revival in different forms, and many scholars consider the urban alleyway as a suitable concept for the creation of small-scale public spaces with a quality of social life and a mix of diverse users, which are vital to the sustainability of the city.[17]

Creativity and Sustainability

Creativity can be understood as a phenomenon in which existing ideas and information are used to produce something new and valuable. According to

Mel Rhodes, the dominant factors of creativity are process, product, person, and place.[18] Similarly, researchers, including Peter Meusburger, Joachim Funke, and Edgar Wunder, are asking such questions as: why are highly creative individuals not evenly distributed over time and space? Why are certain cities and historical periods characterized by great creativity in the visual arts, music, and science whereas others are not? As a result, the development of creativity is seen as not just the achievement of a single person but also as a complex and dynamic set of social and spatial factors.[19]

In comparison, Charles Landry argues that cities' crucial resource is its people and that aspects like human intelligence, emotions, creativity, and innovative ideas are becoming more important than location, natural resources, and access to markets.[20] Combining both approaches, we can argue that it is the complex, dynamic interaction between creative people and their environments which allows us to understand creativity as a dynamic and interactional concept. The relevance of situational, cultural, and contextual principles in this setting is as important as the consideration of time, existing conditions, and chances to gain experiences and solve problems. All of these aspects together allow us to recognize creativity as a valuable concept when it comes to understanding current urban conditions and the state of socioeconomic development of diverse urban settings.

Richard Florida argues that a high quality of urban life helps cities to attract talent and, as a consequence, to grow economically. Furthermore, the growth of a creative class includes outcomes and benefits including new ideas, the high-tech industry, and regional growth.[21] However, if there are no incentives offered by governments or other bodies to support a good quality of urban life, which helps to attract creative people and new talent, it is very unlikely for creative talent alone to be able to cope with problems like economic decline and to be the engine that turns a struggling community into an innovation cluster.[22] Thus, the existence of a creative class does not automatically mean economic revitalization or development as Florida argues. For example, the case of Berlin shows that a city which generates amenities to attract a growing creative class as a result offers a rich cultural and urban life but not enough job opportunities for the growing group of intellectuals. In contrast, others scholars ask for new alternatives to creative capitalism and economically driven creative practices, focusing on new ways to understand and utilize the concept of creativity and creative commons, including not only flexibility and mobility but also precarity, recovery, and accumulation. Guiseppe Cocco and Barbara Szaniecki argue that, out of every success and failure of capitalist processes, new creative practices and identities emerge that are not just aiming to stimulate economic growth; anti-capitalist movements and protests show that creativity and creative capitalism work in diverse ways in different cultures, cities, and economies in the world.[23]

Sustainability is a term that we hear almost every day, but what does it really mean? The words "sustainable" and "sustainability" are not only about us, our culture, and our environment but also about our economies, cities, and countries worldwide, and more. Sustainability can be defined as the ability and capacity to do something, to maintain something, and to sustain ourselves using the resources we have now without damaging the resources and potential resources of the people who will need them in the future. In short, to be sustainable means that an act, activity, or process can continue forever without damaging others.[24] Living sustainably also means to live within our natural environments and cities without harming other people, society, and culture. In the twenty-first century, sustainability is a dominant idea and catchphrase as it is an everyday question whether we should consume food, clothes, energy, and other products whose origins and final destinations we can no longer retrace. Our lifestyles and lives in cities are changing at a fast pace as we are asked to be more flexible, mobile, and competitive in times of rapid globalization.

In such global times, sustainability occurs in almost every context and situation as a great idea, concept, and solution to be applied to everything and anything, but, as a result, sustainability is frequently misunderstood as we do not ask for its original meaning or what we can really achieve by it.[25] Others further claim that sustainability became a concept to justify and sustain first-world lifestyles at a comfort level that we think is ethically appropriate and acceptable while actually knowing that we ignore the most crucial fact: that we base our lives on the resources, abilities, and potential of people who do not belong to the global rich.[26] Instead of just relying on others and their resources, life in cities has to be about the limited resources, existing heritages, and creative potentials to secure a sustainable future for life in cities.

Taken together, both terms—creativity and sustainability—are contemporary buzzwords, wherein we miss the opportunity to explore their real potential and impact once applied. In this context, John Huige states that creativity can contribute to the sustainable transition of the city from one stage to another if we recognize that the transition occurs in cycles. Each phase—whether growth, decline, or recovery—has its own dynamism and appears in many forms, shapes, and networks.[27]

Alternative lifestyles and creative urban practices mirror this dynamism as they reflect the fluidity and flexibility of contemporary urban life and identities. As a result, a growing number of creative commons or communities evolve out of these alternative lifestyles. People are not just forced to migrate but also voluntarily move from one country, city, or community to another to commit themselves, searching for the real meaning of life and/or work as volunteers or artists or in other flexible positions to discover simplicity and forgotten potentials and to use their own creativity.[28] What if more and more people practiced such lifestyles and created such communities in economies and cities that have seen not only economic booms but also declines? Can we see the search for a meaningful everyday life in the city as a chance to recover hidden abilities, potentials, and chances for the future—at best, a sustainable future? The case of Tokyo offers a good example of what happens to a city that has undergone such dynamic changes in recent decades, at the same time looking for alternative approaches to its diverse and complex problems, such as limited resources, shrinkage, and urban decline.

THE CASE OF TOKYO

Having undergone rapid urbanization and modernization in the twentieth century, Tokyo now faces new, complex challenges caused by global change and interregional competition, including an aging and shrinking population and ongoing conflicts over the creation of new office and living space versus the maintenance of long-standing community structures exemplified by traditional forms of inner-city housing.[29] In particular, small-scale, ordinary, public places such as the urban alleyway are being further eroded by the construction of large-scale, high-rise buildings and contested by the emergence of Western-design-oriented public spaces. Responses to some of these challenges include the emergence of new social movements, cultural revivals, and the forging of new, unknown, hybrid lifestyles. These phenomena can be best understood by studying concrete ways in which people (re-)interpret and (re-)occupy small, ordinary places.[30] The central question of this essay is how creative practices and small, urban places shape each other and the contemporary urban landscape of Tokyo to be sustainable in times of urban change.

Tokyo, a city which is rebuilt continuously, has an interesting mix of contradictions, a patchwork of different building types, and scales and trends scattered in a network of small-scale neighborhoods with no street names, framed by larger streets and infrastructures of different eras. Even the tiny plots of land and small-scale, urban spaces are used to build imaginative and creative structures. Especially along the narrow streets, under the railway tracks, and in small-scale, urban voids, we find unique and innovative buildings.[31]

Cotton Mather, P. P. Karan, and Shigeru Iijima state that the "Japanese have not only refined compactness on the landscape, but also they have become [enamored] with the bonsai, the delicate bowl on the dining table, the tiny flower print, [and] the small garden representing the universe [;...] [t]he Japanese landscape mirrors the cultural affinity for smallness, multiple land use, and compactness."[32] This resonates with a statement by Yoshinobu Ashihara, who argues that the Japanese city, while appearing very messy, works remarkably well. The "hidden order" behind the visual chaos is rooted in Japanese geography and culture.[33] Guenther Nitschke gets closer to this idea, arguing that "a space or 'place' is an area defined by 'certain human activities'; constantly, it would move with the activity that gave its definition [; ...] [i]n Japan it is the procession or other event independent of spatial settings that make a place."[34] Thus, small-scale, leftover places are the result of complex processes of social interactions, constantly attracting various invasions of subjective uses, unplanned habitats, and spontaneous activities.[35] My research takes this approach further, arguing that these small-scale, ordinary places and close-knit neighborhoods—such as in the introduced case of Yanaka—attract a new kind of clientele of social entrepreneurs, artists, and other alternative practitioners who are looking for such niches of retreat, ordinary life, and vernacular creativity to establish new lifestyles and creative practices.

Declining Tokyo: Creativity and Sustainability as an Approach to Revitalization

Especially when talking about opposite processes, such as decline and recovery, we should study the case of Tokyo, a city which has been in a spiral of decline for several decades but has been targeted for regeneration, renewal, and alternative ways of growth by the public Tokyo Government, other private developers, and even individuals. In the last few decades, Tokyo has been facing new, more complex urban challenges, especially after a phase of rapid growth, extensive urban sprawl, and large-scale development during the 1980s and 1990s. An increasing inflow of retired people from "the baby boomer generation" and young working people into central Tokyo confronts urban designers with new and complex urban planning issues. These issues either have been left unsolved or have emerged as a result of existing problems like *kakusa shakai*, the increasing economic gap between the rich and poor in rural and urban areas at the local level, or the new urban rich and a growing group of people without secure futures (e.g., the "working poor") at the micro, intra-neighborhood level.[36] These numbers cannot be compared with the level of social and ethnic polarization in American cities like Los Angeles or Chicago. Yet a certain degree and form of social segregation and gentrification can be observed at the neighborhood level, at which the different conflicts are directly felt and visible.[37]

Recently, Tokyo has been utilizing creative policies to encourage the economic development of different areas into creative clusters, attracting a number of new, diverse groups of residents, users, and customers. Yet what happens to areas (urban or rural) whose economies fail to provide the expected outcomes or do not do so in the timeframe set by their leaders? Do these creative clusters continue to strive, do they stagnate, or do they develop in a different direction instead? One central purpose of this essay is to analyze to what extent creativity is: (1) used as a strategic measurement to achieve economic revitalization; (2) seen as a starting point but not considered to be the only factor in this process; or, (3) instead, recognized as the original potential of an area that can be used to support social and economic development. The theoretical starting point of this research is to see the function and meaning of creativity in the process of urban revitalization as a combination of these patterns and as an organic process.

This essay aims to study different processes of creativity and urban practices to show how they affect the term sustainability. The results can help us to understand and develop strategies to revive the city from the inside. This will be done by applying innovative methods, such as urban trailing with specific urban dwellers, for example, the native, the newcomer, and the returnee, to show everyday practices of in/exclusion and the changing role of everyday dwellers.

Living in Yanaka, I developed, over time, deep connections with different local residents, by which I came to understand the ongoing challenges observed and experienced in the community. Following a first phase of direct observation and immersion, I interviewed residents between April and March of 2012 using a snowball principle, starting with one shop/resident, approaching participants with specific questions, and later conducting an urban walk together. Afterwards, I transcribed and translated the interviews, showing the names of participants as pseudonyms, whereby all participants agreed to be part of the research and resulting publication. More specifically, different participants were asked how they experience the neighborhood and life in the city. Some native residents talked about the different positive and negative aspects, drawing on traditional features such as close social ties, friendships, daily gossip among neighbors, trust, and a sense of voluntary obligation. They also talked about how these features have been changing in times of rapid modernization and urban restructuring, resulting in new and sometimes unseen spatial and social realities. Thus, the results of my observations, interviews, and participatory walks allow me to highlight different forms of everyday *in situ* interactions and to provide rich ethnographic insights that bring the different and often unheard voices of the residents to life.[38] Before presenting the interviews, the case of Yanaka will be introduced to set the backdrop for the specific examples and to explain why this district of Tokyo is special.

The Case of Yanaka: From Ordinary Place to Creative Hot Spot

Yanaka district is located in the ward Taito-ku, in eastern Tokyo. 8,282 people lived in the area in 2015. The small, hilly area is known as spurs of the

Musashino plateau and in the east, west, and north is framed by the so-called low-town *shitamachi* quarters of Nezu, Sendagi, and Nippori, where local merchants/craftsmen live. Yanaka was known as *teramachi* (temple town) or "land of temples" during Edo (from the sixteenth to nineteenth centuries). In the hilly area, hundreds of temples, as well as shrines and cemeteries, could be found, especially after 1657 and the Great Meireki Fire, during which many temples were destroyed and then rebuilt in the Yanaka neighborhood.[39]

In-between the temple grounds, feudal lords and their serving samurai lived in small residential quarters, which turned step-by-step during the nineteenth century into quarters of the common folk. As the community of Yanaka survived the Great Kanto Earthquake of 1923 and the Tokyo Air Raids of 1945, the neighborhood still retains the charm of a traditional Edo quarter. The maze of streets, side streets, and alleyways has been growing organically, and many plots along them are diminutive, cramped, and, as a result of subdivision, hidden from streets with major access and often only accessible by foot.[40]

During the year, several seasonal events are held in the neighborhood, as for example in the Yanaka Cemetery Park, famous for its cherry blossoms in spring. The area is famous for its art universities, galleries, and historical buildings. Many of them have been renovated to fit new purposes (a renovated bathhouse is nowadays known as the art gallery Scai the Bathhouse), illustrating that the area is in high demand and blending old and new lifestyles in a unique way, creating an artistic and sophisticated atmosphere. The newest example is a traditional setting of three houses which was supposed be turned into a parking lot but instead opened in May of 2015 under the name Ueno Sakuragi Atari. The complex includes shops, a cafe/beer hall, art studios, and the new Kayaba Bakery.[41]

Yanaka is, throughout Tokyo, known as a special historic district worthwhile to preserve. There is consensus that the area should be protected. Nevertheless, as there were no official preservation laws which could be applied to Yanaka, it is due to the effort of the local community that the district is not yet affected by urban processes such as large-scale, urban renewal projects or high-rise developments inside the historic neighborhood. In 1979, a community center opened in Yanaka to enable residents to engage more in local town planning or so-called *machizukuri* projects. In the 1980s, local initiatives, like the nongovernmental group called *Yanaka Gakkō* (Yanaka School), began to propose ways to deal with processes of urban restructuring, which also started to occur in many inner-city neighborhoods of Tokyo. This approach is known as "community building" *machizukuri*, a process in which the local community can suggest ideas to the local government and other stakeholders to find the best possible compromise.[42]

In 1989, a nongovernmental group called *Yanaka Gakkō* (Yanaka School), a local group made up of residents, professors, and students, was founded to share knowledge and ideas to preserve historical and local resources. Consequently, surveys and other workshops were organized to document historical elements. All members of the community, including residents and children, were invited to festivals, flea markets, and other cultural events—such as the annual art festival Geikoten—to spread local knowledge and engage the community in different activities.

The Geikoten Event, which was established in 1993 and takes place every year during the month of October, invites all community members to participate in the event, whether by exhibiting their artistic work, visiting local stands and galleries, or getting involved in other ways. As many people take part, the whole neighborhood is on the move, visiting local workshops, studios, street stalls, installations, and other open-air spaces. Derelict houses, vacant plots, and other nooks and crannies are used for temporary installations, performances, and art events. Self-produced maps are distributed to visitors to encourage insiders and outsiders to visit the public and private spaces that are open during this special two-week event.

With the establishment of Geikoten, additionally, several facilities, including a gallery displaying local crafts, were opened to help residents to value their own heritage and to show outsiders the rich local culture. Since then, the NGO has started to collaborate with other neighborhood groups to preserve the historical features of the district and to spread the word to other cities.[43] These initiatives aim to draw more attention to public spaces of the community and to ensure that small streets and alleyways remain the basic urban fabric and that typical characteristics are preserved and used as places of everyday life. For this purpose, local community members conduct tours and surveys and use social media to invite people of all walks of life to claim their right to the street and to remember the values of the local culture. Other mainstream media feature the district and its special events, attracting, in return, more visitors and tourists. The main argument is that more public awareness will increase the appreciation, revaluation, and protection of the district's local assets.

Yet a recent survey conducted by the Taito Cultural and Historical Society and the graduate school of the Tokyo University of the Arts in 2012 showed that, out of the 748 traditional wooden building constructed around 1965 and documented in 1986, only 539 still existed in 2002 and even fewer (406) remained in 2012. The major reason might be the Tokyo government is concerned about the safety of these wooden houses and asked house owners to prepare their properties to be more earthquake safe. As a result, some owners decided to reinforce them; however, this was rarely done as it is very cost intensive. Most houses were or are being torn down and replaced with modern, earthquake resistant buildings.[44]

With the rise of local influence, some construction which was planned for the Yanaka area was retracted by the developers, as seen in the case of a proposed car park under the Shinobazu Pond. Not all developers took the local initiatives into account; however, some proposals, such as the Lions Mansion (2000), were realized considering the local *machinami* (townscape), implementing a lower building height and a new architectural design including the urban context.[45]

These features have made Yanaka in the last few decades an increasingly desirable neighborhood, which is nowadays appreciated by young and old people alike who have moved to the area.[46] Yanaka, having in the past escaped high-rise developments typically seen in Tokyo since the 1960s and new urban trends such as gentrification, has been experiencing, since the late 1980s and especially with the fall of land prices since the 1990s, more pressure coming from real estate developers trying to realize new high-rise developments and/or fashionable renewal plans.[47] Even though the efforts of the local community helped to protect major parts of Yanaka, we can nowadays observe a front of high-rise developments along the main street of Kotodoi dori to the south, Shinobazu dori to the west in Nezu, and down to the east in Nippori, forming a typical hard shell around the soft core of Yanaka.[48]

As the number of inhabitants in central Tokyo is rising again, more people are looking for appropriate housing in inner-city areas like Yanaka, shown in figure 6.1. Architects and real estate companies are responding to this trend by increasingly providing new housing in the form of modern high-rise buildings, which are considered to be the mainstream of contemporary, urban, residential developments.[49] As a result, the long-standing, local community with its native residents is facing new challenges of how to integrate the different needs and daily activities of the changing clientele and newcomers within the existing spatial and social structure of the local vicinity. The following interviews mirror, on the one hand, the pressures, conflicts, and challenges that the residents are facing and, on the other hand, their own activities, lifestyles, and creative practices that help to sustain the community.

If we compare the development of Yanaka to that of an urban commons, we find one mutual aim, which is to be open to all and to move the concept of sustainability beyond the conversation about urban gardening.[50] Using the potential of urban commons and the diversity of ordinary places as found in Yanaka can help to create a society and community in which everybody has equal chances to participate, a horizontal organization is practiced, and all can make use of one common resource.

Figure 6.1 The changing urban landscape, seen from Yanaka. *Source*: Photo by the author, 2012.

Examples of Everyday Urban Practices and Creativity

Creativity as a Way of Balancing Tradition and Modernity

Kumai-san, a long-standing resident running her own *tatami* shop in the area of Yanaka, experienced and recognized different kinds of changes affecting her family business as the demand for traditional craft is radically shrinking because the product as such is not desired anymore (people prefer modern apartments and flooring that is easy to maintain and to clean). Furthermore, processing the material requires special skills, making it an expensive item. Asked about her private life and how she makes a living nowadays, Kumai-san answered:

> My father-in-law strives for multiple businesses and is finally successful. These days we do not only *tatami* works but also interior works. However, it is not easy to survive in the shrinking market of this business derived from structural changes. Let's say there used to be many house builders from which we got jobs, but this happens not so often anymore. [...] However, fifteen years ago we opened a gallery, where modern art made by the Geidai Art University is displayed, and so we tried to connect arts with interior design.

Thus, Kumai-san was able to adapt to the changes in her business. She can now make a living running a traditional craft shop and art gallery. The traditional features of the area of Yanaka seem to support her business as different old clients, as well as an increasing number of new customers, visit the area on a regular basis or during seasonal events like exhibitions and local festivals.

As a result of the ongoing gentrification of the neighborhood, several newcomers have moved onto her street, and Kumai-san is interested in developing good communication between her native and new neighbors, as well as with the many visitors who come to the area on weekends. She is sometimes surprised by their spontaneous visits to the shop, which inspire her to find new topics for her exhibitions and ideas for her voluntary work in the local town planning group. Thus, she is not afraid of any possible effects on her daily life as she is welcoming people who are interested in the history of that place and in keeping the area alive. Her small neighborhood became, in recent years, famous for the town walks conducted in Yanaka. Kumai-san even mentioned that, in the morning before our interview, people from the TV station NHK came to visit the area to film the opening of a new shop, and that, the week before our interview, her father-in-law repaired *tatami* mats in the remaining space behind their house, shown in figure 6.2, a feature that many (mainly young) visitors do not know but desire to experience. Thus, creative urban practices of native urban dwellers seem to reflect everyday urban life struggles and chances, influenced by traditional and modern features.

Ordinary Places as a Source of Inspiration and Creativity

Kumai-san's immediate neighbor views the neighborhood in a different way. Akimoto-san is a newcomer to the area and a former student of the nearby university and has just opened a new café together with her partner. Asked why she is not active in the local *machizukuri* groups, she stated that she is too busy during the week running her own business and developing new ideas to attract enough customers and to make a living. In response to whether she likes the traditional atmosphere of the area, she stated:

> I like the area as many people want to visit the different cultural facilities, which secures in turn enough guests for our small café. Yes, the neighborhood has changed in recent years, but I am a newcomer, and, for me, it is important that I can combine work and living.

Being further asked if this was the reason why she moved to the area, she stated that she was attracted by the charm of the old houses and the mix of old and new because many galleries, ateliers, and alternative shops have recently

Figure 6.2 Interviewees in Yanaka. *Source*: Photos by the author, 2012.

opened in the area nearby. She went on to state that she likes the traditional and "old-fashioned pattern of everyday life," but she prefers to keep a good relationship with newcomers to the area as she doesn't feel connected to the old people living nearby. Her neighbor runs a rice-cracker shop; to my question about whether she communicates with him, she answered that she rarely sees him as she mainly visits other facilities like the art gallery on the corner opposite the *tatami* shop.

To compare to Akimoto-san's view, I had a chance to talk to Tanaka-san some days later. Tanaka-san runs the shop next to hers. Asked if he has good communication with his new neighbors, he stated:

> Communication, I would not call it communication. We exchange occasional greetings, but not much more. [...] I am afraid of the current changes going on. And, yes, the new shops are also competitors for me as many visitors to the area are young people who come to Yanaka to enjoy a walk and coffee [...] and sometimes, also buy some souvenirs at my shop, [...] but I am getting old, and my son does not want to take over the place.

The creative changes in the city and growing gentrification process have, on the one hand, positive effects on the spatial and social structure of the area as new people move into the old houses to open new shops, attracting new people to the area. Nevertheless, they also increase existing disparities between natives and newcomers who have different perspectives on the future of the traditional neighborhood. Newcomers to the area are bringing not only new ideas but also new forms of urban lifestyles, which start to dominate and become new urban trends and signs of new forms of gentrification.

Creativity and Nostalgia

As this essay aims to understand the changing perception and recognition of ordinary places, an additional interview and walk—in this case, with a returnee to the area—was conducted to present personal experiences and reflections of people who returned to the neighborhood and have different approaches to creative lifestyles based on their memories of the past.

In the area down the hill in Yanaka, I interviewed Okano-san, who is running a flower shop in an old wooden house (see figure 6.2). It was previously a carpentry shop, which her father rented to somebody until Okano-san returned to the neighborhood ten years ago to take over the place. Talking about her childhood days and how the area has changed over the last thirty years, she stated:

> My family has lived in this house from the time of my father's father. Of course, during my childhood, the locals helped each other. We would have dinner with our neighbors, and they would let us use their bath. I wonder what is left of those times. Doubtless I developed my attraction to old things from this old, working-class neighborhood. Older ladies come to do their shopping at the supermarket nearby with their wicker baskets, which they meticulously repair the minute they get damaged, even slightly. A good quality item you can use for a long time. When I come across these sorts of people, I think they really have style.

This statement shows that Okano-san feels close to, and "rooted" in, the neighborhood, as she has vivid memories when talking about the past. This might have been why she came back to the area as she was longing for the places of her childhood, which may have long disappeared but are still "alive" in her memory and have become again a part of her everyday life (e.g., through the transformation of the house into her shop). She is using the shop as a "place" to meet different people, to follow her own hybrid, alternative lifestyle (mixing professional and private life) as a returnee to the area and to express her creativity and memories through her shop—its interior and the way that she sells her goods. Asked what kind of customers come to her shop, she answered mainly newcomers "who range from residents of new flats in the high-rise buildings to [new] residents of old nice houses, and people who are fond of eccentric products." Although most of her customers are newcomers and many live in the new apartment blocks, she regrets that the area is changing into a neighborhood with mainly high-rise buildings as this also has negative impacts on the area, including a decrease in sunlight and green space. To my question of whether she is active in some kind of local group to support the preservation of the area, she remarked that she is not active because she prefers to lead her alternative life and to run free with her creativity.

Talking about the importance of maintaining public, ordinary places surrounding her house, she explained what role these places play for her:

> Because of my nature, being hurried up, I just prefer to make short-cuts using the alleyways. Alternatively, my nature may be the same as that of children who seek a small space like a box that they like to get into, [...] and I prefer to walk the alley alone in the morning, when everything is empty and quiet, as this reminds me of my childhood days. Nowadays everything has gotten so fast and noisy that I long for that moment I just can hide and escape from my busy, daily life.

This statement shows that she is using leftover urban places as her hideout and her own place of retreat as she is not actively supporting revitalization and preservation attempts but likes to make creative use of the alleyway. This form of hybrid identification with these urban places to some extent reflects her personal situation as a returnee to the area. She was not living in the area during the time that changes, such as emerging high-rise blocks, first became visible. Now back in her neighborhood, she has somehow accepted and internalized the changing urban landscape in following, in her own way, an urban and creative lifestyle. This reaction can be described as a combination of *compensation* (using the urban places for her daily walk to remember), *adaption* (running this kind of shop, she responded to the demand of the new clientele inhabiting and visiting the area, thus allowing her to live here), and *resignation* (as she is not getting active in a local group or similar organization). Finally, it can be argued that her hybrid lifestyle reflects different facets of the changes going on in the area as she, on the one hand, regrets the changes to the neighborhood and, on the other hand, longs for more ordinary places to allow her to reflect on her past, present, and future.

Creative Practices and Sustainable Lifestyles

Walking along the border of Yanaka and Nezu, I entered a shop called Dou-Dou and found a woman in her late thirties standing behind the counter. To answer my question regarding how she started the shop, Ando-san explained that she first sold handmade children's clothing on the Internet and, later, leather goods called "Leprotto (Plot Les)," as well as bags, shoes, hats, and, finally, semi-ordered shirts. She opened the atelier in 2008 along with her husband, who was born in the area and returned to live here in 2003. He sews the goods while she runs the shop. Ando-san discovered this local neighborhood while studying at the Art University near Ueno Park and has since then seen many friends and colleagues start their own small businesses in this area. Her immediate neighbor also runs an atelier for different self-made accessories and bags, but, in contrast to her, she runs her shop independently with her husband.

Asked how she and her husband would describe their lifestyle, Ando-san answered that they both enjoy the calm atmosphere of the neighborhood, which allows them to live a "slow life" and combine work and private life. It was always a small dream of hers to run her own shop, and, to realize this, she mainly relies on the communication and network that she has established with her direct neighbors. Asked whether she is active in the local neighborhood association or long-standing community, she stated that she and her husband take part in different events throughout the year (e.g., the Geikoten in October) but have, so far, not seen the need to become active members of other local groups. She closed by stating that she puts most emphasis on her clients and customers, who are mainly outsiders visiting the area on foot guided by a travel magazine.

Her direct neighbor is also a woman in her thirties, who is running her shop together with a colleague. Their variety of products is larger, and, asked what kind of products they focus on, Tsuda-san answered that they mainly sell on store-made accessories, bags, and alternative fashion goods. Asked why she opened the shop in this area, Tsuda-san stated that Yanaka became more and more famous in the last few years, and a friend told her about the possibility to open a small shop along this narrow lane. She took the chance and opened the first branch in the Tokyo area. She added that this location might be by far the most "attractive" as she thinks that Yanaka reflects the philosophy and style of the shop very well. Thus, she considers the location and urban fabric of the narrow lanes as essential to attract the type of customers who would come to visit and shop in this atelier.

Her acquaintance next door considers the narrow street in a similar way. Yamamoto-san, also a returnee to the area, and a former student of the nearby university, opened a new bagel shop together with her partner at the end of 2008. They are both working hard to run the bakery on top of their main jobs and do not have much time to take part in activities that are not related to their work. Asked if she likes the *shitamachi* atmosphere of the area, she stated that she was attracted by the charm of the old houses and the mix of old and new. Many galleries, ateliers, and alternative shops have recently opened in the area as more young creatives have decided to work and live in the area (see figure 6.3). She went on to state that she likes the traditional and "old-fashioned pattern of everyday life," but prefers to keep a good relationship with newcomers and other alternative people living in the area, as she does not feel connected to the old people living nearby. Eventually asked if her lifestyle supports the maintenance of Yanaka, she stated that she is happy to fulfill her own desire to live a simple, meaningful life which, at the same time, also respects not only the history and traditions but also the possibilities and

Figure 6.3 Shops along Hebimichi (Snake Lane) in Yanaka. *Source*: Photos by the author, 2012.

potential of the area. She considers the fact that she is using all of her energy to run the shop with her husband a reward for the community as she helps to revive and vitalize the area without destroying its historical character.

Some days later, I had a chance to talk to Kawaii-san, who runs a small bakery shop on the same block. She opened her shop at the end of 2009, selling French-style sweets called Florentines—small blocks of baked and candied almonds—first promoted on her online shop. Kawaii-san, who lived near Chicago until a year ago, came back to Tokyo and thought about opening her own shop soon after she arrived and started to live in the local area. Having learned how to produce these small treats, she soon opened her small shop called Atelier de Florentina. She uses only organic ingredients, including almonds, flour, and syrup, which she partially orders from abroad.

Asked why she started to sell French-style sweets in this local neighborhood, Kawaii-san said she likes the sweets and thought it would be a good idea to add something "international" to the area. Furthermore, the sweets are simple to make and do not require much space or time to prepare and bake. Female customers especially like to visit the shop and ask if they can try the different sweets. But, since her sweet shop and the neighboring shops were recently featured in different lifestyle and online magazines, as well as on TV, all kinds of customers have been coming to visit the small lane. The new customers help her to think about the expansion of her assortment and adding a small cafe. Thus, already, this small group of young entrepreneurs is showing how the social landscape of Nezu and Yanaka has been changing in recent years—sometimes for the better and sometimes for the worse.

Discussion

The featured interviews show that we can find different groups of users, residents, and creatives working and living in Yanaka, who we can mainly define as natives, returnees, and newcomers to the area. Natives are, for example, traditional merchants and craftsmen who still make use of the ordinary places like the alleyways as an extension of their workspace, storage facilities, or space for producing/exhibiting some of their work. As well as communicating with neighbors and other merchants, they communicate with visitors to the area who are interested in rediscovering traditional forms of craft and ordering handmade products, and some of whom are even considering moving there. In some ways, this group of residents can be considered living a marginalized lifestyle as they do not spend so much in the local economy but own their houses and sustain their own lifestyle. Yet this lifestyle makes it possible for Yanaka to continue to exist as a close-knit and socially dense community as native people frequent ordinary places for daily walks and local shopping tours and gossip with neighbors. They care for their potted plants, hang their laundry, and follow all kinds of other daily routines that sustain communication, trust, and strong social ties between people of all walks of life.

This cannot prevent the changes seen in many local neighborhoods in Tokyo as, even in Yanaka, the community becomes more fragmented as old homeowners die and newcomers take over their land to build new houses, increasing the gap between native and newcomer, old and young, poor and rich. Those changes are neither offering chances to revive the area using creative approaches, for example, making use of existing structures, nor are they making living in the city more sustainable. Yet sometimes newcomers like Akimoto-san make use of an existing house. She cannot afford to rebuild but rents the house, preserving the structure for new uses and users, even if her cafe is a very unique new business concept (Bossa Nova Cafe). In this way, she is maintaining the old building and inviting all kinds of visitors and, especially, newcomers to discover her cafe and the area, increasing the attraction and gentrification of Yanaka in all its diversity (appreciation, commercialization, and consumption).

Returnees like Okano had the chance to move into an existing building, establishing at the same time a "down-to-earth" flower shop and meeting place for all kinds of people of the community. Her statement that she feels that the area has changed a lot and she wants to preserve Yanaka, even at the small scale, corresponds with the fact that she is not running her shop for the mere purpose of covering her basic living expenses but, especially, to keep the communication with all neighbors alive. This offers the chance to preserve physical and social characteristics of Yanaka, applying a mixed-use concept. This approach is very similar to what people living in urban commons aim for: using the creative and productive potential of a derelict, ordinary place to revive the place for all. The residents and returnees like Yamamoto used existing structures and developed new businesses without being able to live there as the rent prices in Yanaka are far too high to be covered by their family incomes. Yet even these are ways of revitalizing the neighborhood without changing everyday life patterns or access to ordinary places.

Newcomers like artists or other alternative businesses run by Tsuda or Ando appreciate the unique atmosphere of the alleyways, which helps them to express their alternative lifestyles, perform different kinds of subcultures, or to combine work and living. Others open ateliers, galleries, cafes, or small bookshops to attract different kinds of customers, profiting from the popularity of Yanaka, offering products like expensive ceramics, old furniture, or reused kimonos to wear in the form of new fashion designs. Thus, to some extent, artists have caused the gentrification of Yanaka and, therefore, boosted the local and symbolic economy of the neighborhood as more visitors want to experience local food, culture, and other traditional items. At the same time, they also help to revive the neighborhood as more young, working, and well-situated people start to live in Yanaka.

Despite that, we should question if these changes aim to reinforce the sustainability of the neighborhood or not. On the one hand, alternative urban lifestyles and urban places like those found in the case of Yanaka are labeled as more sustainable than others as such an image sells well to attract more urban elites and expatriates, a development that the government of Tokyo especially welcomes. Nonetheless, in reality, alternative lifestyles and creative practices performed by marginalized groups as described above, by making use of existing historical and local resources, offer the incentive and stimulus to start discussing more sustainable ways for the future of an urban community for all.

All in all, I argue that ordinary places like alleyways mirror the rising interest and importance of ordinary places and commons. Situated between the poetics of everyday life and political struggles, ordinary places are allied to urban commons, which were introduced to understand that new urban hot spots, creative urbanism, and the future of sustainability are closely linked to each other.

CONCLUSION

Attempting to explore the potential and activities of marginal, ordinary places as drivers or hinderers of creativity and sustainability in the city, this essay has introduced and discussed specific examples of urban identities which exist and persist within the dynamic urban landscape of contemporary Tokyo. The rising interest and importance of small, ordinary places was analyzed before relating this to their rediscovery as new hot spots, drivers of creative urbanism (with its positive and negative outcomes as a rising number of self-relying communities/organizations and new forms of isolation, displacement, and gentrification), and stimulators to develop creative approaches to sustainability for the revival of the city.

I have argued, in this essay, that creative practices, lifestyles, and acts of reclaiming ordinary places can enhance the social interaction and, in turn, the sustainability of the city. More and more, ordinary places are being discovered to provide living places for diverse work-life concepts, especially for people who do not (want to) fit into the typical, mainstream lifestyles but look for more meaning, commitment, and simplicity in life to make use of their right to reclaim the city. This corresponds with the main task of all actors in the city to provide urban places for all, whether they have the resources to pay for a high-end apartment or not. Ordinary places can provide these in-between spaces as they hold new potentials and chances for the marginalized, who are also part of the city and are actively shaping the city at the grassroots level—an ability that some city dwellers have lost or forgotten how to activate. Thus, to make cities more sustainable, we have first to better understand the creativity of all users and how to support it to make good use of it.

The case of Yanaka shows that ordinary places can stimulate and enhance diverse creative abilities. The alleyways, which are the urban fabric, everyday place, and lifeline of people of all walks of life, foster the "mingling of diversity" and "knowledge spillovers" that Jane Jacobs described.[51] In the streets and alleyways of the city, locals and strangers interact with each other. This interaction is unleashing creativity and innovation; it is important to understand that innovation is a local process starting in one place, in which residents, natives, and newcomers cross-fertilize each other with new ideas. This means that especially large cities like Tokyo need a dense network of resilient, small-scale, and close-knit communities with diverse characters, where the poor and rich, old and young, and local and global interact which each other to make sure that creative capacities are used to set up a common economy and a sustainable future for our cities.

NOTES

1. This research is based on fieldwork conducted in 2012 in Tokyo, supported by the Japanese Society for the Promotion of Science (JSPS) under Grant 09724.

2. Tim Edensor et al., *Spaces of Vernacular Creativity: Rethinking the Cultural Economy* (Abingdon, Oxon, UK; New York: Routledge, 2010).

3. David Harvey, *Rebel Cities: From the Right to the City to the Urban Revolution* (New York: Verso, 2012).

4. Stavros Stavrides, *Common Space: The City as Commons* (London: Zed Books, 2016).

5. For a concise discussion of the change to local grassroots movements and new forms of protest, see Marcus Castells, *Networks of Outrage and Hope: Social Movements in the Internet Age* (Cambridge, UK: Polity, 2012) and João B. M. Tonucci Filho, "Urban Occupations as Urban Commons? Explorations from Belo Horizonte, Brazil" (paper presented at the First IASC Thematic Conference on the Urban Commons, Track 1: Claiming the Urban Commons, Bologna, Italy, November 2015).

6. C. A. Bowers, *Revitalizing the Commons: Cultural and Educational Sites of Resistance and Affirmation* (Lanham, MD: Lexington Books, 2006).

7. Christian Borch and Martin Kornberger, eds., *Urban Commons: Rethinking the City* (Abingdon, Oxon, UK; New York: Routledge, 2015).

8. Elinor Ostrom et al., "Revisiting the Commons: Local Lessons, Global Challenges," *Science* 284, no. 5412 (1999): 278–82, accessed April 8, 2017, http://www.jstor.org/stable/2898207.

9. Martina Löw, "Managing the Urban Commons: Public Interest and the Representation of Interconnectedness," in *Urban Commons: Rethinking the City*, eds. Christian Borch and Martin Kornberger (Abingdon, Oxon, UK; London: Routledge, 2015), 112.

10. There are a variety of commons and ways of interpreting them; next to urban commons, knowledge commons increasingly play a role in the twenty-first century; see Marilyn Strathern, *Commons and Borderlands: Working Papers on Interdisciplinarity, Accountability and the Flow of Knowledge* (Wantage, UK: Sean Kingston, 2004) and Brett M. Frischmann, Michael J. Madison, and Katherine J. Strandburg, eds., *Governing Knowledge Commons* (Oxford: Oxford University Press, 2014).

11. For a detailed discussion of the meaning of cultural commons and how they have changed over time, see Enrico Bertacchini, ed., *Cultural Commons: A New Perspective on the Production and Evolution of Cultures* (Cheltenham, UK; Northampton, MA: Edward Elgar, 2012) and Jaron Rowan Madirolas, "The Creative Industries and the Cultural Commons: Transformations in Labour, Value and Production" (PhD diss., Goldsmiths College, University of London, 2012), http://research.gold.ac.uk/8022/1/CS_thesis_Rowan.pdf.

12. Mary Dellenbaugh et al., eds., *Urban Commons: Moving Beyond State and Market*, Bauwelt Fundamente 154 (Basel, Switzerland: Birkhäuser, 2015).

13. The initiatives observed in urban commons reach a new complexity as diverse users participate; for further discussion, see Peter Blundell Jones, Doina Petrescu, and Jeremy Till, eds., *Architecture and Participation* (Abingdon, Oxon, UK; New

York: Taylor & Francis, 2005) and Paul Jenkins and Leslie Forsyth, eds. *Architecture, Participation and Society* (Abingdon, Oxon, UK; New York: Routledge, 2010).

14. Stavrides, *Common Space*, 18.

15. The increasing consumption and commodification of public places is an urgent issue as people lose their connection to their own intimate space; for a concise discussion, see Bas Hubertus Arnoldus Spierings, "Cities, Consumption and Competition: The Image of Consumerism and the Making of City Centres" (PhD diss., Radboud University Nijmegen, 2006). http://repository.ubn.ru.nl/handle/2066/45763 and Anne M. Cronin and Kevin Hetherington, eds., *Consuming the Entrepreneurial City: Image, Memory, Spectacle* (New York: Routledge, 2008).

16. Anthony Giddens, *The Consequences of Modernity* (Cambridge, UK: Polity Press, 1990).

17. A diverse body of literature and research has focused in recent years on the importance of alleyways and how they can help to increase the sustainability of cities; see Heide Imai, "Sensing Tokyo's Alleyways: Everyday Life and Sensory Encounters in the Alleyways of a City in Transition," in *Everyday Life in Asia: Social Perspectives on the Senses*, eds. Devorah Kalekin-Fishman and Kelvin E. Y. Low (Burlington, VT: Ashgate, 2010), 48–68; Kazuo Usugi et al., *Machi Roji Saisei no Design - Roji ni Manabu Seikatsu Kukan no Saiseijutsu* (Designing for the Revival of a Town with Alleyways) (Tokyo: Shokokusha, 2010); and Toshiyuki Masubuchi, *Rojiura ga Bunka o Umu!: Saigairo to Sono Kaiwai no Hen'yo* (Back Alleys Give Birth to Culture: Changing Narrow Alleys and Surrounding Areas) (Tokyo: Seikyusha, 2012).

18. Mel Rhodes, "An Analysis of Creativity," *The Phi Delta Kappan* 42, no. 7 (1961): 305–10, http://www.jstor.org/stable/20342603.

19. Peter Meusburger, Joachim Funke, and Edgar Wunder, eds. *Milieus of Creativity: An Interdisciplinary Approach to Spatiality of Creativity* (Berlin: Springer Science and Business Media, 2009), 56.

20. Charles Landry, *The Creative City: A Toolkit for Urban Innovators*, 2nd ed. (New York: Earthscan, 2008).

21. Richard L. Florida, *Who's Your City?: How the Creative Economy is Making Where to Live the Most Important Decision of Your Life* (New York: Basic Books, 2008).

22. Enrico Moretti, *The New Geography of Jobs* (Boston: Houghton Mifflin Harcourt, 2012).

23. Guiseppe Cocco and Barbara Szaniecki, eds., *Creative Capitalism, Multitudinous Creativity: Radicalities and Alterities* (Lanham, MD: Lexington Books, 2015).

24. Dennis Rodwell, *Conservation and Sustainability in Historic Cities* (Oxford: Wiley-Blackwell, 2007).

25. M. Ibrahim Khan and M. Rafiqul Islam, *True Sustainability in Technological Development and Natural Resource Management* (New York: Nova Science Publishers, 2007).

26. Paul Kingsnorth, *One No, Many Yeses: A Journey to the Heart of the Global Resistance Movement* (London: Free Press, 2004).

27. John Huige, "Sustainability and the Cultural and Creative Industries," in *Pioneering Minds Worldwide: On the Entrepreneurial Principles of the Cultural and*

Creative Industries, eds. Giep Hagoort, Aukje Thomassen, and Rene Kooyman (Chicago: University of Chicago Press, 2012), 89.

28. Marina van Geenhuizen and Peter Nijkamp, eds., *Creative Knowledge Cities: Myths, Visions and Realities* (Cheltenham, UK; Northampton, MA: Edward Elgar, 2012).

29. André Sorensen, "Building World City Tokyo: Globalization and Conflict over Urban Space," *The Annals of Regional Science* 37, no. 3 (2003): 523.

30. Darko Radović and Davisi Boontharm, eds., *Small Tokyo* (Tokyo: Flick Studio, 2012), 23.

31. Atelier Bow-Wow, *Pet Architecture Guide Book* (Tokyo: World Photo Press, 2002).

32. Cotton Mather, P. P. Karan, and Shigeru Iijima, *Japanese Landscapes: Where Land & Culture Merge* (Lexington, KY: University Press of Kentucky, 1998), 35.

33. Yoshinobu Ashihara, *The Hidden Order: Tokyo Through the Twentieth Century* (Tokyo; London: Kodansha, 1989).

34. Günter Nitschke, *From Shinto to Ando: Studies in Architectural Anthropology in Japan* (London: Academy Editions, 1993), 23.

35. Ash Amin and Stephen Graham especially talk about the importance of unplanned activities in times of globalization, when our everyday life and normal ways of inhabiting space are overregulated with the consequence that local identities and heterogeneity disappear. Ash Amin, and Stephen Graham, "The Ordinary City" *Transactions of the Institute of British Geographers* 22 (1997): 411–29. Also, see Heide Jäger, "Breathing Gaps: Life and Processes in Tokyo's Dynamic Urban Spaces of In-between" (Master's thesis, Faculty of Architecture, Oxford Brookes University, 2005).

36. Stephan Köhn, "Das Zeitalter der Prekarisierungsgesellschaften?: Einige Gedanken zum Diskurs des Prekären in Ost und West," in *Prekarisierungsgesellschaften in Ostasien?: Aspekte der sozialen Ungleichheit in China und Japan*, eds. Stephan Köhn and Monika Unkel (Wiesbaden, Germany: Harrassowitz, 2016), 35.

37. Anthony Fielding, "Social Class Segregation in Japanese Cities: The Case of Kyoto," *Transactions of the Institute of British Geographers* 29, no.1 (2004): 68.

38. Michael Middleton et al., *Participatory Critical Rhetoric: Theoretical and Methodological Foundations for Studying Rhetoric in Situ* (Lanham, MD: Lexington Books, 2015), 35.

39. Paul Waley, *Tokyo: City of Stories* (New York: Weatherhill, 1991), 193.

40. André Sorensen, *The Making of Urban Japan: Cities and Planning from Edo to the Twenty First Century*, Nissan Institute/Routledge Japanese Studies (London; New York: Routledge, 2004).

41. The project Ueno Sakuragi Atari, which offers different spaces, including a space called *mina no roji* (the alleyway for all), invites all kinds of users, including locals, new residents, and visitors. "Ueno Sakuragi Atari: Connecting. Reaching Out," *Ueno Sakuragi Atari*, accessed October 30, 2015, http://uenosakuragiatari.jp/concept_en.html.

42. Yee Sing Teh and Yoh Sasaki, "Rehabilitation Methods and Revitalization Strategies in the Old Inner-City Areas of Rapid Growth Cities in Asia: A Comparison

of Four Cities: Penang, Hanoi, Shanghai, and Tokyo," in *Proceedings of International Symposium on City Planning 2014* (Hanoi, Vietnam), accessed October 10, 2015, http://www.cpij.or.jp/com/iac/sympo/Proceedings2014/35-fullpaper.pdf.

43. Akiko Shibara, "A 10-Year Pathway to *Machizukuri* in Yanaka," *Machizukuri Monthly Magazine*, 2010, 99.

44. Chika Urashima, "Yanaka Inn Transports Tourists Back to Postwar Tokyo," *Asahi Shimbun*, accessed February 19, 2016, http://ajw.asahi.com/article/cool_japan/style/AJ201602190005o.

45. Paul Waley, "Who Cares About the Past in Today's Tokyo?," in *Urban Spaces in Japan: Social and Cultural Perspectives*, eds. C. Brumann and E. Schulz (London: Routledge, 2012), 161.

46. Susan Hirakawa, "Social Movements and the Preservation of Tradition: A Case Study in Urban Tokyo" (Master's thesis, Faculty of Liberal Arts, Sophia University, Tokyo, 2002).

47. André Sorensen, "Neighborhood Streets as Meaningful Spaces: Claiming Rights to Shared Spaces in Tokyo," *City & Society* 21, no. 2 (2009): 220.

48. Peter Popham, *Tokyo: The City at the End of the World* (Tokyo: Kodansha International, 1985).

49. Christian Dimmer and Koen Klinkers, "Downtown Tokyo Revisited: Restructuring and Urban Renaissance," *S+RO* (*Stedebouw & Ruimtelijke Ordening*) 6 (2004): 44–58.

50. Marieluise Jonas and Heike Rahmann, *Tokyo Void: Possibilities in Absence* (Berlin: Jovis, 2014).

51. Jane Jacobs, *The Death and Life of Great American Cities* (New York: Vintage Books, 1961).

BIBLIOGRAPHY

Amin, Ash, and Stephen Graham. "The Ordinary City." *Transactions of the Institute of British Geographers* 22 (1997): 411–29.

Ashihara, Yoshinobu. *The Hidden Order: Tokyo Through the Twentieth Century*. Tokyo; London: Kodansha, 1989.

Bertacchini, Enrico, ed. *Cultural Commons: A New Perspective on the Production and Evolution of Cultures*. Cheltenham, UK; Northampton, MA: Edward Elgar, 2012.

Blundell Jones, Peter, Doina Petrescu, and Jeremy Till, eds. *Architecture and Participation*. Abingdon, Oxon, UK; New York: Taylor & Francis, 2005.

Borch, Christian, and Martin Kornberger, eds. *Urban Commons: Rethinking the City*. Abingdon, Oxon, UK; New York: Routledge, 2015.

Bow-Wow, Atelier. *Pet Architecture Guide Book*. Tokyo: World Photo Press, 2002.

Bowers, C. A. *Revitalizing the Commons: Cultural and Educational Sites of Resistance and Affirmation*. Lanham, MD: Lexington Books, 2006.

Castells, Marcus. *Networks of Outrage and Hope: Social Movements in the Internet Age*. Cambridge, UK: Polity, 2012.

Cocco, Guiseppe, and Barbara Szaniecki, eds. *Creative Capitalism, Multitudinous Creativity: Radicalities and Alterities*. Lanham, MD: Lexington Books, 2015.

Cronin, Anne M., and Kevin Hetherington, eds. *Consuming the Entrepreneurial City: Image, Memory, Spectacle*. New York: Routledge, 2008.

Dellenbaugh, Mary, Markus Kip, Majken Bieniok, Agnes Katharina Müller, and Martin Schwegmann, eds. *Urban Commons: Moving Beyond State and Market*. Bauwelt Fundamente 154. Basel, Switzerland: Birkhäuser, 2015.

Dimmer, Christian, and Koen Klinkers. "Downtown Tokyo Revisited: Restructuring and Urban Renaissance." *S+RO* (*Stedebouw & Ruimtelijke Ordening*) 6 (2004): 44–58.

Edensor, Tim, Deborah Leslie, Steve Millington, and Norma M. Rantisi. *Spaces of Vernacular Creativity: Rethinking the Cultural Economy*. Abingdon, Oxon, UK; New York: Routledge, 2010.

Fielding, Anthony. "Social Class Segregation in Japanese Cities: The Case of Kyoto." *Transactions of the Institute of British Geographers* 29, no.1 (2004): 64–84.

Florida, Richard L. *Who's Your City?: How the Creative Economy is Making Where to Live the Most Important Decision of Your Life*. New York: Basic Books, 2008.

Frischmann, Brett M., Michael J. Madison, and Katherine J. Strandburg, eds. *Governing Knowledge Commons*. Oxford: Oxford University Press, 2014.

Giddens, Anthony. *The Consequences of Modernity*. Cambridge, UK: Polity Press, 1990.

Harvey, David. *Rebel Cities: From the Right to the City to the Urban Revolution*. New York: Verso, 2012.

Hirakawa, Susan. "Social Movements and the Preservation of Tradition: A Case Study in Urban Tokyo." Master's thesis, Faculty of Liberal Arts, Sophia University, Tokyo, 2002.

Huige, John. "Sustainability and the Cultural and Creative Industries." In *Pioneering Minds Worldwide: On the Entrepreneurial Principles of the Cultural and Creative Industries*, edited by Giep Hagoort, Aukje Thomassen, and Rene Kooyman, 85–91. Chicago: University of Chicago Press, 2012.

Imai, Heide. "Sensing Tokyo's Alleyways: Everyday Life and Sensory Encounters in the Alleyways of a City in Transition." In *Everyday Life in Asia: Social Perspectives on the Senses*, edited by Devorah Kalekin-Fishman and Kelvin E. Y. Low, 48–68. Burlington, VT: Ashgate, 2010.

Jacobs, Jane. *The Death and Life of Great American Cities*. New York: Vintage Books, 1961.

Jäger, Heide. "Breathing Gaps: Life and Processes in Tokyo's Dynamic Urban Spaces of In-between." Master's thesis, Faculty of Architecture, Oxford Brookes University, 2005.

Jenkins, Paul, and Leslie Forsyth, eds. *Architecture, Participation and Society*. Abingdon, Oxon, UK; New York: Routledge, 2010.

Jonas, Marieluise, and Heike Rahmann. *Tokyo Void: Possibilities in Absence*. Berlin: Jovis, 2014.

Khan, M. Ibrahim, and M. Rafiqul Islam. *True Sustainability in Technological Development and Natural Resource Management*. New York: Nova Science Publishers, 2007.

Kingsnorth, Paul. *One No, Many Yeses: A Journey to the Heart of the Global Resistance Movement*. London: Free Press, 2004.

Köhn, Stephan. "Das Zeitalter der Prekarisierungsgesellschaften?: Einige Gedanken zum Diskurs des Prekären in Ost und West." In *Prekarisierungsgesellschaften in Ostasien?: Aspekte der sozialen Ungleichheit in China und Japan*, edited by Stephan Köhn and Monika Unkel, 1-36. Wiesbaden, Germany: Harrassowitz, 2016.

Landry, Charles. *The Creative City: A Toolkit for Urban Innovators*. 2nd ed. New York: Earthscan, 2008.

Löw, Martina. "Managing the Urban Commons: Public Interest and the Representation of Interconnectedness." In *Urban Commons: Rethinking the City*, edited by Christian Borch and Martin Kornberger, 109–26. Abingdon, Oxon, UK; London: Routledge, 2015.

Masubuchi, Toshiyuki. *Rojiura ga Bunka o Umu!: Saigairo to Sono Kaiwai no Hen'yo* (Back Alleys Give Birth to Culture: Changing Narrow Alleys and Surrounding Areas). Tokyo: Seikyusha, 2012.

Mather, Cotton, P. P. Karan, and Shigeru Iijima. *Japanese Landscapes: Where Land & Culture Merge*. Lexington, KY: University Press of Kentucky, 1998.

Meusburger, Peter, Joachim Funke, and Edgar Wunder, eds. *Milieus of Creativity: An Interdisciplinary Approach to Spatiality of Creativity*. Berlin: Springer Science and Business Media, 2009.

Middleton, Michael, Aaron Hess, Danielle Endres, and Samantha Senda-Cook. *Participatory Critical Rhetoric: Theoretical and Methodological Foundations for Studying Rhetoric in Situ*. Lanham, MD: Lexington Books, 2015.

Moretti, Enrico. *The New Geography of Jobs*. Boston: Houghton Mifflin Harcourt, 2012.

Nitschke, Günter. *From Shinto to Ando: Studies in Architectural Anthropology in Japan*. London: Academy Editions, 1993.

Ostrom, Elinor, Joanna Burger, Christopher B. Field, Richard B. Norgaard, and David Policansky. "Revisiting the Commons: Local Lessons, Global Challenges." *Science* 284, no. 5412 (1999): 278–82. Accessed April 8, 2017. http://www.jstor.org/stable/2898207.

Popham, Peter. *Tokyo: The City at the End of the World*. Tokyo: Kodansha International, 1985.

Radović, Darko, and Davisi Boontharm, eds. *Small Tokyo*. Tokyo: Flick Studio, 2012.

Rhodes, Mel. "An Analysis of Creativity." *The Phi Delta Kappan* 42, no. 7 (1961): 305–10. http://www.jstor.org/stable/20342603.

Rodwell, Dennis. *Conservation and Sustainability in Historic Cities*. Oxford: Wiley-Blackwell, 2007.

Rowan Madirolas, Jaron. "The Creative Industries and the Cultural Commons: Transformations in Labour, Value and Production." PhD diss., Goldsmiths College, University of London, 2012. http://research.gold.ac.uk/8022/1/CS_thesis_Rowan.pdf.

Shibara, Akiko. "A 10-Year Pathway to *Machizukuri* in Yanaka." *Machizukuri Monthly Magazine*, 2010.

Sorensen, André. "Building World City Tokyo: Globalization and Conflict over Urban Space." *The Annals of Regional Science* 37, no. 3 (2003): 519–31.

———. *The Making of Urban Japan: Cities and Planning from Edo to the Twenty First Century*. Nissan Institute/Routledge Japanese Studies. London; New York: Routledge, 2004.

———. "Neighborhood Streets as Meaningful Spaces: Claiming Rights to Shared Spaces in Tokyo." *City & Society* 21, no. 2 (2009): 207–29.

Spierings, Bas Hubertus Arnoldus. "Cities, Consumption and Competition: The Image of Consumerism and the Making of City Centres." PhD diss., Radboud University Nijmegen, 2006. http://repository.ubn.ru.nl/handle/2066/45763.

Stavrides, Stavros. *Common Space: The City as Commons*. London: Zed Books, 2016.

Strathern, Marilyn. *Commons and Borderlands: Working Papers on Interdisciplinarity, Accountability and the Flow of Knowledge*. Wantage, UK: Sean Kingston, 2004.

Teh, Yee Sing, and Yoh Sasaki. "Rehabilitation Methods and Revitalization Strategies in the Old Inner-City Areas of Rapid Growth Cities in Asia: A Comparison of Four Cities: Penang, Hanoi, Shanghai, and Tokyo." In *Proceedings of International Symposium on City Planning 2014*. Hanoi, Vietnam. Accessed October 10, 2015. http://www.cpij.or.jp/com/iac/sympo/Proceedings2014/35-fullpaper.pdf.

Tonucci Filho, João B. M. "Urban Occupations as Urban Commons? Explorations from Belo Horizonte, Brazil." Paper presented at the First IASC Thematic Conference on the Urban Commons, Track 1: Claiming the Urban Commons, Bologna, Italy, November 2015.

"Ueno Sakuragi Atari: Connecting. Reaching Out." *Ueno Sakuragi Atari*. Accessed October 30, 2015. http://uenosakuragiatari.jp/concept_en.html.

Urashima, Chika. "Yanaka Inn Transports Tourists Back to Postwar Tokyo." *Asahi Shimbun*. Accessed February 19, 2016. http://ajw.asahi.com/article/cool_japan/style/AJ201602190005.

Usugi, Kazuo, Hitoshi Aoki, Kazuro Iseki, and Satoshi Okamoto. *Machi Roji Saisei no Design—Roji ni Manabu Seikatsu Kukan no Saiseijutsu* (Designing for the Revival of a Town with Alleyways). Tokyo: Shokokusha, 2010.

van Geenhuizen, Marina, and Peter Nijkamp, eds. *Creative Knowledge Cities: Myths, Visions and Realities*. Cheltenham, UK; Northampton, MA: Edward Elgar, 2012.

Waley, Paul. *Tokyo: City of Stories*. New York: Weatherhill, 1991.

———. "Who Cares About the Past in Today's Tokyo?" In *Urban Spaces in Japan: Social and Cultural Perspectives*, edited by C. Brumann and E. Schulz, 148–66. London: Routledge, 2012.

Chapter 7

Coexistence, Conflict, and Governance in Multi-Ethnic Districts

Two Case Studies in the Municipalities of Padua and Venice, Italy

Claudia Mantovan

Cities have always been places where strangers and all sorts of people rub shoulders with one another. Being poles of attraction for individuals and groups in search of opportunities makes them ideal places for the coexistence of people coming from different social and cultural backgrounds, with different life experiences behind them.[1] But processes underway in recent decades have tended to alter this coexistence of different people, who are increasingly at risk of living no longer side-by-side, but on separate islands with no bridges linking them together.

This tendency derives from changes that have been taking place in Western society ever since the second half of the 1970s, and that are attributable to the ongoing processes of globalization, a phenomenon accompanied by a marked increase in international migratory movements. Since the turn of the new millennium, virtually every developed nation in the world has become a country of immigration.[2] The migratory flows of the last thirty years have evolved not only in quantitative, but also in qualitative terms, with an increase in such elements of differentiation as country of origin, gender, generation, human capital, social status, and occupational skills. This adds to the complexity of society in a manner hitherto unknown, even to the point of inducing some well-known scholars to speak of the advent of "super-diversity"[3] and of a "society of strangers."[4]

Globalization also coincides with increasing levels of social exclusion and polarization. The acceleration of modern capitalism and the fiscal crisis of the welfare state model have given rise to a dramatic increase in the inequalities in the distribution of incomes and assets in every industrialized country, and especially in the United Kingdom and the United States.[5] Starting in the 1990s, the gradual dismantling of the welfare state has also coincided almost everywhere with an unprecedented growth in states' investment in their means of repression, a process that Loïc Wacquant describes as the passage from a social state to a "penal state."[6] The State is no longer able to fully exert its sovereignty over the economic dynamics, which have become global, so it restricts itself to maintaining conditions of "law and order" functional to the reproduction of capital and to legitimizing the current mechanisms for the distribution of wealth.[7] It is precisely due to the pruning back of social support programs that once succeeded in containing hardship, poverty, and social conflict that the main focus of political action has been shifting from the fight against insecurity (i.e., against the conditions that threaten social and economic security) to the fight against unsafety (i.e., the insecurity triggered by criminality).[8] The political emergence of the demand for security against criminality and the extensive criminalization of poverty[9] are structurally related to the governance of new processes of social exclusion, becoming the most effective way to "naturalize" the imposition of new social models of exclusion.[10]

In most countries of the Western world, these new theoretical approaches to the management of deviance and criminality have been translated into "zero tolerance" policies, or the so-called "actuarial" policies based on situational prevention and on the repression of a behavior that, though not a crime (or only a minor offence), is considered undesirable and gives citizens the impression of living in a degraded environment. The repressive action has consequently focused on beggars who make a nuisance of themselves, streetwalkers, drunks and drug addicts who hang out in public places, the homeless, and so on.[11]

The repercussions of the global changes briefly outlined above are having visible effects on urban life, posing new challenges for those responsible for governing our cities. Now that it is no longer possible to confine otherness to faraway places, or to reabsorb within the social body those individuals who are now considered permanently "wasted lives,"[12] we are seeing that the impossibility of keeping the marginalized and the "different" out of our urban city centers is prompting the construction of material and symbolic walls within the city to defend its inhabitants from the threat, or supposed threat, posed by such people.[13] The tendency underway in Western cities—and especially in the United States (where we often see the most extreme version of tendencies that are visible in Europe too)—is to organize urban space according to a genuine "ecology of fear,"[14] dividing it into ghettos for the poor and "gated communities" for the rich.[15] The universal dimension of the public space is eroded; it comes to be perceived as a dangerous place where people passing through are assumed to be hostile simply because they are unfamiliar. Immigrants and the marginalized are held responsible for much of the sense of insecurity generated by the structural changes underway in our society, facilitated by the criminalization and stigmatization of migrants and ethnic minorities in the political public debate and in the media,[16] and this is giving rise to phenomena of urban conflict. These trends run the risk of threatening the realization of comprehensive sustainability within cities, which includes a realization of social justice and equality in society.

These dynamics are typical of modern cities but are somehow exacerbated in districts near railway stations. The areas around the station of any medium-to-large city often have many of the same characteristics: a strategic position in terms of the local communication and transport routes, large flows of people in transit, and the availability of certain low-threshold services. Hence, they have the usually sizeable presence (and visibility in public spaces) of migrants, social drop-outs, and more or less blatant drug-dealing and/or streetwalking activities. These areas are often described by the local media as "degraded" and unsafe, and it is not unusual for episodes of urban conflict to occur due to the concomitant presence of social groups with contrasting needs (regarding the usage of these public spaces, for instance), which has prompted the creation of "citizens' committees." Focusing our research on these particular areas makes it easy to identify tendencies and problems that affect Western society generally (albeit less severely), such as the resident population's growing complexity and diversity in terms of their country of origin, social status and lifestyles, the increasing social exclusion, the fragmentation of society, and the construction of material and symbolic boundaries between different groups.

For the study presented here,[17] we judged it as interesting to analyze two districts containing the railway stations of two different cities, Padua and Mestre (Mestre is part of the municipality of Venice, but, unlike the islands of the city's historical center, it is on the mainland). To be more precise, in Padua, we considered the Stazione and Arcella urban units, and, in Mestre, the so-called Piave district. The areas considered in the two case studies lie less than forty kilometers apart in northeastern Italy and have numerous elements in common, particularly as concerns the rapid demographic changes that have taken place in the last twenty years and the problems reported by residents and traders. Some interesting differences emerge between them, however, one of which concerns the profoundly different approaches taken by their local authorities in an effort to deal with the problems of these "critical" districts, even though the two administrations shared the same political orientation (at the time of our study, they both belonged to the center-left).

This study aimed mainly to analyze (1) the problems encountered by the different urban populations living in the areas examined (residents, workers, and people who neither live nor work in the district, but who frequently visit its public spaces) and the reasons for any conflict between them; (2) the action taken in the districts (be it organic or only extemporaneous) by the various actors mobilized to improve the quality of life in the area, that is, local authorities, nonprofit and profit-making organizations, immigrants' associations, and citizens' committees. In addition to the content of their interventions, we also analyze whether their schemes were implemented separately or coordinated by any of the various actors involved, for which reasons, and with what results. By comparing the results in both cities, the objective of the latter part was to highlight the different ways that subjects tackle the challenges of today's urban life, in an attempt to identify which more closely approaches the realization of comprehensive sustainability in cities.

DECONSTRUCTING THE PROTEST AGAINST THE DISTRICTS' "DEGRADATION": THE MULTI-DIMENSIONALITY OF THE PROBLEMS EXPERIENCED BY RESIDENTS AND TRADERS

In both case studies, some of the residents and traders complained that their districts had become unsafe and "degraded." This malcontent has prompted the creation of "citizens' committees," which protest against what they see as the deterioration of their area; they feel abandoned by the institutions, and the media describe their districts in horrific terms as an epicenter of criminal activity. Such forms of resident mobilization, which have increased considerably in Italy since the 1990s, focusing especially on issues of safety and the environment,[18] are partly due to the decline of the traditional channels of social and political representation and mediation affecting the country as a whole. As mentioned above, the "safety" issue is the way in which society today interprets the problems deriving from the changes underway as a consequence of globalization and the welfare policy crisis. The decline of politics has led to it gradually being substituted by a "moralist" type of logic, a debate in which a confrontation between the "good guys" and the "bad guys" has replaced the conflict between the social classes.[19] The creation of residents' committees to protest against the micro-criminality on their streets thus becomes a way to attract the attention of the local administrators—especially since the introduction of direct elections for Italy's local mayors in 1993, which has led to citizens making increasing demands on them.

If we analyze the problems raised by the people interviewed for our study in more depth, it soon becomes clear that they have very little to do with criminality proper. Instead, they tend to confirm the importance of factors identified in the relevant sociological literature as being related to the perception of insecurity. Among such factors, what stand out are the so-called "environmental and social incivilities":[20] what people complain about most is the presence of individuals perceived as deviant in relation to their own standards for the care and maintenance of public spaces and the coexistence of people using them. The residents object to noisy groups of people sitting and drinking beer on the pavements, abandoning their rubbish on the ground, relieving themselves in public, begging, sleeping in doorways or shadowy corners, and so on. The people we interviewed complained about phenomena of genuine micro-criminality too (streetwalking, drug dealing), not because they saw those behaviors as a source of real physical or economic danger to themselves, but because they listed them among the previously mentioned "undesirable" types of behavior. It is worth mentioning at this point that none of the committee members interviewed had personally been the victims of any criminal actions in their districts; and they said that their own area was actually not as dangerous as the local media claimed, or as it seemed to their acquaintances who lived elsewhere.

Another important element that emerged among the problems reported by residents and traders was the poor quality of the urban space. Both the areas analyzed have a high density of residential and other buildings that have saturated the physical space, leaving little room for people to meet and socialize. This is particularly true of Arcella in Padua, a district described as "claustrophobic," where the lack of public spaces where young people (and immigrants in particular) might meet is the reason why some spaces are put to a variety of uses (like the car park outside a supermarket used as a skate park). This increases the young immigrants' visibility in the district, however, and perception of them as a "disturbing" presence, adding to the original residents' sense of insecurity, especially among the elderly, who are particularly numerous in the districts analyzed in both Padua and Mestre (they account for about one-third of the resident indigenous population).

The areas analyzed also suffer from a lack of trust in the institutions and from weak social relations, additional factors that have a bearing on how secure people feel. As regards the institutions, there is evidence of a generalized perception of the local authority's inaction and/or inefficiency in dealing with the problems that the locals experience on a daily basis. It is important to draw a distinction here, however, because this problem was felt much more in Padua than in Mestre, thanks to the type of action being taken in the latter district, discussed below, which is reducing the distance between the local institutions and some of the citizens. As for the social relations, the two districts analyzed are profoundly fragmented socially, are highly heterogeneous in terms of ethnicity, age group, and social class, and have a rapid turnover of residents. These features give rise to a sort of "mosaic" of groups and spaces, often consisting of "tiles" that are spatially juxtaposed but disconnected from one another in relational and communicational terms. The physical proximity of different social classes or groups does not necessarily reduce their social distance, and it may even dilate the gap between them.[21] It is not enough to be neighbors in order to establish relations, unless they are based on a common identity, and this also depends on people's occupations, family ties, geographical origins, or economic relationships that can be shared and can serve as a reference, because "there is nothing more intolerable than the physical proximity (experienced as promiscuity) of people who are socially very separate."[22]

Particularly when it comes to places for socializing, we find evidence in our two case studies of a "rupture" between the original residents and workers on the one hand and the immigrant residents and city users on the other. The former withdraw to the private spaces of their own and their friends' homes and a handful of restaurants and bars. The latter, as mentioned earlier, occupy the district's public spaces in order to meet, socialize, and exchange information. The original residents' withdrawal into their private spaces is only partly because of the presence of "undesirables" in the district's public spaces. It is largely due to changes in the ways in which Western societies tend to socialize and to their district's demographics. Many of the people interviewed also emphasized that these districts lack appropriate meeting places, such as community centers for the elderly.

In the end, we could say that the main factor behind the "crisis" taking place in these districts, as experienced by a part of their inhabitants, can be broken down to powerful processes of change that have overwhelmed them in the last two decades. These changes mainly relate to the macro-variables discussed above: the acceleration and globalization of international migratory movements and an ever-greater social exclusion. As already mentioned, these phenomena are showing their visible effects in our cities, with increasing numbers of migrants arriving from various countries and people living in the margins of society, especially in the parts of the city that afford them more opportunities, such as the areas around the railway stations forming the object of our study. Cities thus provide the stage where the effects of globally originated changes are enacted, and this poses a major challenge to the city administrations. For a start, they have to respond to calls for the "removal" of the migrants, the marginalized, and the "deviants" from the public spaces coming from the members of citizens' committees, who see themselves as the district's legitimate users, entitled to lay down the rules of behavior for the other urban populations, often simply by virtue of their prolonged presence in the area as residents or traders. These dynamics are not new. The importance of the temporal dimension in processes of community closure and its use as a symbolic resource in the exclusion of others were already brought to light many years ago by Norbert Elias and John Scotson.[23]

The citizens' demands are simply impossible to meet, however, using the means available to city governments alone. This is self-evident given the "global" causes behind the phenomena described, as frequently emphasized by the local administrators themselves and by the exponents of the police forces of the two municipalities, who we interviewed or heard speak at public meetings.

Their explanations have focused on the fact that the type of behavior that the citizens complain about is impossible to repress for legal reasons (prostitution is not a crime, and Roma people arriving from Romania are EU citizens and entitled to circulate freely, etc.). The authorities often complain that the weapons at their disposal are blunt, and they call for changes to the legislation that would enable them to take more appropriate action (such as revisiting the current law on prostitution). It has also been emphasized that mere situational prevention is not really effective; it only has a "displacement effect," shifting the problems and the undesirable presences elsewhere, often only temporarily.

It is the pressure exerted by the citizens' committees on the exponents of the local institutions to deal with these problems (an impossible task in many ways) that lies behind the latter's adoption of policies that are often purely "symbolic" (administrative orders are issued, the police make their presence felt, etc.). Their purpose is merely to provide reassurance, and their declared objective is only to take action on the citizens' "perceived security." The effect of such measures is actually counterproductive: as the annual report drawn up by the operators of the *Accoglienza invernale* (Winter welcome) project for the homeless in Padua have emphasized, the intensification of the police checks in the area around the city's railway station in recent years has had the effect of making the low-threshold service-providers lose contact with the most needy, including the migrants with no documents and the Roma, with a negative fallout not only on these people's security, but also on the safety of the autochthonous citizens. It is in nobody's interests, not even of the collectivity at large, to increase the presence of individuals with all sorts of problems—including health issues—who find no answers or solutions of any kind in the territory. Measures involving an intensification of the police's presence have completely failed to improve the "objective" safety of the townspeople, or even their "perceived" security. The promoters of citizens' committees whom we interviewed said themselves that they sometimes had the opposite effect, raising the sense of social alarm and people's perception of living in a "state of siege."

Faced with these same scenarios, the two municipal administrations in Padua and Venice have adopted very different policies. We now go on to see in more detail what action they have taken in the two cities and what effects they have had.

MESTRE: A "KALEIDOSCOPE" OF SCHEMES COORDINATED BY THE LOCAL AUTHORITY

On May 19, 2006, some residents of the Piave district in Mestre called a meeting at the community center in via Sernaglia to complain about their area's problems. What triggered their concern was the fact that the pedestrian zone in San Francesco square had become a meeting place for people with drug problems, but the issues they discussed were broader, including the rapid demographic and commercial transformation underway in the district as a result of the rising numbers of immigrants. The tone of the meeting was "inflamed": the citizens protested loudly, calling for a stronger presence of police in the district.

In response, the Venice local authority tried to enlist these individuals and involve them directly in organizing action to deal more constructively with the problems that they reported. In particular, the then-alderman for social policies appointed the local authority's ETAM *Animazione di Comunità e Territorio* (Community and Territorial Animation) unit to take action in the Piave district. As an ETAM operator explained, entrusting an action of this type to a service that forms part of the local administration is an unusual approach on the Italian scene,[24] which makes it possible to take action with a greater efficacy and continuity:

> Our service is part of the Venice Local Authority's social policies department. We deal with community animation. Among other things, we promote the citizens' participation in the life of the city [...] We're probably the only institutional service concerned with community animation in Italy, which goes to show a particular characteristic: [...] the fact that the Venice local authority decided already at the end of the 1980s to create a service of this type and to make it a permanent part of its staffing plan, with its own educators. (Radiana Gregoletto, ETAM educator)

What this operator told us confirms the findings of a previous study comparing the work of street educators in Venice and Padua, and that is the fact that, unlike what happens in Padua (and many other Italian municipalities), the work done on the streets of Venice is a service that forms an integral part of a local authority service sector; it is implemented by about 100 educators working for the local authority's various social services. This has several important positive effects, one of which is to ensure the loyalty and autonomy of the operators, and another concerns the continuity of their activities.[25]

Soon after the above-mentioned, angry meeting, two ETAM educators were assigned to a project for the Piave district that was begun in 2006. Consistently with the community animation approach, these educators promoted activities that were to be organized by small groups of residents. This gave birth to the *Gruppo di lavoro Piave* (Piave working group), which currently has as its spokesperson one of those angry people who helped to organize that first meeting and became a member of a committee established to protest the problems in San Francesco square. This working group has chosen to remain informal and is loosely composed of about fifteen people, almost all of them autochthonous district residents with a medium-to-high formal education.

In cooperation with ETAM, the working group has embarked on what might be defined as an "integrated prevention" scheme with a distinct focus on "community prevention."[26] It organizes a number of micro-activities and micro-projects that all serve the general purpose of promoting mutual understanding, social harmony, and cohesion in the Piave area. The group's actions range from improving the urban furniture to social animation to economic requalification.

At the "heart" of the measures implemented by the Piave working group and ETAM, there are, above all, actions that focus on sociocultural animation and the promotion of intercultural encounters. In this setting, there have been both one-off projects and schemes that have become a stable and/or regular feature of community life. Among the latter, one of the flagship undertakings is undeniably the multi-ethnic choir *Voci dal Mondo* (Voices from the world), created in 2008 by one of the ETAM educators, and consisting of approximately fifty members from both the autochthonous and the immigrant populations. Other activities have become well established too, like the "barter markets" for children and adolescents up to thirteen years old, organized every third Saturday afternoon of the month in the San Francesco square pedestrian zone, or the *cena di quartiere* (street party), which has been organized in June in the same square for five years now. As for the more extemporaneous community animation schemes, these have been implemented partly by bringing events normally organized elsewhere in the city to the Piave district and, vice versa, by "taking" the Piave road into the heart of Mestre with an artistic installation called *Via Piave in Piazza* set up in Mestre's central Piazza Ferretto and representing the Piave road's commercial businesses.

Another aspect of the complex and multifaceted kaleidoscope of measures for the Piave district concerns the work done on the topic of communication, perceptions, and representations, undertaken mainly through the medium of the figurative arts. A periodical titled *Le voci di via Piave* (Voices of the Piave road), first published in October of 2007, aims to challenge the distorted and alarmist pictures of the district painted by the local press, making space for other voices that are usually not represented in the mass media. The periodical also reports on the activities and the public debates organized by the Piave work group in an effort to "nurture a new way to think about the Piave road" (Roberta Zanovello, ETAM educator). Various public meetings have been organized on the topics of security and on the changes underway in the district. Several photography exhibitions have been held, as well, showing old and new citizens of Mestre and the Piave area, also with a view to providing food for thought on the social phenomena underway in the district. This is a crucial area of intervention partly because a recent study on coexistence and conflict in districts of immigration showed that the local community's ability to produce coherent narrations about the area in which they live is an important integrating factor. There is "a specific social capital (proportional to the degree of dynamism and the propensity for cooperation of the political actors and associations in the district) that becomes manifest in the capacity to combat the distorted representations of intergroup relations in the district, and to convey authentic representations successfully, to the townspeople and further afield."[27] In today's society of communication, this "narrative autonomy" is a key factor of social cohesion and is also crucial to the realization of comprehensive sustainability, given that the latter depends not just on practices but also on representations, culture, and conceptual work.

The fact that so many coordinated and coherent schemes have been implemented in the Piave district is largely thanks to the local authority's role as "director" and coordinator through its ETAM unit. On several occasions during our interviews, the founders of the Piave working group mentioned how important the ETAM unit's support had been to the success of their undertakings.

The action that is being taken in the Piave district has also had an effect on the immigrants' participation (or some of them at least). Consistently with the picture seen elsewhere in Italy,[28] so too in Mestre, the immigrants have been forming associations, mainly consisting of members of the same nationality. Although it is not easy for them to find the time and the resources they need to spend on their participation, there are nonetheless clear signs of a network of associations of immigrants active in the area and of the public visibility of several of their leaders, particularly those who cooperate with ETAM and the Piave working group. It is, therefore, safe to say that one of the effects of the projects begun in the Piave district has been to pinpoint those immigrants with a stronger desire to commit themselves to the territory, to join forces and develop a network, and to emerge on the public stage.

Several representatives of a citizens' committee called *Un impegno per la città* (A commitment to the city)—established in October of 2005 by a number of Piave district residents and traders living or working nearest to the railway station to protest against the area's "degradation"—have quite often been involved in activities promoted by the Piave working group. Relations between these two groups of citizens feature a certain degree of ambivalence, and they are sometimes frankly critical of each other. The Piave working group accuses the committee of fomenting alarmism by emphasizing only the negative aspects of the area and of giving priority to their protest rather than to a more constructive commitment. The committee retaliates, saying that the Piave working group tries to gloss over the district's problems, partly because it works so closely with the local authority. But there are also examples of these groups interacting and cooperating with one another. In fact, the active presence of the local authority in the district assured by the ETAM has another, crucial function: it builds a bridge between the townspeople and the institutions—an important factor in times like the present, when the former have trouble finding appropriate channels for communicating with the public administrations because of the decline in the traditional mechanisms of mediation and representation (such as the political parties and the trade unions). As mentioned in the previous paragraph, and already illustrated in other studies (see Petrillo's investigation on the citizens' committees of the 1990s in Genoa),[29] the citizens' protests and calls for the police to deal with matters that, strictly speaking, have nothing at all to do with their safety are really a way of voicing their frustration because they feel abandoned by the institutions and lack other, more appropriate references. As mentioned earlier, a loss of trust in the institutions is one of the variables that increases people's perception of insecurity. The work done by the ETAM educators thus also involves listening to the citizens (and especially to those taking a more confrontational stance), making them feel that the institutions take an interest in their needs and helping them put their problems in the right perspective.

PADUA: THE SPONTANEOUS AND FRAGMENTARY ACTIVATION OF THE PRIVATE SOCIAL SECTOR

Unlike the situation in Mestre, the local authority governing the city of Padua has, in a sense, preferred to ride the wave of its citizens' discontent. In crucial areas like the districts containing the railway station, it has adopted a policy based on a massive use of administrative orders, and on an intensification of the police's presence, even with cooperation from the Army. Other measures for the area around the railway station have focused merely on urban requalification while much less attention seems to have been paid to investing in social schemes. If anything, the latter have been even further reduced in recent years: the operators working on the *Accoglienza invernale* project for the homeless, managed by an association called *Noi—Famiglie padovane contro l'emarginazione e la droga* (Us—Paduan families against marginalization and drugs), have reported, for instance, that abandonment of a street education scheme in 2008 has had the effect of reducing the number of people in difficulty being taken into care.

The Paduan administration invests little in social policies and tends (like many other Italian local authorities) to delegate social activities to the third, not-for-profit sector. In doing so, it relinquishes the chance to have a role in coordinating and orienting these activities, and that is why any action taken in Padua to deal with the numerous problems of the areas around the railway station has been scarcely systematic. There is no global project of the kind adopted in Mestre. The local authority has limited itself to acting as a partner in projects promoted by others. It is because of the local institutions' evasion of their responsibilities and/or inefficacy (something that all our interviewees mentioned) that many other actors have become involved in efforts to deal with the problems of the district around the railway station—and with more or less democratic nuances. In other words, there are signs of a number of schemes being implemented by various committees and associations to "occupy" the public spaces: some have aimed to include all the people gravitating around the station; others have been seen as a sort of "war" waged against other people to regain possession of the public spaces rather than an effort to improve relations between all the parties concerned.

The large-scale project promoted by Banca Etica and undertaken by the Associazione Mimosa in cooperation with the University of Trieste, which goes by the name of *La città partecipata* (The shared city), was certainly a scheme of the former type. This project consisted of action research conducted in the square outside Padua's railway station. Alongside Banca Etica and the Associazione Mimosa, various private and public subjects were partners in the project, including Padua's local authority, the Confesercenti (confederation of traders in commerce and tourism), the Italian State railways, several hotels in the area, the Fondazione Cariparo, and the Cooperativa Adriatica (a supermarket chain that financed the project, together with Banca Etica). The project involved activities intended to satisfy the demands of local residents and traders (analyzed as part of the research), which were mainly to restore the district's image and to organize more cultural and social events in the area. The project succeeded in having much less impact and territorial visibility than the projects undertaken in the Piave district in Mestre, however, and this seems to be at least partially due to the fact that it was not the local administration that was promoting and coordinating the project in Padua. In an interview, an operator of the Associazione Mimosa made it clear that social processes need a "director," and it is difficult to achieve this without the decisive contribution of the political decision-makers:

> What is needed for the area around the railway station is a sort of "control room," someone to govern the changes that are already underway and cannot be stopped [...]. We are unable, physically and economically, to guarantee certain processes [...] because it's not just that we need money, we also need decisions, a political management of the matter. [...] It is a rather shortsighted approach to manage things without a founding concept, a final idea about how we aim to handle an area, a district [...] and not only from an architectural standpoint. [...] When it comes to social issues, Padua abounds with associations and voluntary workers and, because it is so rich and the third sector has provided numerous solutions, this may have led the politicians to sit back instead of developing their

> own ideas and choosing a course of action; perhaps it is because there is a lot of wealth on this side that the other side has kept to the sidelines. If you look at Venice, it is rather the other way around: the local authority is very strong from the social standpoint, and there is a third sector that is probably only about a third of the size of ours, or even less. (A. P., Associazione Mimosa)

The final words of this interview seem particularly interesting and not only because the interviewee spontaneously draws a comparison to the situation in the municipality of Venice. The hypothesis that emerges is that it is because of a very strong third sector that the public administration has failed to act as a farsighted "guide" in orienting the processes of change underway in its territory, leading them in a direction that will promote social cohesion and interaction between all members of the population (traders, residents, people attending the public spaces), safeguarding the rights of all the parties concerned. In actual fact, other studies that we have conducted in areas where we identified the same phenomenon have given us the impression that it is the local authority's evasiveness/inefficacy in governing the social phenomena in its area that has induced the third sector to take a leading role. In other words, faced with unresponsive political decision-makers, lay organizations and Catholic voluntary groups see themselves as being invested with the task of providing solutions and services. That is why, in what we defined a while back as the *local immigration field* in Treviso and Vicenza, Caritas and the trade unions have come to act as coordinators and promoters far more than in Venice, where this position was already occupied by the local authority.[30] Returning to the present research, it, therefore, seems reasonable to say that the need for sociocultural animation in public spaces is strongly felt by the local populations of Mestre and Padua alike, but in the latter city it is the private social sector that is trying to meet this need and in a manner that is unavoidably more fragmented and without an organic goal.

For instance, several groups have taken action independently to organize activities designed to revitalize the public spaces in the area behind Padua's railway station too. To give a couple of examples, several traders organized an event called *Viale Arcella in festa*, a sort of trade fair with activities for children, and a group of opposition councilors on the local district council arranged a *reading* near the Borgomagno flyover.

Various *comitati di cittadini antidegrado* ("antidegradation" citizens' committees) have also sprung up on both sides of the railway station. These groups all consist exclusively of Italian residents and traders (see our comments above on this type of organization), and they are rather self-centered, each concerned only with their own particular part of the district. In addition to their sole interest in their own, very limited neighborhoods, another reason for such fragmentation lies in the diversity of their strategies: several of the spokespeople for the most proactive of these committees (those doing more than just protesting) have told us that they do not always approve of the others' methods. Their activities include drawing up petitions, taking action to improve awareness, and organizing public meetings. But the committee spokespeople interviewed were also loudly demanding that the institutions take steps to implement a commercial and social revitalization of their areas. In some cases, they themselves promoted community animation schemes, though usually only for the benefits of the autochthonous population. This is the case, for instance, of the burraco card game tournament promoted by the *Comitato Stazione* (Railway Station Committee) in 2011 and held on the so-called "boulevard" outside Padua's railway station. As the brochure describing the event explained, the tournament was seen as "the last card against degradation," and its purpose was to induce "drunks and drug dealers who hang out and sleep here day and night" to move away from the area, and "to draw the attention of the Paduan people to this area (characterized by a huge explosive potential) for the organization of cultural and recreational events that would have a positive return in terms of the area's social requalification."[31] The goal, therefore, was not to facilitate communication and interaction between the different members of the community gravitating around the railway station (between Italians and immigrants, for instance), but to repossess this public space for "Paduan people" (where Paduan "naturally" means only the autochthonous population, the only people considered as legitimate citizens), and to oust any other "disturbing" occupants.

In the districts bordering the railway station in Padua, we thus see a number of separate situations in which groups of individuals take action independently to deal with what they identify as the area's main problems, and their objectives are often rather heterogeneous and uncoordinated. What is missing is a "director" in the form of a local authority capable of intercepting, adding value, creating a network, and focusing the energies of those who are taking action, or who are potentially willing and able to do so. Among the latter, for instance, there are the immigrants' associations. The picture that emerged from our research is one of immigrant associations that have withdrawn behind closed doors, becoming scarcely visible in the public sphere. This seems to be partly attributable to the failure of the local institutions (the district councils and, to some degree, the local authority too) to take effective steps to establish genuine communication channels and interfaces with the exponents of the variegated world of the immigrants' self-organization. Even the relations between the *Razzismo Stop* (Stop Racism) association (an emanation of the "Disobbedienti" political movement active in supporting migrants, who appreciate it enormously) and Padua's administrators have always been less cooperative than those between Venice's administrators and the *Caracol* cooperative (which likewise has its roots in the left-wing-occupied social centers). The lagoon city's administration often involved *Caracol* in organizing activities in the area of its social policies, whereas the association in Padua was not invited to take part in the *Accoglienza invernale* project nor in any other social inclusion scheme. Even the citizens' committees find it quite difficult to establish communication channels with Padua's local government.

CONCLUSION

The comparison drawn between the two cases considered in this study provides some useful food for thought in light of our reconstruction of the elements of conflict between the different urban populations living in the districts concerned and of our analysis of the work done by public and private actors to deal with the problems identified.

On the first aspect, the outcomes of the project are consistent with other reports in the scientific literature on the topic. Almost every study on the issue of insecurity has reached the conclusion that people's fear of crime does not stem from any direct experience of violence and criminality.[32] It has already been demonstrated that the perception of insecurity depends more on factors that have nothing at all to do with crime, such as the quality of the urban spaces in which people live, as well as their social vulnerabilities, social networks, and trust in others and in their institutions. Regarding the lack of care for public spaces, it is particularly obvious that an important factor that contributes to making people feel insecure and abandoned by the institutions relates to a whole array of common "incivilities." They are a symptom of peoples' difficult relations with one another, and they interfere with the chances of living in harmony alongside individuals who have other habits and customs, and who seem to be unable to accept rules that others see as fundamental to a civil coexistence.[33] This issue is closely linked to the matter of interpersonal trust: one of the sources of insecurity studied by ethnomethodology is the violation of communication rules, which leads to a suspension of the trust that a person places in another as a dependable and competent agent, consequently undermining elements that are normally taken for granted in our daily lives and introducing existential anxieties that take the form of suspicion and hostility.[34]

It is, therefore, hardly surprising that analyses on the motives behind the sense of insecurity and the complaints of citizens living in districts of the type described here (which all feature communities with a marked ethnic heterogeneity and very different social classes) consistently acknowledge the enormous weight of what is perceived as a violation of the rules of coexistence.[35] As one of the best-known voices among the scholars of urban insecurity in Italy puts it, "what concerns people in the vast majority of Italian cities is not the violent, bloody crime, but the cracks appearing in the essential dependability of daily life."[36] The emergence of this issue is certainly partly attributable to the growing *super-diversity* of Western societies. The new influx of immigrants characterized by numerous elements of internal differentiation clearly makes negotiating "rules of civil coexistence" more complex and increasingly gives people the feeling that they are living in an unpredictable, unfamiliar world.

These impressions are confirmed by the interviews that we conducted with the spokespeople of the "antidegradation" citizens' committees, who wanted those individuals whom they saw as undesirable "removed" from their district. Here, again, we can see the important contribution of environmental and social incivilities to the townspeople's sense of insecurity and demand for action. This call for the "removal" of certain undesirable types, often associated with a request for the closure of poles of attraction for such people (such as the soup kitchens in Padua or the so-called "ethnic" shops), is partly due to changes underway in the model of social control and to the consequent criminalization of poverty (as discussed in the introduction of this essay). The outcome is a mainstream way of thinking that associates the whole grey area of incivilities and "urban disorder" with the criminal domain. Worse still, as a result of this process, numerous social problems (such as begging) are attributed a penal valence by linking them with the issue of urban degradation and antisocial behavior.[37]

In this scenario, however, the citizens' committees (or the more proactive among them at least) have some elements in common, and one of these is the strong call for the institutions to revitalize the districts around the railway station commercially and socially. This request goes to show, once again, that these committees are driven not only by a sense of insecurity, but also (and possibly even more strongly) by other needs, such as the need to retrieve a sociality of proximity. It is fundamentally important for public decision-makers to hear these voices and respond to these needs—also, to prevent such self-appointed organizations from drifting in an increasingly antidemocratic direction. As Tamar Pitch writes:

> The decline of the mass political parties, of their forms of association and participation, has left a void in which various different, more or less ephemeral forms of association take shape. What the latter lack, however, are reliable channels for communication and mediation with the local and national institutions of political power. I suggest that the curvature taken by many of these mobilizations therefore depends largely on this ephemeral relationship with the institutions, on the limited authoritativeness of the institutions, and even on the instrumental use made of the former by the latter. It is up to the institutions to provide a procedural framework capable of protecting individual rights

> against moral majorities, while making room for negotiation and confrontation between cultural diversities. This means that citizens' mobilizations can be prevented from having privatizing consequences if their organization and commitment proceed along clearly regulated lines that assure representation and transparency. So it is not a matter of demonizing these mobilizations—quite the opposite. They bear witness to a desire for civic participation, which is essential nowadays to produce an inclusive governance. It is more a matter of focusing this desire for protagonism and participation towards non-privatizing, inclusive, "democratic" goals.[38]

In order to promote a sustainability that achieves the objectives of social justice and equality in today's cities, we can, therefore, see the huge importance of a local authority capable of orienting and guiding the processes of change in the territory it governs, of listening to all the organized actors (even the more conflictual), and of ensuring their involvement, in order to focus their energies on the objective of creating an inclusive city where the rights of every inhabitant are respected.

These considerations are sustained by the findings of the second part of our study, conducted to produce a map of the public and private actors that have been committed to improving the quality of life in the districts forming the object of our research. What emerges from our analysis is that, faced with much the same problems, the administrations in Padua and Mestre have adopted two very different policies, which could be briefly described using the formula "delegated participation" as opposed to "guided participation." The local authority in Padua has basically opted to ride the wave of protest voiced by a part of its citizens, implementing a policy based on law and order and urban requalification while making no effort to take action organically on the numerous problems in the areas around the city's railway station. The administration in Mestre, on the other hand, has sent its own specialized service for the mediation of conflict and community animation into the Piave district. Its operators work in cooperation with a group of residents to accompany and guide the processes of change underway in the district, also intercepting the more conflictual members of the community and striving to involve them in proactive and constructive schemes. The resulting "kaleidoscope" of measures is having a number of positive effects on the territory. For a start, the autochthonous citizens' loud appeal for the public spaces of their district to be "repossessed," brought back to life with fun, cultural events, and socially and economically requalified has been amply satisfied in Mestre but left unfinished in Padua. In addition, the Venetian local authority's "directorship" has also had a positive effect on the development of a network of undertakings and on ensuring that they are inclusive and promote intercultural relationships. The local authority's proactive approach and guidance have thus succeeded in promoting a governance that adds value and generates a network to sustain all the actors operating in the territory—not only the Piave working group but other subjects too, such as the associations of immigrants and the citizens' committees.

Padua is quite a different story, in which the local authority plays no such part. We find a number of schemes being implemented by committees and associations to "retrieve" the public spaces, some of which aim to include all the people who gravitate around the railway station (such as the *Città partecipata* project coordinated by the Associazione Mimosa) while others take the form of a war to be waged against outsiders to win back the public spaces, making no attempt to facilitate relations between all the parties involved. From this point of view, the comparison between the "antidegradation" burraco card-playing tournament organized by the Stazione di Padova committee and the annual district dinners organized in Mestre's Piave district offers an emblematic example of the difference between the two cities' approaches: the former is an elitist initiative intended only for the autochthonous population, the latter an event designed to promote intercultural relationships.

What emerges from our research is the importance of a local administration capable of orienting and guiding the processes of change and of promoting and coordinating a network of public and private actors in the various policy sectors—an aspect that has been emphasized by numerous other scholars too.[39] Contrary to claims advanced by the rhetoric of subsidiarity that governance implicates, attributing less importance to the public actor when constructing public goods, we can, therefore, say that the importance of the public body tends to become even greater on the governance stage, and this is because the functions of guidance and coordination that subsidiarization requires tend to increase.[40] As Carlo Pisano writes in a recent contribution on the "governance of security," this is the reason why several authors have proposed the concept of "State anchored pluralism."[41] Used in a normative sense, this concept refers to the role of "anchor" that the State should continue to play in the production of security goods. In other words, the State is not just a "node" among many others within a "network" of actors involved in the local governance of security. It has a specific, fundamental role inasmuch as it has the responsibility and the capacity to coordinate all the various actors involved in the network.[42]

But serving in this role of "director" requires a vision and clear political and social goals that (as in the Venetian approach) must assuredly be to create an inclusive city where the rights of all the parties concerned are respected. As Alessandro Baratta suggests in one of his well-known writings,[43] we ultimately need to stop talking about the "right to security" of the few, and to reiterate the need for the "security of the rights of everyone" instead. If security is a private good, it will be in short supply and must be competed for, which is why the topic of security for a city's inhabitants "implicates staging a perennial conflict between victims and offenders."[44] We, therefore, need the political forces to gradually delineate and govern security as a public good, moving toward "producing a greater security for everyone, and especially for those (the most defenseless) whose rights have been less well safeguarded."[45]

To reach these goals, social policies and urban policies must proceed in synergy. Concerning the latter, in particular, several highly authoritative scholars have said that, in order to limit the tendency for disaggregation and the construction of material and symbolic boundaries in modern cities, we need public policies and projects that aim to create spaces where "different" people can come together and experience a genuine exchange by participating in shared practices to restore the public space on a local level.[46] Architectural and urban planning projects should, therefore, protect and cultivate *mixophilia*, creating and disseminating open, attractive, and hospitable public spaces that all categories of people will be tempted to visit and share because "the art of living peacefully and contentedly with differences, and of benefiting from the consequent variety of inputs and opportunities, is becoming the most important ability that citizens need to learn and practice."[47] While fear of contamination and exposure to differences has generated more or less deliberate segregations, our town centers nonetheless remain the only places where remedies to these ills can be adopted, the only places where the related risks can be converted into individual and collective opportunities on a daily basis, because they are always great social laboratories, the ideal places for experimentation and contamination.[48]

NOTES

1. Georg Simmel, *Die Großstädte und das Geistesleben* (Dresden: Petermann, 1903).
2. Douglas S. Massey, "La ricerca sulle migrazioni nel XXI secolo," in *Stranieri in Italia. Assimilati ed esclusi*, edited by Asher Colombo and Giuseppe Sciortino (Bologna: Il Mulino, 2002), 25–49.
3. Steven Vertovec, "Super-diversity and Its Implications," *Ethnic and Racial Studies* 30, no. 6 (2007): 1024–54, doi:10.1080/01419870701599465.
4. Ash Amin, *Land of Strangers* (Cambridge, UK: Polity Press, 2012).
5. Thomas Piketty, *Le capital au 21e siècle* (Paris: Seuil, 2013).
6. Loïc Wacquant, *Les prisons de la misère* (Paris: Éditions Raisons d'Agir, 1999).
7. Zygmunt Bauman, *Globalization: The Human Consequences* (New York: Columbia University Press, 1998); and Lucia Re, "Le trasformazione delle politiche di controllo sociale nell'era della globalizzazione," *Dei delitti e delle pene* 8, no. 1–2–3 (2001): 141–62.

8. Zygmunt Bauman, *In Search of Politics* (Cambridge, UK: Polity Press, 1999). We refer, here, to Bauman's well-known tripartite interpretation of the concept of "security": the Polish sociologist distinguishes between *certainty* (security of a cognitive type, which has to do with the predictability and intelligibility of the surrounding environment and our place in it), *security* (that has to do with each individual's social and occupational condition, and depends on the stability and reliability of their world), and *safety* (a physical type of security that relates to the threat to an individual's person and possessions).

9. Loïc Wacquant, *Punishing the Poor: The Neoliberal Government of Social Insecurity* (Durham, NC: Duke University Press, 2009).

10. Massimo Pavarini, "Introduzione. L'aria della città rende (ancora) liberi? Dieci anni di politiche locali di sicurezza," in *L'amministrazione locale della paura. Ricerche tematiche sulle politiche di sicurezza urbana in Italia*, ed. Massimo Pavarini (Rome: Carocci, 2006), 11–64.

11. Emilio Santoro, "Castigo e delitto," in *Invito allo studio della società*, ed. Pier Paolo Giglioli (Bologna: Il Mulino, 2005), 69–100.

12. Zygmunt Bauman, *Wasted Lives. Modernity and Its Outcasts* (Cambridge, UK: Polity Press, 2004).

13. Zygmunt Bauman, *Fiducia e paura nella città* (Milano: Mondadori, 2005).

14. Mike Davis, *Ecology of Fear: Los Angeles and the Imagination of Disaster* (New York: Metropolitan Books, 1998).

15. Derek S. Hyra, *The New Urban Renewal: The Economic Transformation of Harlem and Bronzeville* (Chicago: University of Chicago Press, 2008).

16. Marcello Maneri, "Si fa presto a dire 'sicurezza'. Analisi di un oggetto culturale," *Etnografia e ricerca qualitativa* 6, no. 2 (2013): 283–309.

17. This article summarizes the main findings of a research project titled *La partecipazione di autoctoni e migranti alla vita della città come fattore di sicurezza urbana: due casi studio nei Comuni di Padova e Venezia* (Participation of the autochthonous and immigrant populations in city life as a factor of urban security: two case studies in the municipalities of Padua and Venice), designed and conducted by Claudia Mantovan and presented by Prof. Giuseppe Mosconi for a competition for projects of excellence 2009/2010 organized by the Fondazione Cariparo, which selected and funded the project. The research was conducted from January 2011 to January 2014, based on a composite method: review of background data on the districts under analysis; press review of the daily newspapers *Il Mattino di Padova* and *La Nuova Venezia*; ethnographic observation in the public spaces and retail shops in the districts; fifty semi-structured interviews with politicians, spokespeople from the citizens' committees, representatives of the autochthonous associations and cooperatives, and of the associations of immigrants; analysis of the documentation produced by the local administrations, district councils and other organizations considered. The full results of the research were published in Claudia Mantovan and Elena Ostanel, *Quartieri contesi. Convivenza, conflitti e governance nelle zone Stazione di Padova e Mestre* (Milano: FrancoAngeli, 2015).

18. Donatella Della Porta, *Comitati di cittadini e democrazia urbana* (Soveria Mannelli, Italy: Rubbettino, 2004).

19. Tamar Pitch, *Contro il decoro. L'uso politico della pubblica decenza* (Rome-Bari: Laterza, 2013).

20. Leonardo Chiesi, "Le inciviltà: degrado urbano e insicurezza," in *La sicurezza urbana,* ed. Rossella Selmini (Bologna: Il Mulino, 2004), 129–40.

21. Jean-Claude Chamboredon and Madeleine Lemaire, "Proximité spatiale et distance sociale. Les grands ensembles et leur peuplement," *Revue française de sociologie* 11 (1970): 3–33.

22. Pierre Bourdieu, "Effets de lieu," in *La misère du monde,* ed. Pierre Bourdieu (Paris: Seuil, 1993), 259 (my translation).

23. Norbert Elias and John Scotson, *The Established and the Outsiders* (London: Sage, 1994).

24. The Italian welfare system is traditionally characterized by a high degree of delegation to third sector organizations. Tiziana Caponio, "Policy Networks and Immigrants' Associations in Italy: The Cases of Milan, Bologna and Naples," *Journal of Ethnic and Migrations Studies* 31, no. 5 (2005): 931–50. doi:10.1080/13691830500177891.

25. Annalisa Butticci et al., "Vigili di quartiere ed educatori di strada: Padova e Venezia," in *L'amministrazione locale della paura. Ricerche tematiche sulle politiche di sicurezza urbana in Italia,* ed. Massimo Pavarini (Rome: Carocci, 2006), 141–225.

26. On the various types of prevention, see Rossella Selmini, "Le attività di prevenzione: una premessa teorica," *Quaderni di Città Sicure* 20 (2000): 37–52.

27. Ferruccio Pastore and Irene Ponzo, eds., *Concordia Discors. Convivenza e conflitto nei quartieri di immigrazione* (Rome: Carocci, 2012), 303 (my translation).

28. Claudia Mantovan, "Immigration and Citizenship: Participation and Self-organisation of Immigrants in the Veneto (North Italy)." *Forum Qualitative Sozialforschung/Forum: Qualitative Social Research* 7, no. 3 (May 2006). http://www.qualitative-research.net/index.php/fqs/article/view/140/307; and Claudia Mantovan, "Cohesion without Participation: Immigration and Migrants' Associations in Italy," *Patterns of Prejudice* 47, no. 3 (2013): 253–68, doi:10.1080/0031322X.2013.797172.

29. Antonello Petrillo, *La città delle paure. Per un'archeologia dell'insicurezza urbana* (Avellino, Italy: Elio Sellino, 2003).

30. Claudia Mantovan, *Immigrazione e cittadinanza. Auto-organizzazione e partecipazione dei migranti in Italia* (Milano: FrancoAngeli, 2007).

31. Luca Siragusa, "Torneo di Burraco," brochure, Comitato Zona Stazione di Padova, 2011.

32. Hugues Lagrange. "La peur à la recherche du crime," *Déviance et société* 17, no. 4 (1993): 385–417.

33. Chiesi, "Le inciviltà."

34. Harold Garfinkel, *Studies in Ethnomethodology* (Englewood Cliffs, NJ: Prentice Hall, 1967).

35. See, for instance, Pastore and Ponzo, *Concordia Discors*; Séverine Germain, "Le retour des villes dans la gestion de la sécurité en France et en Italie," *Déviance et Société* 36, no. 1 (2012): 61–84, doi:10.3917/ds.361.0061; Dino Giovannini and Loris Vezzali, eds., *Sicurezza, coesione sociale e immigrazione. Prospettive teoriche e analisi di un caso* (Milano: Unicopli, 2011); Rossella Selmini, "Il punto di vista

dei comitati di cittadini," *Quaderni di Città Sicure* 11 (1997): 77–94; Barbara Giacomozzi and Rossella Selmini, "Nuovi attori: censimento dei comitati di cittadini operanti in Emilia-Romagna sulla sicurezza," *Quaderni di Città Sicure* 5 (1996): 211–19; and Sally Engle Merry, *Urban Danger: Life in a Neighborhood of Strangers* (Philadelphia: Temple University Press, 1981).

36. Giandomenico Amendola, "La paura e la vita quotidiana," in *Insicuri e contenti. Ansie e paure nelle città italiane*, ed. Giandomenico Amendola (Naples, Italy: Liguori, 2011), 13 (my translation).

37. Rossella Selmini, "Origine, sviluppi ed esiti delle politiche di governo locale della criminalità nell'Italia contemporanea," in *Dalle città all'Europa. Strategie di sicurezza urbana*, ed. Franco Corradini (Reggio Emilia, Italy: Nuova Prhomos, 2014), 23–40.

38. Tamar Pitch, "Sono possibili politiche democratiche per la sicurezza?," *Rassegna Italiana di Sociologia* 42, no. 1 (2001): 153–54. doi:10.1423/2561 (my translation).

39. See, for instance, Germain, "Le retour des villes"; Matteo Bassoli, "La governance locale: alcuni aspetti teorici," in *La governance del territorio. Partecipazione e rappresentanza della società civile nelle politiche locali*, eds. Matteo Bassoli and Emanuele Polizzi (Milano: FrancoAngeli, 2011), 15–37; Emanuele Polizzi, "Nodi e prospettive della governance all'italiana," in *La governance del territorio. Partecipazione e rappresentanza della società civile nelle politiche locali*, eds. Matteo Bassoli and Emanuele Polizzi (Milano: FrancoAngeli, 2011), 211–20; Luigi Bobbio, *Amministrare con i cittadini. Viaggio tra le pratiche di partecipazione in Italia* (Soveria Mannelli, Italy: Rubbettino, 2007); Pitch, "Sono possibili politiche democratiche"; and Filippo Barbera, "Le politiche della fiducia. Incentivi e risorse sociali nei Patti Territoriali," *Stato e Mercato* 63, no. 3 (2001): 413–50, doi:10.1425/472.

40. Polizzi, "Nodi e prospettive."

41. Ian Loader and Neil Walker, *Civilizing Security* (Cambridge, UK: Cambridge University Press, 2007).

42. Carlo Pisano, "La governance della sicurezza. Dinamiche partecipative e ruolo dell'autorità," in *La governance del territorio. Partecipazione e rappresentanza della società civile nelle politiche locali*, eds. Matteo Bassoli and Emanuele Polizzi (Milano: FrancoAngeli, 2011), 149–69.

43. Alessandro Baratta, "Diritto alla sicurezza o sicurezza dei diritti?," in *La bilancia e la misura. Giustizia, sicurezza, riforme*, eds. Stefano Anastasia and Mauro Palma (Milano: FrancoAngeli, 2001), 22–36.

44. Pavarini, "Introduzione," 38 (my translation).

45. Ibid., 38.

46. Bauman, *Fiducia e paura*; Manuel Castells, *La città delle reti* (Venice, Italy: Marsilio, 2004); Boaventura De Sousa Santos, *Democratizzare la democrazia* (Troina, Italy: Città Aperta, 2003); Patrick Le Galès, *European Cities: Social Conflict and Governance* (Oxford: Oxford University Press, 2003); and David Harvey, "Frontiers of Insurgent Planning," *Plurimondi* 1, no. 2 (1999): 269–86.

47. Bauman, *Fiducia e paura,* 33 (my translation).

48. Gabriella Turnaturi, "La città," in *Invito allo studio della società*, ed. Pier Paolo Giglioli (Bologna: Il Mulino, 2005), 183.

BIBLIOGRAPHY

Amendola, Giandomenico. "La paura e la vita quotidiana." In *Insicuri e contenti. Ansie e paure nelle città italiane*, edited by Giandomenico Amendola, 1–24. Naples, Italy: Liguori, 2011.

Amin, Ash. *Land of Strangers*. Cambridge, UK: Polity Press, 2012.

Baratta, Alessandro. "Diritto alla sicurezza o sicurezza dei diritti?" In *La bilancia e la misura. Giustizia, sicurezza, riforme*, edited by Stefano Anastasia and Mauro Palma, 22–36. Milano: FrancoAngeli, 2001.

Barbera, Filippo. "Le politiche della fiducia. Incentivi e risorse sociali nei Patti Territoriali." *Stato e Mercato* 63, no. 3 (2001): 413–50. doi:10.1425/472.

Bassoli, Matteo. "La governance locale: alcuni aspetti teorici." In *La governance del territorio. Partecipazione e rappresentanza della società civile nelle politiche locali*, edited by Matteo Bassoli and Emanuele Polizzi, 15–37. Milano: FrancoAngeli, 2011.

Bauman, Zygmunt. *Fiducia e paura nella città*. Milano: Mondadori, 2005.

———. *Globalization: The Human Consequences*. New York: Columbia University Press, 1998.

———. *In Search of Politics*. Cambridge, UK: Polity Press, 1999.

———. *Wasted Lives. Modernity and Its Outcasts*. Cambridge, UK: Polity Press, 2004.

Bobbio, Luigi. *Amministrare con i cittadini. Viaggio tra le pratiche di partecipazione in Italia*. Soveria Mannelli, Italy: Rubbettino, 2007.

Bourdieu, Pierre. "Effets de lieu." In *La misère du monde,* edited by Pierre Bourdieu, 249–62. Paris: Seuil, 1993.

Butticci, Annalisa, Francesco Faiella, Giuseppe Mosconi, Alvise Sbraccia, and Francesca Vianello. "Vigili di quartiere ed educatori di strada: Padova e Venezia." In *L'amministrazione locale della paura. Ricerche tematiche sulle politiche di sicurezza urbana in Italia*, edited by Massimo Pavarini, 141–225. Rome: Carocci, 2006.

Caponio, Tiziana. "Policy Networks and Immigrants' Associations in Italy: The Cases of Milan, Bologna and Naples." *Journal of Ethnic and Migrations Studies* 31, no. 5 (2005): 931–50. doi:10.1080/13691830500177891.

Castells, Manuel. *La città delle reti*. Venice, Italy: Marsilio, 2004.

Chamboredon, Jean-Claude, and Madeleine Lemaire. "Proximité spatiale et distance sociale. Les grands ensembles et leur peuplement." *Revue française de sociologie* 11 (1970): 3–33.

Chiesi, Leonardo. "Le inciviltà: degrado urbano e insicurezza." In *La sicurezza urbana,* edited by Rossella Selmini, 129–40. Bologna: Il Mulino, 2004.

Davis, Mike. *Ecology of Fear: Los Angeles and the Imagination of Disaster*. New York: Metropolitan Books, 1998.

De Sousa Santos, Boaventura. *Democratizzare la democrazia*. Troina, Italy: Città Aperta, 2003.

Della Porta, Donatella. *Comitati di cittadini e democrazia urbana*. Soveria Mannelli, Italy: Rubbettino, 2004.

Elias, Norbert, and John Scotson. *The Established and the Outsiders*. London: Sage, 1994.

Garfinkel, Harold. *Studies in Ethnomethodology*. Englewood Cliffs, NJ: Prentice Hall, 1967.

Germain, Séverine. "Le retour des villes dans la gestion de la sécurité en France et en Italie." *Déviance et Société* 36, no. 1 (2012): 61–84. doi:10.3917/ds.361.0061.

Giacomozzi, Barbara, and Rossella Selmini. "Nuovi attori: censimento dei comitati di cittadini operanti in Emilia-Romagna sulla sicurezza." *Quaderni di Città Sicure* 5 (1996): 211–19.

Giovannini, Dino, and Loris Vezzali, eds. *Sicurezza, coesione sociale e immigrazione. Prospettive teoriche e analisi di un caso*. Milano: Unicopli, 2011.

Harvey, David. "Frontiers of Insurgent Planning." *Plurimondi* 1, no. 2 (1999): 269–86.

Hyra, Derek S. *The New Urban Renewal: The Economic Transformation of Harlem and Bronzeville*. Chicago: University of Chicago Press, 2008.

Lagrange, Hugues. "La peur à la recherche du crime." *Déviance et société* 17, no. 4 (1993): 385–417.

Le Galès, Patrick. *European Cities: Social Conflict and Governance*. Oxford: Oxford University Press, 2003.

Loader, Ian, and Neil Walker. *Civilizing Security*. Cambridge, UK: Cambridge University Press, 2007.

Maneri, Marcello. "Si fa presto a dire 'sicurezza'. Analisi di un oggetto culturale." *Etnografia e ricerca qualitativa* 6, no. 2 (2013): 283–309.

Mantovan, Claudia. "Immigration and Citizenship: Participation and Self-organisation of Immigrants in the Veneto (North Italy)." *Forum Qualitative Sozialforschung/Forum: Qualitative Social Research* 7, no. 3 (May 2006). http://www.qualitative-research.net/index.php/fqs/article/view/140/307.

———. *Immigrazione e cittadinanza. Auto-organizzazione e partecipazione dei migranti in Italia*. Milano: FrancoAngeli, 2007.

———. "Cohesion without Participation: Immigration and Migrants' Associations in Italy." *Patterns of Prejudice* 47, no. 3 (2013): 253–68. doi:10.1080/0031322X.2013.797172.

Mantovan, Claudia, and Elena Ostanel. *Quartieri contesi. Convivenza, conflitti e governance nelle zone Stazione di Padova e Mestre*. Milano: FrancoAngeli, 2015.

Massey, Douglas S. "La ricerca sulle migrazioni nel XXI secolo." In *Stranieri in Italia. Assimilati ed esclusi*, edited by Asher Colombo and Giuseppe Sciortino, 25–49. Bologna: Il Mulino, 2002.

Merry, Sally Engle. *Urban Danger: Life in a Neighborhood of Strangers*. Philadelphia: Temple University Press, 1981.

Pastore, Ferruccio, and Irene Ponzo, eds. *Concordia Discors. Convivenza e conflitto nei quartieri di immigrazione*. Rome: Carocci, 2012.

Pavarini, Massimo. "Introduzione. L'aria della città rende (ancora) liberi? Dieci anni di politiche locali di sicurezza." In *L'amministrazione locale della paura. Ricerche tematiche sulle politiche di sicurezza urbana in Italia*, edited by Massimo Pavarini, 11–64. Rome: Carocci, 2006.

Petrillo, Antonello. *La città delle paure. Per un'archeologia dell'insicurezza urbana.* Avellino, Italy: Elio Sellino, 2003.

Piketty, Thomas. *Le capital au 21e siècle*. Paris: Seuil, 2013.

Pisano, Carlo. "La governance della sicurezza. Dinamiche partecipative e ruolo dell'autorità." In *La governance del territorio. Partecipazione e rappresentanza della società civile nelle politiche locali*, edited by Matteo Bassoli and Emanuele Polizzi, 149–69. Milano: FrancoAngeli, 2011.

Pitch, Tamar. "Sono possibili politiche democratiche per la sicurezza?." *Rassegna Italiana di Sociologia* 42, no. 1 (2001): 137–58. doi:10.1423/2561.

———. *Contro il decoro. L'uso politico della pubblica decenza*. Rome-Bari: Laterza, 2013.

Polizzi, Emanuele. "Nodi e prospettive della governance all'italiana." In *La governance del territorio. Partecipazione e rappresentanza della società civile nelle politiche locali*, edited by Matteo Bassoli and Emanuele Polizzi, 211–20. Milano: FrancoAngeli, 2011.

Re, Lucia. "La trasformazione delle politiche di controllo sociale nell'era della globalizzazione." *Dei delitti e delle pene* 8, no. 1–2–3 (2001): 141–62.

Santoro, Emilio. "Castigo e delitto." In *Invito allo studio della società*, edited by Pier Paolo Giglioli, 69–100. Bologna: Il Mulino, 2005.

Selmini, Rossella. "Il punto di vista dei comitati di cittadini." *Quaderni di Città Sicure* 11 (1997): 77–94.

———. "Le attività di prevenzione: una premessa teorica." *Quaderni di Città Sicure* 20 (2000): 37–52.

———. "Origine, sviluppi ed esiti delle politiche di governo locale della criminalità nell'Italia contemporanea." In *Dalle città all'Europa. Strategie di sicurezza urbana*, edited by Franco Corradini, 23–40. Reggio Emilia, Italy: Nuova Prhomos, 2014.

Simmel, Georg. *Die Großstädte und das Geistesleben*. Dresden: Petermann, 1903.

Siragusa, Luca. "Torneo di Burraco." Brochure, Comitato Zona Stazione di Padova, 2011.

Turnaturi, Gabriella. "La città." In *Invito allo studio della società*, edited by Pier Paolo Giglioli, 159–83. Bologna: Il Mulino, 2005.

Vertovec, Steven. "Super-diversity and Its Implications." *Ethnic and Racial Studies* 30, no. 6 (2007): 1024–54. doi:10.1080/01419870701599465.

Wacquant, Loïc. *Les prisons de la misère*. Paris: Éditions Raisons d'Agir, 1999.

———. *Punishing the Poor: The Neoliberal Government of Social Insecurity*. Durham, NC: Duke University Press, 2009.

Chapter 8

The Erosion of the Cultural Commons and the Possibilities of Participatory Urbanism

Public Art in New Orleans, Detroit, and Port-au-Prince

Joseph Donica

Public art almost always has to struggle with the contradictions of its own definition. The public is rarely consulted on art built in its name even though the public has to contend with that art's meaning on a daily basis while moving through the city. Terms like "democracy" and "participation" are thrown around by architects, artists, and designers. But publics are rarely confused regarding the irony of the use of those terms since those publics give consent neither to the space in which it is built nor the design of the art built in their name. The lack of participation by the public in competitions for public art mirrors the lack of representation that many communities have in the urban spaces that they use. Many urban centers have conducted design competitions in which candidates' submissions are chosen without any information about the designers' identities being given to the judges. Two of these stand out as extremely contentious in the history of public art in the United States: the Vietnam Veterans Memorial in Washington, D.C., designed by Maya Lin; and the 9/11 Memorial and new World Trade Center site in New York City. A design competition provides a nod to concepts like democracy and participation, but competitions like these serve to reveal the inherent contradictions and complexities of these terms and their interactions with the contemporary urban fabric, made up of powerful agencies and firms and the publics who are left to deal with the built environment that those institutions decide to construct. So, the "public" of public art often acts as a screen to mask private interests. In this essay, I address the challenges that public space has faced—especially after disasters—and what public art in three cities has done to respond to those challenges and reinvigorate the concept of shared public space.

The recent reframing of the idea of a public—as something other than everyone who uses a space—has its roots in urban planning after the destruction that World War II brought on Europe and the reignited interest in cosmopolitan civic values to act as a guard against state violence on its citizens. After World War II, the use of modern sculpture for memorials seemed out of step with the public's renewed interest in shared civic values. And not until the Vietnam Memorial did the public see a place for modern sculpture for memorials—the reason why Lin's memorial looms large over any discussion of contemporary memorials.[1] Toward the end of the twentieth century, art's relationship to its own site of creation, as well as to those who interact with it, became more integral to its meaning and importance. Lin's memorial was important because it stood on the National Mall. Art placed in a museum, in many ways, is art without a context. The physical permanence and connection to public space is what gives public art its particularly strong political and social possibilities. While public art since the 1970s has seen interventions such as participatory art, community art, anti-monuments, and activist art, memorials have not benefited from the evolution that other public art forms have. My focus here on public memory arises from by belief that few shared experiences have as much political promise as memory does. "'Public' situates shared memory where it is most salient to collectives, in constituted audiences, positioned in some kind of relationship of mutuality that implicated their common interests, investments, or destinies, with profound political implications," Greg Dickinson, Carole Blair, and Brian L. Ott argue in their volume on public memory.[2] This essay is about those political implications.

The stakes in the politics of public memory are so high because they shape the spaces that we share, and spaces shape the very concepts that we use to understand our relationships to others with whom we share those spaces. It is an anxious process, indeed. The often-heated controversies over memorials constructed in the public's name reveal the public's anxiety over more than art, though. On the one hand, publics have demanded memorials to major losses. The cases of the Vietnam Memorial and the newer World War II Memorial in D.C. illustrate this demand. On the other hand, there is ambivalence in the public's collective feelings toward what is or what should be remembered at such sites. In 1974, Douglas Davis claimed, "[t]he twentieth century has been obsessed by the dream of a public art."[3] The same year, Amy Goldin sharpened that reality: "I think the reason there is so little genuine public art today is our own disbelief in the reality of the public world itself."[4] This disbelief in the public world, the world of participation and contending views of the future of urban space, is felt today whenever art is installed in cities. If public art is to be more participatory, then publics will have to provide strategies to challenge this disbelief if public space is to have any real connection to the future of the interactions that occur between citizens in urban spaces. With the culture of crisis that defined the past decade and a half—Iraq, Katrina, the 2008 financial crisis, the 2010 Haiti earthquake, and the terrorist attacks in Paris and Brussels—came a crisis of memory that raised questions about the democracy of memory. Who has a right to memorialize loss? What spaces are acceptable for the expression of that memory? And what "memory work" should those spaces secure for future generations?

Public memory of such events has been overtaken by corporate interests and states that are allied with those interests. The rebuilding of places affected by monumental disasters, such as New Orleans, has largely been left up to corporations and without consultation with the publics who inhabit them. We only need to look at the privatization of the New Orleans public schools for evidence of this move toward the corporatized American city in the wake of disaster. Anthropologist Donald Nonini has considered the effects that private interests have had on the global commons in the past three decades as we have seen an erosion of the commons globally. He states at the outset of his collection, which studies the breakdown of the global commons, "[w]hat is now at stake at this point in world history is control over 'the commons'—the great variety of natural, physical, social, intellectual, and cultural resources that make human survival possible."[5] Nonini breaks down the concept of the commons into categories: natural resource commons, social commons, intellectual and cultural commons, and species commons (the human body, including organs and gene sequences). Public art projects fall into the category of the intellectual and cultural commons and as such "are non-rival goods; that is, one person engaged in learning a scientific method or playing music does not diminish the learning or pleasure of other people simultaneously doing the same."[6] One may argue that public space available for public art is limited and therefore a rival good, but this essay highlights three sites at which this rule has been challenged—New Orleans, Detroit, and Port-au-Prince. These three cities, all rich in their histories of public art and participation in public space, have seen an abnormally high erosion of their cultural commons after physical and economic disasters. Thus, in these cities, there emerges a connection between participation and the cultural commons. Nonini states the existence of such commons, in fact, does not rely on rivalry and competition but on participatory action. He says, "[o]ne can argue that intellectual and cultural resources can be created and regenerated only through social exchange and sociability."[7] The point that Nonini makes about intellectual and cultural commons being "central to sustaining the cultural heritage of ethnic and national minorities"[8] is a central function of public art and public space in New Orleans, Detroit, and Port-au-Prince. All three cities have long histories of populations who have seen incredible oppression, physically and economically. New Orleans, Detroit, and Port-au-Prince have also stood as symbols of the economic inequality of urban centers in North

America due solely to the imposition of the private sphere on the public in efforts to rebuild after disasters. But the publics in all three cities also have a heritage of pushing against the encroachment of neo-capitalist interest in public space.

These cities have shown us, throughout their histories, that the commons can be radically rethought in favor of more access to the public if the political will is there. Nonini's work challenges the assumption that the tragedy of the commons presupposes. Garrett Hardin's famous explanation of William Forster Lloyd's early concept of the tragedy of the commons assumes that all resources held in common will always be depleted because individuals acting rationally will act in their own self-interest.[9] Hardin's assumption is dangerous in cities like New Orleans, Detroit, and Port-au-Prince, where economic structures long susceptible to privatization exist. However, Nonini's work has an important emphasis that challenges this assumed tragedy. His work suggests that participation in the commons is not a zero-sum game. One's use of the commons should not deplete the commons so that another has limited access to the resources of the commons. Therefore, the commons *should* by definition be a sustainable repository of resources, ideas, and culture. With the increasing globalization of cities that ties them to international markets and trade cycles, the cultural commons face a significant crisis. When the sustainability of the physical spaces of cities is threatened, so too is their culture. But there are alternatives. In their volume *Alternatives to Economic Globalization*, John Cavanagh and Jerry Mander offer ten principles for sustainable societies.[10] The principles suggest that something has been lost in the globalized city and that something should be reclaimed. To Cavanagh and Mander, both members of the International Forum on Globalization, that which has been lost is the efficacy of the commons to make local life better for those who share local commons. They blame "corporations and global bureaucracies that increasingly have served these corporations" for this erosion.[11] And specific aspects of the cultural commons that are public in name are "virtually privatized" because the government has reduced funding. They use the example of PBS in the United States, which has had to advertise for private funding in the face of budget cuts from the government. Cavanagh and Mander make the apt point that the political will to protect parts of the commons from market forces has yet to be harnessed.[12]

It remains to be seen if such political will could be harnessed given the present system of corporate funding of political campaigns and the lobbying system. Private interests and global business depend on the commons being limited and unsustainable. This presents two major problems for participatory urbanism. First, the past thirty years have seen erosions of not only public space but also the entities and institutions tasked with their protection, such as unions, social welfare programs, regulatory agencies, and publicly funded urban development agencies. Second, as I have already stated, there is an increasing use of the word "public" to describe projects and urban development that have little input from the public. A buzzword now that indicates such a shift is the "public-private" partnerships that universities and governments use as they have come to rely on corporate funding for almost all of their urban development. Where else does one need to look for evidence of the erosion of the commons?

In these two primary challenges, we see a site emerge where public art intersects with the idea of the sustainable city. The recent Paris climate talks have put public discussions of sustainable urban life back at the center of concern for urbanists and residents alike. As essential as the agreement among two hundred countries to reduce emissions is for the continued use of the commons by cities' residents, the participation in such talks had nothing to do with residents or citizens, the ones who will bear the biggest brunt of regulations and who will foot most of the bill. While the talks begin to address the problems of climate change, the United States' role in them is problematic in two ways. First, they do not come close to one hundred percent renewable energy that many activist organizations like Greenpeace have called for. And President Obama has stated that the controlling interests in the talks are corporate and private as these are the entities with the most to lose. Tim McDonnell, writing for *Mother Jones*, claims, "[t]he fossil fuel industry is bank rolling the Paris climate talks."[13] He looks to the corporate sponsors of the talks as evidence that the deal signals ominous things to come for urban dwellers. Obama has framed a narrative about creating a world safe for our children and grandchildren, but the reality on the ground is one of profits for private interest being lost in the race to sustainability.[14] So, even those grand initiatives using the rhetoric of the commons have contributed to their erosion. City parks, public space, and even space set aside for public art have long suffered from this erosion. These sites are patrolled and monitored during the hours that they operate, and those hours of operation are rarely controlled by the public for whom the rhetoric of public space is named.

The conflict arises between those who build public space and the rhetoric of the public good, as Nonini explains, "[w]hen governments subject already existing commons systems to monitoring and regulation."[15] In a way, this is government treating publicly held property as private property. Only, in this situation, the property owner does not have primary rights to the land. This essay examines this conflict over the commons at the site of public art because these sites have grown even more fragile as private interests have used disasters like a hurricane, a financial crisis, and an earthquake as opportunities to erode the commons and place their control in the hands of a few while still using the rhetoric of the public square as a justification for its erosion. Whether the commons "are renewed or not is a political question," and, in New Orleans, Detroit, and Port-au-Prince, that question has never been farther from an answer.[16]

If public art lies at the intersection of the social and intellectual commons (therefore, is subtractive and non-subtractive to these commons), then it becomes—and indeed is—the site of major conflict. Such conflicts arise from the competing interests not only in the spaces where public art is built and displayed but also over the meaning made in these spaces. In almost every instance of the announcement of new public art in a major U.S. city, there has been intense controversy. Three aspects of contemporary public art intensify the debates whenever a design is unveiled: (1) its disconnect from the past; (2) its lack of didacticism; and (3) its failure to embody the views of the public in its design and content.[17] However, public art in the form of memorials necessarily has an immediate relationship to the past. A memorial looks to the past and tells a public how to remember a history of loss. This is contentious because controlling public memory is a dangerous proposition. Many memorials take years just to plan in order to avoid these controversies. But somewhat more controversial because of its subtlety is the memorial's relationship to the future. Public memorials (I include the Heidelberg Project in this category) have a looking-back-looking-forward nature to them that communicates a dichotomous message: don't forget, but move on. This strange injunction communicated through public space is complex because it seeks to encapsulate the values of an entire city's, if not a country's, values. Public memorials are controversial because they incite conversation over what the public values. Public art is symbolic, and this symbolism can create real rifts within communities' imaginations. While the artists and designers who construct public art may intentionally obscure such symbolism, the public generally insists that art should mean something. And there is a sense that that meaning must be political.

But politics, like memory, is highly susceptible to interpretation. In his work on public sculpture, Michael North points to the elasticity and power of the symbolism of public art. "The names and myths attached" to public art, he argues, "often endure longer than" the art itself.[18] This anonymity of contemporary public art results not from the representation of the public but, rather, as North argues, from "[a]n expression of laws above and beyond any collection of people." This type of art responds "to nothing outside itself."[19] North's argument is certainly helpful in thinking about how the public makes sense of the art and public spaces built in its name. But his argument also raises an important question of whether the public *should* respond to nothing outside itself. Below, I present three cities for which public art challenges the non-referential art that North has observed. The art in New Orleans, Detroit, and Port-au-Prince are almost nothing but references to their surroundings and to the loss with which the public, for whom the art has been made, must live.

But referencing something more than itself is a complex task for public art. Nowhere was this more obvious in recent history than with the 9/11 Memorial. Joel McKim, in his article that provides a reading of New York's 9/11 Memorial in the light of Giorgio Agamben's aesthetic theory, offers a perspective on commemorative public art that moves beyond both didacticism and therapeutic "working through" in order to produce a space that is pure communication—a space that offers its only lessons and only comfort in and through language."[20] Public art has great potential in an age where publics are increasingly aware of the inequalities that define their political identities because public art eludes the museum and the class identifications that come along with it. But this is not a necessary elusion. Designers can communicate inequality through public art just as much as a museum can. The work of the designer in participatory art is to leave meaning open for interpretation yet still guide the public away from messages that shut down participation. North suggests a seemingly simple hybrid: "nonreferential art with a message."[21]

There are several examples of such nonreferential art that even invites limited participation by the public. Beyond the examples of participatory public art that I examine below, there are several—more "official"—examples of this participatory art. Both the Oklahoma City Bombing Memorial, as well as the Pentagon 9/11 Memorial, which derives its basic idea from the Oklahoma memorial, invites the public to sit and to contemplate. The Oklahoma memorial does this with chairs representing the number lost, and the Pentagon memorial does so with benches. Spontaneous memorials are more in line, though, with the type of participation that invites both the public's expressions of loss, as well as their critique of social systems that often lead to that loss. We need only remember the spontaneous memorials that filled corners of Manhattan after 9/11 with flowers, notes, and other items, to imagine the power that participatory art can have. Many cultural critics, though, have raised questions about what the extended existence of such participation in public spaces would look like.

By highlighting some of these possible spaces in New Orleans, Detroit, and Port-au-Prince, one of my goals in this essay is to show how crises open up certain spaces for the public to use in more participatory ways and with more active strategies for securing participation in them. After a few decades of art built in the public's name without their consent, in order to move public art into a more radically participatory realm, it is time for a critique of such art. "Given the market's near total saturation of our image repertoire, [...] artistic practice can no longer revolve around the construction of objects to be consumed by a passive bystander," Claire Bishop says of the emergence of public art since the early 1990s. She calls, instead, for "an art of action." She goes on to ask for an art that takes steps to "repair the social bond."[22] The focus on the social aspect of art in Bishop's work comes from a trend in "advanced art of the last decade" focusing on collectivity and not the individual personae of the artist.[23] She sees two roles for participatory art as a catalyst for social change. First, there is an opposition in the art to the material demands of the market. And, second, there is a collaborative bent to it that avoids self-interest. Because of this "twofold gesture of opposition and amelioration," Bishop states, "there can be no failed, unsuccessful, unresolved, or boring works of participatory art, because all are equally essential to the task of repairing the social bond."[24] The post-crisis spaces of New Orleans, Detroit, and Port-au-Prince, spaces whose identities have been transformed by crisis, have redefined the concept of participation through commemorative art. If participatory art has the ability to repair broken social bonds, then the relationship between the art and the communities that interact with them on a daily basis needs close attention.

It is no surprise that, in urban environments, "the sharpest equalizing claims have come from socially excluded groups, who have often imagined an alternative social order."[25] The memorialization of loss in the three communities presented here offers ways of thinking about public art that respond to the needs that Bishop highlights. Built into the contemporary urban fabric is a longing for a public art that repairs social bonds. The structures of memory in these communities communicate a porousness of the term "public" and can, in part, reflect the fragility of the communities themselves. Traditional memorials (think of the 9/11 memorial, the World War II Memorial in D.C., or the Oklahoma City bombing memorial) often act as screens that conceal troubled pasts and ask—sometimes insist—that communities move on from loss. But a different type of memorial has emerged in the past decade that works against concealment and toward repair. In a late capitalist age, the primary form of participation within traditional democratic structures is built on capitalist exchange. City dwellers interact with each other as they buy goods and services at corporately owned stores, restaurants, bars, and theaters. But parts of New Orleans, Detroit, and Port-au-Prince have found a point of solidarity in their rejection of participating in this exchange. All three cities have rich histories of seeking alternative ways of interacting with the built environment and reestablishing the sociality that crisis interrupted.

EXISTENCE AFTER DISASTER: THE NINTH WARD MEMORIAL IN NEW ORLEANS

After Hurricane Katrina, many public art projects and memorials invited participation by creating spaces at the memorials to those lost in the storm that people could use as sites of mourning but also for daily enjoyment and collective political action. The participation at these sites of memory for New Orleans's residents went further than many recent memorials to disasters because they circumvented a strictly political interpretation and provided an openness that allowed local citizens to interpret the disaster themselves. Depoliticizing memorials is one strategy for increasing participation, but this is still limited participation. The strategies for creating participatory art that emerged after Katrina move us closer to the art that Bishop calls for in that such art is radically participatory by being conceived by citizens who create spaces for other citizens to mourn in, enjoy, use to critique social structures, and even destroy. Thus, participatory public art is a kind of art that implicates citizens in its creation.

The Katrina Memorial in the Lower Ninth Ward in New Orleans was built by the local community from local materials. As figure 8.1 shows, the memorial isn't aesthetically edgy but does, through its open design, ask the community to interact with it and thus be a part of the Ninth Ward's recovery. Hurricane Katrina opened up possibilities for public art in New Orleans. On the one hand, it opened up the possibilities for local artists and designers to create a new aesthetic for a city so rooted in the past. And, on the other hand, it presented the challenge of connecting that troubled past to the possibility of a future for the city's communities whose continued existence most of the

Figure 8.1 Ninth Ward Hurricane Katrina Memorial, New Orleans, Louisiana. *Source*: Photo courtesy of David Tulloch.

nation questioned. New Orleans, to this day, has collectively sent one message to a nation that thinks it is still under water: we are still here. This is the message that New Orleans needed to communicate to the nation after the storm, and the one that it had the biggest problem accomplishing.

Most public art gives a sense of permanence. This aspect is communicated primarily through materials like concrete, steel, granite, and limestone. The Ninth Ward Katrina Memorial, designed by David Lee, makes use of inexpensive materials (pavers, wooden beams and chairs, and concrete blocks) and communicates a seriousness of intent that also evokes the makeshift memorials that appeared all over the city in the wake of the storm. The memorial sits at the foot of the Ninth Ward Bridge in the middle of a street. Because of its location and its red and sky-blue colors, the memorial forces itself and its message on residents. The memorial is a plaza that is, perhaps, not that useful as a place to interact with because of its location in the street, but the message of the memorial is not lost on viewers. The partially built walls of a house invoke rebuilding. And, if that was not clear enough, then a sign in the window of the house reads, "I am coming home! I will rebuild! I am New Orleans!" (see figure 8.2).

The ability to find one's identity through a phrase like "I am New Orleans" in a public space that is so wrought with loss is an important feature of any public space that is to be, in any way, participatory. But the memorial goes further in its attempts to allow residents to identify with the loss. The empty chairs are an interesting choice for the memorial. They simultaneously invoke the loss of people who will not be able to make use of them and seem waiting to be filled by returning and new residents. Chairs and benches have been used to powerful effect in memorials. The chairs of the Oklahoma City memorial to those lost in the bombing of the federal building and the benches of the Pentagon 9/11 Memorial both use sitting spaces to communicate not only loss but also contemplation. Those memorials fail to address social repair, though, because neither looks at a message beyond the memorial itself. Participatory commemorative art has to do more than ask the viewer to interact with it physically. The Ninth Ward Memorial opens its spaces for the viewer to interact with, but it also communicates a message that goes beyond an ambiguous imperative to "not forget." Forget what? Those lost? The social relations that led to the loss? The loss of city space? The Ninth Ward Memorial is clear: "I am coming home! I will rebuild! I am New Orleans!"

TOWARD AN INTERACTIVE PUBLIC ART: THE HEIDELBERG PROJECT IN DETROIT

The Ninth Ward Memorial accomplishes much of what Bishop asks for participatory art to do in repairing social bonds. But it is not as physically

Figure 8.2 Ninth Ward Hurricane Katrina Memorial, New Orleans, Louisiana. *Source*: Photo courtesy of David Tulloch.

or conceptually complex as the Heidelberg Project in Detroit. This project goes much further toward a participatory art of repair than the Ninth Ward Memorial in that it invites residents or even tourists to become a part of the project. One instance of this participation is the "Yellow House" portion of Heidelberg, shown in figure 8.3, which has a box of markers outside where visitors could sign their names on the house in which local residents still lived. The rest of the project, begun in the late 1980s, is a memorial of sorts to the slow economic disaster that left Detroit in ruins. Begun by Tyree Guyton on Detroit's east side in Guyton's childhood neighborhood, Heidelberg has stood as a symbol of urban participation and the dangers of that participation. Gillian Jakab of the *Michigan Daily* describes the project:

> What has come to be known as the Heidelberg Project is two city blocks of ever-evolving, art-adorned houses, found object sculptures and interior urban-folk museum spaces that have stood as playfully as they have steadfastly against the blight of inner-city decay. It is named for Heidelberg Street, in Detroit's McDougall-Hunt neighborhood, the block where it began and on which Guyton grew up.[26]

Figure 8.3 The Yellow House, Heidelberg Project, Detroit, Michigan. *Source*: Photo by the author.

In its recent history, twelve acts of arson have left much of the project in charred ruins. And, over its history, the city has bulldozed parts of the project. MaryAnn Wilkinson, curator of exhibits at the University of Michigan's Museum of Art, says of Heidelberg: "It's an outdoor dynamic project that changes constantly. Not only because of Tyree's work, but because it gets torn down sometimes or it gets burned down sometimes."[27] Reflecting the change of its surrounding environment is one of the most effective functions of participatory urban art.

Heidelberg reminds residents and visitors that physically interacting with the city can change the way that your senses experience the city. Heidelberg also reminds them, though, that participation is not just about building but also about destroying. As already stated, through its thirty-year existence, individuals burned portions of the structures down. So, Heidelberg opens questions of what we want when we say we want the democracy that participatory urbanism provides. Community, in the sense of a local organizing principle, has guided citizen participation in Detroit as much as in New Orleans. The working-class communities of Detroit since the beginning of its economic decline have found alternatives to the social and economic structures that led to that decline. These alternatives, for better or worse, have been found within the church, local neighborhood organizations, and the social structures of drug distribution. Alternatives to the economic globalization that in many ways left the city to die, some reparative and others destructive, are reflected and even enacted by the alternatives imagined by the city's art. This collective action reminds Detroit's residents of those alternatives because it has been so contentious with arsonists wanting to burn it down, the city wanting to bulldoze it or put it in an indoor gallery, and the global art community's skeptical fascination with the project.[28] However, just as Detroit stands for something in the national imagination, so too does Guyton's art stand for something: the enduring possibility of public space to operate and thrive outside of, and in spite of, the erosion of the culture commons.

Detroit's troubles began years ago. We know that now, but the city's problems have been the subject of nationwide public discussion only in the past decade. A slow economic disaster destroyed the city, leaving in its wake physical destruction similar to Katrina's effects on the Gulf Coast. And, as New Orleans does, Detroit holds a place in the mythology of the nation in ways that are similar to New Orleans. Detroit stands for something. As New Orleans stands as the country's most un-American city, Detroit, for much of its history, stood as the country's most American city. It serves many roles in the nation's imagination. It is simultaneously a place that the rest of the nation distances itself from by thinking that Detroit's problems would never spread to other areas. But the role of Detroit in the nation's imagination is different than New Orleans's because Detroit, for many years, was the symbol of middle-class American prosperity. Much of the coverage of Detroit's economic decline has focused on this one point that seems to frighten others in the middle class. While the city's economy and once immense wealth arose from the car companies that it sheltered, Detroit has had a thriving arts scene for decades. The controversies over governor-appointed city manager Kevyn Orr's plans to sell some of the Detroit Institute of Art's paintings brought Detroit's art scene to the public's attention once again. And the situation raised serious questions about the idea of public art since the DIA is one of the only truly public art museums in the country; it has been owned and run by taxpayers since its founding. But the public and underground art scene in Detroit is also thriving. The Heidelberg Project is an excellent example of public art that seeks radical solutions to repairing broken social bonds while also asking serious questions about participation as the basis for reparative art.

The east side neighborhood where The Heidelberg Project is located was hit particularly hard by the downturn. It is an urban area that has seen drastic change. However, reflecting change should be a function of public art and a way to read the history of a city. Heidelberg's vision is clearly outlined on its website: "The theory of change [...] begins with the belief that all citizens

[...] have the right to grow and flourish in their communities. [...] A community can redevelop and sustain itself from the inside out, by embracing its diverse cultures and artistic attributes as the essential building blocks for a fulfilling and economically viable way of life."[29] Heidelberg is a space that offers actual solutions to economic problems in its neighborhood by asking the public to participate in its creation. But the project also opens up the possibility for radical participation. There are hundreds of installations that make up the project that include a variety of materials and found objects, including shoes (see figure 8.4), electronics, glass bottles, garbage cans, and discarded materials from the city's crumbling infrastructure. While the materials and location of the Heidelberg Project suggest randomness and a DIY style, the arrangements of the installations are reminiscent of the formal layout of an art museum.

Viewers participate in the project in a number of ways. The project is spread out over city blocks where residents still live. So, the viewer automatically becomes a part of a living community that is rebuilding through art. The installations, some funny and some overtly political, are arranged so that the viewer can walk around and on them, touch them, and rearrange them. Other

Figure 8.4 An installation of shoes, Heidelberg Project, Detroit, Michigan. *Source*: Photo by the author.

Figure 8.5 View of the Polka Dot House, Heidelberg Project, Detroit, Michigan. *Source*: Photo by the author.

houses such as the "Polka Dot House," shown in figure 8.5, are residences but are open to visitors to roam through. The Heidelberg Project opens up the idea of participation in public art by involving citizens in its creation. But the project also leads us to consider that radical participation has to include the public's ability to destroy that art, as well. The skeletons of two of the houses that were burned are still visible and have been repurposed as art installations. The acts of arson and the repurposing of the remains after them stand as examples of participatory art that is open to change just as the community around it is.

RESPONDING TO THE WORLD: TELE GHETTO, PORT-AU-PRINCE, HAITI

In 2010, the residents of Port-au-Prince, Haiti were confronted with the worst natural disaster that the city had experienced. The death toll was estimated to be two hundred thousand. In the face of incomprehensible loss, the city was faced with the situation of moving on without the resources to do so. Much of the relief came from sources outside of the country. However, as was the case in New Orleans and Detroit, the impact of the disaster was exacerbated by forces outside of Port-au-Prince residents' control. Port-au-Prince's physical and economic infrastructure were so eroded that full recovery from such a disaster seemed unlikely. Six months after the earthquake, almost none of the rubble had been removed, and much of the city still looked as if the earthquake had happened the day before. Recovery and repair has happened in a decentralized way in the city as residents of small communities seek to commemorate the overwhelming loss while trying to build self-governing communities in the face of an emerging decentralized governing apparatus.

The Tele Ghetto art project exemplifies the dual attempt at commemoration and rebuilding. It was created in 2009 and has served a number of roles in the community, including a critique of the lack of governance in Haiti, as well as a way to commemorate the loss of the earthquake. Tele Ghetto was first named the Ghetto Biennial and was meant to be a periodic art festival. It has since become a permanent installation in one of Port-au-Prince's most physically ravaged neighborhoods. What Tele Ghetto means to many of the residents is a way for them to define their communities and recovery for themselves. With the onslaught of media coverage of Haiti after the earthquake, many residents lost any means to participate in the narratives arising from international media. Romel Jean Pierre, cofounder of the project, describes how he walked around the installations with a plastic bottle made to look like a camera in order to ironically subvert the media's attention. He says, "[w]e keep a bottle, a real bottle, and cut this and give it form into a camera. And we have created Tele Ghetto. It's the voice of persons who don't have the international media for a talk. It's the voice of the people who stay away from society."[30]

Tele Ghetto's installations reflect the history of this area of the city and give a nod to the future while offering innovative art that has been recognized by museums around the world. Whether this attention by the established art world is welcome by the founders or not, the project pushes the boundaries of what participatory art seeking to repair social relations can do. The art of Tele Ghetto is made almost completely of found objects, and much of it comes from the rubble of the earthquake. Many of the pieces reflect the country's deep history of voodoo practices while others are lively or funny, and some show the deep pain still felt in the city by using the skulls and bones of those who died in the earthquake.

Tele Ghetto received international attention through the work of Paris-based artist JR. JR's project *Inside Out* featured Tele Ghetto. A truly transnational art installation, *Inside Out* allows individuals around the world to mail their photos of someone's face to JR's studio, where he increases their size and then prints out and mails the enlarged reproduction back to the individual, who then pastes the extremely large photo on public walls. The project was taken up by many around the world. Particularly interesting is the effect that the installation had on Tele Ghetto. Tele Ghetto was already a creative enclave where residents see themselves as rogue artists. But the lively art throughout has an undercurrent of pain. Many of the sculptures there are made from objects found after the earthquake, as well as human skulls found in the wreckage. The photos taken by Tele Ghetto residents and pasted up on the walls act as a reminder that art after disaster isn't just a way to remember the disaster but a means for allowing citizens to take back and struggle over the spaces that they have difficulty participating in since those places are completely changed.

Tele Ghetto was featured in *Inside Out: The People's Art Project*, a documentary that covered the different urban spaces transformed by JR's project. JR announced the project in 2011 when he was awarded the TED Prize at a TED conference in Long Beach, California. The documentary begins with JR's call on the TED stage for "the creation of a global participatory art project with the potential to change the world."[31] Despite the overstated rhetoric, JR's project poses interesting questions for communities considering the potential for transformative, participatory art projects. Communities all over the world have participated in it, and many controversies have erupted when the pictures have been pasted on public buildings and sacred places. The pictures have appeared on the Israeli West Bank Barrier, on trailers on the Sioux reservations in South Dakota, and in public plazas in Shanghai. When the cofounders of Tele Ghetto submitted their photographs of Port-au-Prince residents making funny faces, they hoped to show the positive change happening in Haiti despite the overwhelming crisis. A project like *Inside Out*, while temporary and ephemeral, is one of the most effective participatory art projects happening today because it allows communities to ask their own questions about identity and recovery and to determine their own solutions to what that recovery looks like.

PARTICIPATORY URBAN DESIGN REDUX: IT STARTS WITH SHARED GOVERNANCE

If recovery of urban space after disaster and its renewal as sustainable space is to have any impact on the lives of cities' residents, the governing structures that led to such disasters—or at least exacerbated them—need to be based on sustainable models as well. Political scientist Archon Fung's idea of empowered participatory governance can be translated into a participatory urban aesthetic theory by looking to, and modeling projects that seek to, not only avoid understanding disasters within existing political structures but also give spaces over to citizens to make spaces that work for them—a practice that begins with the belief that citizens know best how spaces will work for them. The Ninth Ward Memorial, the Heidelberg Project, and Tele Ghetto are made up of the scrappy materials of their surroundings, but they offer powerful expressions of the ways in which identities, as well as cities, can be shaped by rethinking what it means to participate in the spaces and surfaces of the city by individual citizens acting collectively. And all three of these projects have incited debate, which is the starting point for participatory urbanism. They offer a set of strategies for thinking about the city that does not reach out for assumed publics but seeks participation from potential publics—publics who are seeking a space for participation.

Fung has focused his work on strategies for allowing citizens to have more direct participation in the systems that govern their daily lives. His emphasis on *empowered participation*, which values transparency from the top down on issues of governance and urban planning, allows citizens "substantial and equal opportunities to participate directly in decisions that affect them."[32] Giving residents a stake in these decisions seems like a basic model for recovery along participatory lines, but it is rarely used and certainly wasn't in New Orleans, Detroit, and Port-au-Prince after their disasters. Participation is about the kind of sustainable democracy that a city wants. And democracy is about empowerment of each resident no matter how messy or complex that might end up being. Fung's idea of community democracy is empowered because the policy flows directly from community discussion groups who determine what officials implement. Community groups do not act in an advisory role but as actual policy makers. Of course, along with transparency, Fung's method emphasizes decentralization, which frees citizens "to imagine innovations that depart from conventional wisdom and routine."[33] There is a way to translate Fung's empowered participation into an aesthetic theory, as well as into a method for communities to plan public art projects. The communities of the Ninth Ward, Detroit, and Port-au-Prince do not confront democracy as an abstract idea but as a living practice that they try to improve on a daily basis. In Fung's words, "If you want to find democracy's heroes, look in the streets."[34]

In the past decade, we have seen no lack of attempts at democracy in the streets in the United States from the continued anti-war movement, to the grassroots organizing in New Orleans after Katrina, to Occupy Wall Street, and now to Black Lives Matter. But we have also seen an increasing take-over of public space by memorials, which at times have little to do with the public's memory of events. Erika Doss explains why we have seen such a surge in memorial building in the past decade and a half. In *Memorial Mania*, she makes the case that "memorials [...] are archives of public affect [...] that are embodied in their material form and narrative content."[35] She explores the reason why we build so many memorials today—especially in the United States. Doss wants us to think about the anxieties that the contemporary memorial mania represents: "who and what should be remembered," "particular narratives about the nation and its publics," and the demand by citizens for representation and respect.[36] Memorials play a role in shaping "perceptions of social order, national identity, and political transition."[37] She contends that, as opposed to the Habermasian idea of a public sphere "in which rational citizens exchange ideas and come together in shared progressive actions, contemporary American public life is especially marked by emotional appeals and affective investments."[38] Richard Gilman-Opalsky sees more hope for the public sphere than Doss in that he thinks it does more than merely host communication. It can also "work to form a collective political will and [...] bring this to bear on other actors."[39] Opalsky's work can help us think through sites discussed here because he emphasizes the transgressive potential of public spheres. He argues, "[t]ransgressive public spheres have been formed by structurally disadvantaged people and/or people excluded for ideological reasons or reasons of misrecognition."[40] This misrecognition of excluded people can then become the impetus or motivation for collective governance that starts with the memory of the experience of social inequality that these people share. In New Orleans, Detroit, and Port-au-Prince, these memories have been translated into unofficial sites of memory, but these sites of memory also serve to incite such political collectivity. Transgressive public spheres, as Opalsky goes on to write, "always implicate some kind of collective identity. [They] require the regrouping of disparate collectivities and the recasting of political identities against exclusivist group identities."[41] Such identification with the group can become the basis for new modes of participation.

Participatory urbanism is about organizing urban life as it exists and taking into account citizens' reality as the only reality on which to base urban interactions. But participatory urbanism is also about imagining alternative futures. Those like Jon Forester see the political possibilities of urban planning that are responsible, engaged, and organized toward democratic practice on the ground in a way that opens up real alternatives in urban planning in the face of unequal power differentials between communities and those who have the power to force communities into set spaces.[42] Places like New Orleans, Detroit, and Port-au-Prince are ready for alternatives and alternatives that include existing networks of publics. Urban planners and theorists have discussed the networked city since the late 1980s. We have begun to see the complex ways in which urban communities harness that networked quality and understand it in its multiple applications—mobile technologies, governing practices, and special configuration.

Alternative public art can reflect the many alternative modes of social interaction that crises initiate within communities. The ability of national memorials to rally citizens around a set of shared values is dubious at best. Citizenship is a contested category just as art and cultural values are. Urban space is increasingly used less as a static place and more as, in the words of Manuel Castells, "a space of flows."[43] Such spaces "reconceptualize new forms of spatial arrangements under the new technological paradigm."[44] The architecture of cities, its public art included, gives symbolic form to political realities. How, James Donald asks, "are social forces and relations crystallized in the fabric, institutions, and encounters of the city?"[45] If public art is to remain open to ways of remembering the past that fundamentally question participation within the existing democratic structures of urban spaces, then it has to relate that past to alternatives for future governing structures. David Harvey, the urbanist and longtime proponent of creating democratic, urban spaces, calls this type of participation "[t]he right to the city," echoing Henri Lefebvre's 1967 book. Harvey advocates for "a more ecologically sensitive mode of city living." And the right to the city is nothing short of a completely "alternative vision of the city."[46] The right of residents to the spaces where their lives happen, Harvey emphasizes, "[d]oes not arise primarily out of various intellectual fascinations and fads [. ...] It primarily rises up from the streets, out from the neighborhoods, as a cry for help and sustenance by oppressed peoples in desperate times."[47] Citizens' ability to have a right to their city after major disasters like a hurricane, earthquake, or slow economic decline begins with strategies that protect public space as both a symbol of participation and an actual path to participation. This dual-participatory strategy is the very basis for sustainable cities. Urban development and the increasing takeover of cities by private interests pose a physical and existential threat to residents who increasingly have little recourse to defend the spaces that they inhabit. While we may not see a full return to strong public participation in city governance and development in the next few decades, there are spaces of hope—such as the ones presented here—that provide a map to what the sustainable city can be. As Harvey points out, "[t]he right to the city is an empty signifier. Everything depends on who gets to fill it with meaning."[48]

So, I come back to the idea—or the model—of the cultural commons as a sustainable repository of resources, ideas, and culture. We tend to discuss the idea of sustainability only in terms of the physicality of cities. This is an important aspect. The fields of energy production, waste management, and urban design must keep a focus on radical physical designs for sustainability. However, physical models for sustainability are only as strong as the assumptions behind them and the consensus of the public for whom sustainable cities should be built. How can we build or even conceive of such participation by the public in the processes that build these cities when even the ideas that sustainability is grounded in have been co-opted by private interests? In his work on the erosion of intellectual and cultural commons, Nonini decries their enclosure as a result of "private intellectual property rights."[49] The same private interests that have enclosed the intellectual commons "require the use of urban space but reject responsibility for polluting it or paying for its remediation."[50] Nonini describes this abuse of the commons as "weardown." He explains this erosion as part "of the material world on which capitalism depends for its own existence."[51] What role, in the face of this weardown of the material world, does public art and even public space have to play without becoming part of the weardown itself? If, as Nonini says, the "weardown of the commons takes place when corporations and allied states cut the internalized costs of capitalist production and neglect to take measures to reproduce the conditions of production," then public art in places like New Orleans, Detroit, and Port-au-Prince can reclaim those spaces that corporations and allied states have little or no interest in repairing and rebuilding.[52] This rebuilding by citizens after disaster has come to define participatory urbanism in places like these three cities, where divestment of capital has long characterized their histories of decline. Nonini states emphatically that "[t]he protections and regeneration of the [. . .] commons [. . .] fall[s] on those who are still connected to the commons on an everyday basis" and that "social movements [. . .] are now posing and will continue to pose major threats to corporations' savage 'business as usual' and to the oligarchic corporate states that support them."[53]

While New Orleans has seen a reinvestment and even co-opting of its public spaces and services by corporations, there are still spaces that hold hope for truly participatory urban interaction and diffusion of rampant inequality. Detroit and Port-au-Prince's futures are still unclear. The political infrastructure in Michigan has little interest in bringing back a Detroit that looks anything like its former self. And Haiti's political system has been plagued by corruption and mismanagement for decades. The infrastructure and industry that once held Detroit together has certainly crumbled, as have its great symbols of participatory urbanism like the Heidelberg Project, but the remnant citizenry there continues to inspire a history of hope for such radical spaces. Port-au-Prince is certainly caught in the crosshairs of devastating poverty, disease, and lack of almost all necessary infrastructure for citizens to thrive. But spaces like Tele Ghetto demonstrate the ultimate promise of participatory public art—art that is by, about, and for the people who use it on a daily basis.

NOTES

1. Harriet Seine and Sally Webster, "Critical Issues in Public Art," *Art Journal* 48, no. 4 (1989): 288.

2. Greg Dickinson, Carole Blair, and Brian L. Ott, eds., *Places of Public Memory: The Rhetoric of Museums and Memorials* (Tuscaloosa, AL: University of Alabama Press, 2010), 6.

3. Douglas Davis, "Public Art: The Taming of the Vision," *Art in America*, May–June 1974, 84.

4. Amy Goldin, "The Aesthetics of the Ghetto: Some Thoughts about Public Art," *Art in America*, May–June 1974, 34.

5. Donald M. Nonini, "Introduction: The Global Idea of 'the Commons,'" in *The Global Idea of 'the Commons'*, edited by Donald M. Nonini (New York: Berghahn Books, 2007), 1.

6. Ibid., 1.

7. Ibid., 7.

8. Ibid., 11.

9. John Baden and Douglas S. Noonan, *Managing the Commons* (Bloomington, IN: Indiana University Press, 1998), 76.

10. John Cavanagh and Jerry Mander, eds. *Alternatives to Economic Globalization: A Report of the International Forum on Globalization* (San Francisco: Berrett-Koehler, 2004), 79–103. The principles are new democracy, subsidiarity, ecological sustainability, common heritage, diversity, human rights, jobs and employment, food security and safety, equity, and a precautionary stance toward technological innovation. While the title of their collection might seem to challenge globalization entirely, the aim of their project is to suggest ways globalization can happen without necessarily eroding the commons.

11. Ibid., 109.

12. Ibid., 121.

13. Tim McDonnell, "The Fossil Fuel Industry Is Bankrolling the Paris Climate Talks," *Mother Jones*, December 22, 2015, accessed January 2, 2016, http://www.motherjones.com/environment/2015/12/climate-change-summit-paris-cop21-fossil-fuels-sponsors/.

14. Andrew Winston, "What Business Leaders Need to Know about the Paris Climate Conference," *Harvard Business Review*, last modified December 1, 2015, accessed December 26, 2015, https://hbr.org/2015/12/what-business-leaders-need-to-know-about-the-paris-climate-conference/.

15. Nonini, "Introduction," 4.

16. Ibid., 6.

17. Michael North, *The Final Sculpture: Public Monuments and Modern Poetics* (Ithaca, NY: Cornell University Press, 1985), 18.

18. Ibid., 27.

19. Ibid., 34.

20. Joel McKim, "Agamben at Ground Zero," *Theory, Culture, & Society* 25, no. 5 (2008): 83.

21. North, *Final Sculpture*, 39.

22. Claire Bishop, *Artificial Hells: Participatory Art and the Politics of Spectatorship* (London: Verso, 2012), 11.

23. Ibid., 12.

24. Ibid., 13.

25. Richard Johnson, "Equality," in *New Keywords: A Revised Vocabulary of Culture and Society*, eds. Tony Bennett, Lawrence Grossberg, and Meaghan Morris (Oxford, UK: Blackwell, 2005), 109.

26. Gillian Jakab, "Make It Greater: The Heidelberg Project's 30-year Legacy," *The Michigan Daily*, October 14, 2015, accessed January 2, 2016, https://www.michigandaily.com/section/arts/30-year-anniversary-tyree-guyton%E2%80%99s-heidelberg-project-celebrated-umma-copy/.

27. Ibid.

28. Ibid.

29. *The Heidelberg Project*, last modified January 2, 2016, accessed February 1, 2016, http://www.heidelberg.org/.

30. *TeleGhetto: Guerilla Media in Haiti*, last modified August 2, 2015, accessed August 2, 2014, http://www.teleghetto.com/.

31. *Inside Out: The People's Art Project*, last modified January 2, 2016, http://www.insideoutproject.net/.

32. Archon Fung, *Empowered Participation: Reinventing Urban Democracy* (Princeton, NJ: Princeton University Press, 2004), 4.

33. Ibid., 5.

34. Ibid., ix.

35. Erika Doss, *Memorial Mania: Public Feeling in America* (Chicago: University of Chicago Press, 2010), 13.

36. Ibid., 2.

37. Ibid., 10.

38. Ibid., 15.

39. Richard Gilman-Opalsky, *Unbounded Publics: Transgressive Public Spheres, Zapatismo, and Political Theory* (Lanham, MD: Lexington Books, 2008), xv.

40. Ibid., xiv.

41. Ibid., xv.

42. Joel Forester, *Planning in the Face of Power* (Berkeley, CA: University of California Press, 1988), 162.

43. Manuel Castells, *The Informational City: Information, Technology, Economic Restructuring, and the Urban Regional Process* (Oxford, UK: Blackwell, 1989), 146.

44. Ibid., 146.

45. James Donald, "City," in *New Keywords: A Revised Vocabulary of Culture and Society*, eds. Tony Bennett, Lawrence Grossberg, and Meaghan Morris (Oxford, UK: Blackwell, 2005), 35.

46. David Harvey, *Rebel Cities: From the Right to the City to the Urban Revolution* (London: Verso, 2012), ix.

47. Ibid., xiii.

48. Ibid., xv.

49. Nonini, "Introduction," 14.

50. Ibid., 15.

51. Ibid., 15.

52. Ibid., 17.

53. Ibid., 21.

BIBLIOGRAPHY

Aguilar, Louis. "12 Heidelberg Fires, All a Mystery." *Detroit News*, February 20, 2015. Accessed January 2, 2016. http://www.detroitnews.com/story/news/special-reports/2015/02/20/heidelberg-arsons-detroit-hard-solve/23769375/.

Baden, John, and Douglas S. Noonan. *Managing the Commons*. Bloomington, IN: Indiana University Press, 1998.

Bishop, Claire. *Artificial Hells: Participatory Art and the Politics of Spectatorship*. London: Verso, 2012.

Castells, Manuel. *The Informational City: Information, Technology, Economic Restructuring, and the Urban Regional Process*. Oxford, UK: Blackwell, 1989.

Cavanagh, John, and Jerry Mander, eds. *Alternatives to Economic Globalization: A Report of the International Forum on Globalization*. San Francisco: Berrett-Koehler, 2004.

Davis, Douglas. "Public Art: The Taming of the Vision." *Art in America*, May–June 1974, 84–85.

Dickinson, Greg, Carole Blair, and Brian L. Ott, eds. *Places of Public Memory: The Rhetoric of Museums and Memorials*. Tuscaloosa, AL: University of Alabama Press, 2010.

Donald, James. "City." In *New Keywords: A Revised Vocabulary of Culture and Society*, edited by Tony Bennett, Lawrence Grossberg, and Meaghan Morris, 32–35. Oxford, UK: Blackwell, 2005.

Donica, Joseph. "The Heidelberg Project." 2014. JPEG file.

Doss, Erika. *Memorial Mania: Public Feeling in America*. Chicago: University of Chicago Press, 2010.

Forester, Joel. *Planning in the Face of Power*. Berkeley, CA: University of California Press, 1988.

Fung, Archon. *Empowered Participation: Reinventing Urban Democracy*. Princeton, NJ: Princeton University Press, 2004.

Gilman-Opalsky, Richard. *Unbounded Publics: Transgressive Public Spheres, Zapatismo, and Political Theory*. Lanham, MD: Lexington Books, 2008.

Goldin, Amy. "The Aesthetics of the Ghetto: Some Thoughts about Public Art." *Art in America*, May–June 1974, 23–45.

Harvey, David. *Rebel Cities: From the Right to the City to the Urban Revolution*. London: Verso, 2012.

The Heidelberg Project. Last modified January 2, 2016. Accessed February 1, 2016. http://www.heidelberg.org/.

Inside Out: The People's Art Project. Last modified January 2, 2016. http://www.insideoutproject.net/.

Jakab, Gillian. "Make It Greater: The Heidelberg Project's 30-year Legacy." *The Michigan Daily*, October 14, 2015. Accessed January 2, 2016. https://www.michigandaily.com/section/arts/30-year-anniversary-tyree-guyton%E2%80%99s-heidelberg-project-celebrated-umma-copy/.

Johnson, Richard. "Equality." In *New Keywords: A Revised Vocabulary of Culture and Society*, edited by Tony Bennett, Lawrence Grossberg, and Meaghan Morris, 109–12. Oxford, UK: Blackwell, 2005.

McDonnell, Tim. "The Fossil Fuel Industry Is Bankrolling the Paris Climate Talks." *Mother Jones*, December 22, 2015. Accessed January 2, 2016. http://www.motherjones.com/environment/2015/12/climate-change-summit-paris-cop21-fossil-fuels-sponsors/.

McKim, Joel. "Agamben at Ground Zero." *Theory, Culture, & Society* 25, no. 5 (2008): 83–103.

Nonini, Donald M. "Introduction: The Global Idea of 'the Commons'." In *The Global Idea of 'the Commons'*, edited by Donald M. Nonini, 1–25. New York: Berghahn Books, 2007.

North, Michael. *The Final Sculpture: Public Monuments and Modern Poetics*. Ithaca, NY: Cornell University Press, 1985.

Seine, Harriet, and Sally Webster. "Critical Issues in Public Art." *Art Journal* 48, no. 4 (1989): 287–90.

TeleGhetto: Guerilla Media in Haiti. Last modified August 2, 2015. Accessed August 2, 2014. http://www.teleghetto.com/.

Tulloch, David. "Ninth Ward Katrina Memorial." 2008. *Blogspot.com*. Accessed August 2, 2016.

Winston, Andrew. "What Business Leaders Need to Know About the Paris Climate Conference." *Harvard Business Review*. Last modified December 1, 2015. Accessed December 26, 2015. https://hbr.org/2015/12/what-business-leaders-need-to-know-about-the-paris-climate-conference/.

Part III

REPRESENTATIONS OF SUSTAINABILITY AND THE CITY

Chapter 9

The Ambivalence of Noise

Requiem for Fossil Fuels *by Bruce Odland and Sam Auinger*

Alexander Kleinschrodt

Leading up to the surprisingly successful Paris Climate Conference (COP21) in December of 2015, the notion of decarbonization appeared in public discourse. It represents the idea of a complete overcoming of the world economy's dependence on oil, coal, and other not-renewable fuels.[1] While the term decarbonization did not make it into the multilateral Paris treaty, it would still be a consequence of its full implementation that our societies will be subject to massive change, not only in an economic sense but also when it comes to the sensuous experience of our lifeworlds. Those developments provide a new context for a piece by O+A, a duo formed by sound artists Bruce Odland and Sam Auinger, called *Requiem for Fossil Fuels*. It dates back to 2007,[2] and, in the meantime, the conditions for the reception of the piece have definitely changed: the turn away from fossil fuels has not yet become a reality, but a prospect on which a wide range of stakeholders now agrees.[3]

With *Requiem for Fossil Fuels*, Odland and Auinger are able to articulate an ambiguous perspective on such a transformation. While their critical standpoint toward fossil-fueled societies with only a minor interest in keeping up with the demands of sustainability is obvious, the interpretation of *Requiem* cannot be restricted to an unambiguous message like "we have to get rid of all this and be super-green!" Instead, as this essay tries to demonstrate, the meaning of the piece is generated out of a negotiation of losses and promises, fascination and irritation, as well as impressions of beauty and deterioration. In doing so, the focus of attention lies on the modes of representation used in *Requiem*, which are specific to the medium of sound and the means of sound art, especially when linked back again to traditional musical practices.

The first part of the essay sums up discourses about sound, the city and sustainability, in which *Requiem for Fossil Fuels* itself is situated. This is followed by an in-detail description of the piece, which—because of its original formal and compositional features—is of particular importance. With this background, an interpretation of *Requiem* is developed which identifies elements of ambivalence on several levels of the piece, whose connection sheds light on the characteristic, even quite unique standpoint articulated by Odland and Auinger.

SOUND STUDIES AND URBANITY

It may still be unexpected to some that problems of sustainability and climate change can be addressed in the medium of sound, be it musical sound or other. The discipline of sound studies[4] and its forerunners[5] have discovered a long tradition of what is called a primacy of the visual and, going along with it, a marginalization of sound and hearing in Western culture—a diagnosis shared by Odland and Auniger, who claim that the "overwhelmingly visual nature of our economy and culture is undeniable."[6]

From the early days of sound studies, the interest in the aural properties of human settlements and built environments was constantly present among scholars active in this field, be it in the sense of villages in the countryside[7] or of modern urban agglomerations.[8] In close connection with this research itself, the field of urban sound art emerged and became a new field of studies. With their artworks often placed as interventions in urban spaces, sound artists highlight the artificial character and constructedness of urban surroundings, also pointing at the essential role of sound in this process, as sound studies scholars Gascia Ouzounian and Sarah Lappin declare: "[t]hrough the lens of urban sound art, we argued, the city itself could be understood as being composed, and recomposed, through sound."[9] This epistemic quality can also be found in the work of Odland and Auinger. In addition, they try to substitute this metaphoric "lens" with a more specific "hearing perspective."[10] With this concept they aim at an integral understanding of culture, and the focus on sound is expected to form a basis for it: "we learned about our culture by listening not only to the sounds it is proud of—the music, the signal tones, the vast charged hum of the economy—but also to the unintended hums and buzzes, the energy lost as sound vibrations as we burn fossil fuels and turn them into every aspect of modern life."[11] Their approach aims at a reversal of the usual attitude toward such by-product sounds, which does not concede them any significance.

The outdoor sound recordings made by Odland and Auinger for the use in their *Requiem for Fossil Fuels* have mostly been conducted in New York City.[12] For an artwork dealing with the age of fossil fuels, this is not the most obvious choice. After all, New York can at least not be cited as a model of the so-called automotive city or *autogerechte Stadt*,[13] to use the term coined by the German city planner Hans Berhard Reichow.[14] A rectangular grid of roads, which is a highly characteristic and well-known trait of Manhattan, was criticized by Reichow, just because it is, as he thought, not suitable for allowing traffic to flow unhindered. This may illustrate that New York City with its center on Manhattan Island is lacking a major attribute of modernity from the perspective of postwar city planners.[15] But still, there is another, more general interpretation of New York City, which makes it possible to establish a connection to the concept of an age of fossil fuels. From this perspective, New York and, again, especially Manhattan is considered to have been a laboratory of modernity. This view is summed up tellingly by architect Rem Koolhaus: "[e]specially between 1890 and 1940 a new culture (the Machine Age?) selected Manhattan as laboratory: a mythical island where the invention and testing of a metropolitan lifestyle and its attendant architecture could be pursued as a collective experiment in which the entire city became a factory of man-made experience, where the real and the natural ceased to exist."[16] By the combination of these ascriptions with the reference to a machine age (which up to now necessarily was also an age of fossil fuels) makes obvious some discursive links between the city and the conditions of modernity. Koolhaus goes even further, claiming that Manhattan "is the 20th century's Rosetta Stone,"[17] which means that it is a key location, that has to be examined in order to fully understand modernity.

CONVERGENT LAYERS OF SOUND

Requiem for Fossil Fuels is deliberately placed at a point where the before-mentioned discourses—sound, the city, and modernity—intersect. To achieve this, Odland and Auinger opted for a postmodernist approach: they developed a unique adaptation of an older genre of music, in this case the Catholic requiem mass. In this way, their work and their concerns are embedded within a musical tradition ranging from the First Viennese School (with Mozart's requiem mass as the most prominent example) to Krzysztof Penderecki and beyond. Starting with *Introitus* and *Kyrie* and closing with *Agnus Dei* and *Communio*, the piece by Odland and Auinger features the full Latin text of the requiem mass, which is performed by a quartet of singers (soprano, alto, tenor, bass). They follow a precisely notated score[18]; the music is mostly composed in a tonal idiom and does not contain complex rhythmic constellations so common in contemporary music. In many ways, it resembles vocal counterpoint traditions in Renaissance music—not just in the sense of a generic allusion and because of the use of a vocal quartet. Odland and Auinger, both trained as composers before becoming known as sound artists,[19] also integrated compositional techniques associated with this style.

Frequently, melismatic singing is employed, in which several notes or whole phrases are sung on one syllable of the given text (as can be heard immediately in the *Introitus* part). Although most musicologists would argue that this is just some compositional device for setting a text to music, which in the first place carries no meaning, it can hardly be denied that this kind of technique is in itself for many (Western) listeners associated with a sacred ambiance, especially when it is executed by a solo voice.[20] Furthermore, the vocal part of *Requiem for Fossil Fuels* integrates the concept of imitation, meaning that a phrase provided by a first voice is repeated by a second one and, possibly, passed on to a third and fourth. The fugue, embodied in the works of Johann Sebastian Bach, is the most common example of a musical form shaped by this device, but it already became an important feature in the architecture of polyphonic vocal music during the Renaissance period, structuring not just single sections but, for example in the case of Josquin des Prez, whole mass compositions. Odland and Auinger use this technique on several stages of their *Requiem*, most prominently at the end of the *Sanctus*. In the *Agnus Dei* section, it can also be heard, but here it is realized in another way: the singers are told to react to each other by repeating a given phrase while they can choose when exactly to do so; the score describes this as "to echo freely."

The vocal layer of *Requiem* is almost constantly confronted with outdoor sound recordings. This second layer is not a permanently fixed accompaniment like a tape recording would be[21]; instead its formation is controlled via software[22] and can be shaped by a performer during the performance in real-time. In the performances of *Requiem* by O+A this is done by Auinger, while Odland acts as conductor. The layer of outdoor sound recordings consists of two components: on the one hand, there are short and distinct sound events, among them horns and buzzing alarm sounds, single vehicles and the footsteps of commuters on the streets. At least this part of *Requiem for Fossil Fuels* reminds one of another well-known New York composition: Steve Reich's *City Life* (1995) also comprises field recordings made in this city. In the case of Reich, only very short fragments like a car horn or the cry of a street vendor have been used, and these samples from and of the city (in the musical sense but also in the literal one) are incorporated into the textures of his minimalist instrumental music. In a performance of *City Life*, the outdoor sound recordings are provided by three musicians playing sampling keyboards, and the sounds have to be played like a musical phrase at an exactly defined moment. In *Requiem for Fossil Fuels*, Sam Auinger as performer is free to bring in sound events in a more improvisational manner, although generally their appearance is roughly defined by a time code in the score.

However, the soundscape recordings, which form the second component of the *Requiem*'s outdoor sound layer, would not be compatible with Steve Reich's music. These are longer field recordings representing a whole acoustic panorama recorded on a certain point in time and space. In a performance of the *Requiem* those recordings are played back in real-time; they have neither been cut nor is the temporal development of the recordings altered in any other way, compared to the noises which one immediately could have experienced in the situation where the recordings were made.[23]

Bluntly speaking, this component of the piece could be described as a continuous din or hum, which may be experienced as stressful and annoying by some listeners. But apparently, these noises are not completely stable and monotonous: their microstructure is constantly changing during the course of the recordings. In addition, every section of *Requiem* is based on another soundscape recording, therefore providing a sense of *varietas* (diversification) that Renaissance music theorist Johannes Tinctoris regarded as necessary for an appealing piece of music. Furthermore, the soundscape recordings have been modified when it comes to their sonic composition. Characteristic for the approach chosen by O+A is the fact that this did not happen with the help of electronic equipment in a studio post-production setting.[24] Actually, the microphones used for the field recordings were mounted into so-called tuning tubes. Because of their particular length, these tubes possess a certain resonance frequency, resulting in an emphasis of those components in the surrounding noises which correspond with this frequency: in this way, they seem to produce a musical tone out of the noise they are exposed to.[25]

The crucial aspect of *Requiem for Fossil Fuels* is manifest in the structural connections O+A develop between the layers of their piece, between the seemingly contradicting spheres of sublime vocal music—carrying with it all the connotations of high art, aesthetic autonomy, inwardness, and transcendence—and the noisy reality of New York City. The by now tuned din of the city soundscape is actually more than a mere background to the vocal music. Odland and Auinger hint at this point by referring to the underlying recordings as *cantus firmi*. With this denomination (meaning "preassigned chants"), Odland and Auinger once more allude to Renaissance vocal music, where a *cantus firmus* is a preexisting melody, which is sung by one of the voices and treated as an invariant element of a piece of music. With the definition of a *cantus firmus*, a kind of harmonic grid is established, which the other vocal parts have to follow. Much the same happens in *Requiem for Fossil Fuels*, where the four singers adjust to the tones provided by the modified soundscape recordings.[26] In this way, the soundscapes become an integral part of the musical composition. They have been promoted to a point of reference for the singers, which is characterizable by a pitch just as the singer's tones are. But remarkably, there is still a surplus, because the soundscape recordings do not dissolve completely in a musical structure with its own and exclusive logic; they can still be heard as a reference to everyday life.

The convergence of the two layers of sound is taken further still. In *Requiem*, a lot of direct interaction between the singers and the recordings is enacted. Since the layer of outdoor sound recordings, as I mentioned before, is not fixed, a certain kind of musicianship is involved in its creation under the conditions of a live performance: the shorter sound events are triggered by Sam Auniger sitting at an audio workstation, and some of these sounds, for example, a slamming noise in the first minute of *Introitus*, are introduced to function as a cue for the singers, who react to the sounds by starting with a phrase from the score. However, there is not only a temporal coordination established between the layers of the composition. Just as the tuning tube tones set the key for the parts of *Requiem*, some pitches which can be heard in the shorter sound events are adopted by the vocal quartet. The most obvious instance where this is happening can be observed in the *Dies Irae* part. Here the sound of a large truck is made audible, starting as a weak hum and performing a big crescendo up to fortississimo (fff).[27] At the dynamic peak, the truck driver obviously steps on the brake pedal, because the slowing down of the truck can clearly be traced acoustically via the squeaking, which is produced by the friction of brake discs and brake pads and which is coming to the fore acoustically in this moment. The score records this sound with the pitch of b'—the same note with which the following alto solo starts. Actually, the singer already begins with the tone while the truck's brakes are still audible, so that for a moment the technical noise and the vocal sound are hardly distinguishable from each other. This musical coordination of pitches (or, technically speaking, perfect consonance) is realized in an even more sophisticated manner in the *Offertum* part. Here the singers follow the sound of passing airplanes, which were captured in a tuning tube recording. Because of the doppler effect,[28] the frequency of the planes' engines constantly declines. The singers retrace these noises by performing long glissando phrases in unison with the recorded sound. This continuous gliding between pitches is structured by a change of the sung syllable of the Latin requiem text every time a tone of the scale is passed.

ECO ART AND A CHRISTIAN FRAMEWORK

In recent debates about the arts and sustainability, one question is frequently asked: is there a surplus of artistic practices with environmental concerns, a specific capacity which is not redundant to governmental action or preexisting scientific and societal discourses?[29] Naturally, this kind of self-questioning arises from the doubts of many people involved in the arts—artists, as well as theorists—if their work can by any chance be a contribution to building a more sustainable society. If this is denied, a consequential conclusion would be to act like German sculptor Charlotte Posenenske, who produced artworks with Marxist ideas in the background throughout the 1960s. But since she was more and more dissatisfied with the apparently lacking possibilities of art to directly tackle social inequalities, she finally refused to be an artist any longer in 1968, began to study sociology, and became active as a welfare worker. A positive answer, however, would imply that the arts could be more than a mere mirror to the "state of things" and are able to bring in a fundamentally different perspective or mode of action.

At this point, the artistic concept of *Requiem for Fossil Fuels* gains further relevance (though admittedly its performance will probably not result in direct social action), because in my opinion the piece is unique not just in its form and composition, but also because it articulates the problems of a fossil-fueled society in a very specific way. To me, this has to do with the idea of ambivalence. To make this a bit clearer, I have to refer to Harald Welzer, a social psychologist, who is a very influential figure in the German debate about sustainability and the prospects of a great ecological transformation. In interviews he makes no secret of his fascination for the confidence with technological progress characteristic for the 1960s—which he himself witnessed as an adolescent—and especially for fast sports cars.[30] However, today Welzer is one of the loudest critics of German automobile culture, which he condemns as a disastrous fetish. I also discovered this contradiction in connection with air travel when it comes to myself: constantly trying to optimize my carbon footprint and being quite a bit irritated that the COP21 treaty of all things just excluded air traffic from the carbon reduction aims, I am still fascinated by the technology of airplanes and hardly willing to leave the visitors deck of the airport once I am there. So, my feelings toward this phenomenon can legitimately be called ambivalent.

From this perspective, *Requiem for Fossil Fuels* can be understood as representing an ambivalent position toward modernity and technological progress. Because of this, the piece by Odland and Auinger cannot easily be dismissed as well-meant eco art, which has nothing more to offer than an already familiar statement or observation, that has to be affirmed, and, at least, the comfortable feeling of belonging to "the good ones" by doing so.[31] Anyway, it is striking that it seems not as easy to identify common eco-stereotypes in the medium of sound as it is in other media: with respect to the visual realm, for example, the polar bear on a melting ice floe comes to mind, which failed to become an icon of climate change. Instead, the deplorable creature is now usually addressed to illustrate a kind of visual communication, which is not able to depict the complexity of global environmental concerns, since the consequences of global warming can be expected to be by far more extensive. Likewise, in the field of verbal discourse, many stories about humankind and nature seem to have been told already. Of course this observation should not imply that there is a limit to poetic invention when it comes to ecological topics, but it is aimed at the apparently limited repertoire of structuring basic narratives used in this context so far. Ecocritic Scott Slovic has shown how deeply American nature writers in the twentieth century are dependent on traditional narratives, although they doubtlessly adopt them for great benefit. Slovic identified the modes of "rhapsody" (a praise of a primeval or future harmony in nature) and "jeremiad" (a warning of a decline or catastrophe) as basal rhetoric means,[32] demonstrating how a widely read author like Rachel Carson made use of narratives corresponding to those categories.[33]

With their choice of a form contesting the boundaries between music and sound art, Odland and Auinger opened up several possibilities to go round such terrain. A closer look at the traditional function of a Catholic requiem mass is already suitable to develop this idea further, because its meaning is twofold—or ambivalent. On the one hand, the requiem mass is meant to bring up a form for mourning a recently deceased person, channeling individual affects and embedding them into the ritual framework of Catholic belief; the element of loss predominates in this sense. On the other hand, the crafts and powers of musical composition help to transport a Christian interpretation of death, the conviction that death represents not the absolute end, but that the immortal soul lives on in an afterworld. The Swiss theologian Hans Küng sums up the Christian message in the following way: "[b]eing Christian means trying to realize a humanism, which is able to cope with not just the positive things, but also all the negative things like misery, guilt, senselessness, [and] death, based on an imperturbable trust in God."[34] Although the Christian idea of the Last Judgment is very present in the requiem text, I would like to stress that this should not lead to placing an artwork like *Requiem for Fossil Fuels* anywhere near an apocalyptic narrative, which for the last decades served "as a standard feature of environmentalist polemic."[35] Apocalypse, one could say, is—at least in its secular meaning[36]—about total annihilation, and this is not the perspective articulated in Odland and Auinger's *Requiem*. They clearly depict the end of a particular culture or era, but this cannot be imagined as "Untergang des Abendlandes" (a "decline of the West" in the sense of Oswald Spengler); the customary culture is not erased but is in an act of transforming itself. One should keep in mind that a requiem mass needs a congregation willing to mourn someone and, in a way, overcome death. Transferred to *Requiem for Fossil Fuels*, this implies a world society, jointly but not without grief, disbanding itself from fossil fuels, motivated by the belief that it is possible to abandon the route leading to a truly apocalyptic scenario, which the climate scientist Hans Joachim Schellnhuber described as "burning ourselves" (*Selbstverbrennung*).[37] With Küng, this could perhaps be seen as the realization of a universal humanism—or, as he puts it, *Weltethos* (world ethic)—probably informed less religiously and more ecologically.

So, if, at this stage of the discussion, one is urged to name the concrete subject of *Requiem for Fossil Fuels*, then the answer may be: transition. In the Christian belief of a merely transitory existence on earth and an after-worldly eternity, like in the quest for a sustainable modernity, the word "transition" represents a passing from one state to the next. The emphasis is, with this term, placed on the process of transcending, neither at a starting point nor at a definitive destination. While the aim of a decarbonized society foreshadowed in *Requiem* would probably have been described as utopian some decades ago, the possibilities for relating *Requiem* to the realm of utopia have changed since the time of its premiere, as I mentioned in this essay's beginning.

At this point, I would like to develop the concept of ambivalence further, now with respect to sound. First of all, the city noise, which forms the basis of *Requiem*, has to be considered. Apparently, the English word "noise" has an aspect of ambivalence to it. The *Oxford English Dictionary* defines "noise" as a "sound, esp. a loud or unpleasant or undesired one." "Especially" means that "noise" or "a noise" can also function as a neutral term—although this is not the most frequent way of using the word—with roughly the same, wide range of meanings like "sound," which encompasses all phenomena perceivable by the ear.[38] It is also not generally a contradiction to speak of pleasant noises.[39]

This linguistic consideration can be extended toward the phenomena denoted by the word "noise." The noises of a fossil-fueled society, by-products of combustion processes like they are presented in *Requiem for Fossil Fuels*, are neither "good" nor "bad" if singled out and regarded just for themselves. They can still be interpreted as impressive, powerful, and pleasant.[40] It would not be plausible for automobile companies to employ specialists for designing and improving the sound of cars[41] if this category of sounds had to be qualified as "unpleasant" under all conditions and in all situations. Nevertheless, there is certainly a problematic dimension: the overabundance and ubiquity of these noises, leading to a soundscape which can be experienced as less and less nuanced and transparent. But, even independent from all value judgments, these particular noises can also be heard as traces of a culture with a priority on economic growth, necessarily marginalizing other constituting powers of a pluralistic, succeeding society.

In the *Sanctus* part of *Requiem for Fossil Fuels*, the soundscape recordings feature, as Odland and Auinger tell us, "helicopters hovering, landing and taking off with their cargo of captains of industry on a helipad in the East River near Wall Street." By this provocative gesture—combining the most tranquil and transcendent part of the requiem mass with the sounds of a massive machine used in combat, "filtered first through a tuning tube, then through digital filtering so that only harmonious intervals remain"[42]—the ambivalence described above is demonstrated in a nutshell: on the one hand, as Odland was quoted as saying after *Requiem*'s premiere, "using all that oil to transport one or two people [in a helicopter] is a glorious display of power."[43] During the *Sanctus* part, the singers, certainly not by chance, intonate the words "pleni sunt coeli et terra gloria tua"—"heaven and earth are full of Thy glory." On the other hand, the control over this power, treated like a sanctuary itself, is perhaps not wisely used by present societies. The pervasion of the Manhattan soundscape by the helicopters bears witness to this while both tendencies of interpretation remain constantly present for the listener.[44] The framing of these sounds in an artwork, their manipulation via tuning tube and software, and their fusion with the particular message of a requiem mass make this ambivalence more evident.

PARALLELS, PAST AND FUTURE

In an attempt for a conclusion, I would like to situate *Requiem for Fossil Fuels* in a larger poetic context, building a bridge to a particular artwork dating from the first heyday of a global environmental movement around the 1970s. Actually, I believe that there are certain parallels between *Requiem for Fossil Fuels* and the film *Koyaanisqatsi* (1982) by director Godfrey Reggio. This film is composed solely of impressions of rural and urban environments, which are shown either in time lapse or slow motion. The images are accompanied by minimal music composed for the film by Philip Glass. In the history of its reception, the famous but also criticized film was often the object of a shortened interpretation, claiming that it was a manifesto against modernity, while the pure and unharmed nature was represented there as an ideal. This does not correspond, in an obvious manner, with statements by Reggio himself, who said about his intention behind the film: "[t]hese films [*Koyaanisqatsi* and its sequels] are meant to provoke, they are meant to offer an experience rather than an idea [...]. It is up [to] the viewer to take for herself, what it is that [this] means. For some, it's an environmental film; for some, it's an ode to technology."[45]

A closer look at the film itself also provides evidence for this. Most striking is that the film, with its sometimes sublime imagery, features no voice-over commentary, possibly explaining what there is to see and providing an authorial interpretation of how to see it, as is common in documentary films about nature or environmental topics. So the viewer is actually, like Reggio suggests, left alone with him- or herself, having to decide what to make of the audiovisual material. However, it is rather unlikely that a viewer's primary reaction to the film will be solely disaffirmation of what is shown because the fascinating quality of many sequences can hardly be denied. Especially the visual "flow" of the accelerated shots, particularly of nighttime traffic, have to be mentioned here, which became a kind of trademark device for the film. Still, there are some parts of *Koyaanisqatsi* that make it difficult to neglect negative associations completely, like the repudiating facades of a nuclear power plant or the long-lasting, slow-motion shot of a jumbo jet directly approaching the viewer in a frightening way.

From a musicological perspective, it may also become plausible that a biased interpretation, overemphasizing a one-sided positive or negative reading of the film, will lead to some contradictions. The music by Philip Glass, with its sometimes very fast repetitions of certain melodic patterns, clearly complements and heightens the rhythmic qualities of the imagery. It can even be said that these inherent qualities of the images are accentuated via the superimposition of music, which is, in this respect, similar to the visual part of the film. The recipient is invited to discover this similarity, and, in this way, the film as a whole gains a structural density and complexity, which the visual layer alone could not have achieved. If, however, the tendency to a negative reading prevails, then it is this complexity which has, as a consequence, to be interpreted as perplexing, overwhelming, and simply "too much of it all." It would be precisely Glass's music, then, which, as a kind of catalyst, ensures the viewer's disaffirmation of modernity as it is shown by Reggio in his film,[46] with the sole purpose of expressing bustle, overabundance, and, finally, collapse.

It should not be forgotten that there actually is an interpretation of postwar, avant-garde music from the period before Glass that is about declaring that the *raison d'être* of modern art music is to reject any attempt for a positive identification with itself. In his *Philosophy of New Music*, Theodor W. Adorno writes, aimed at the twelve-tone music of Arnold Schönberg: "[a]ll the darkness and guilt of the world it [music] has put upon itself. All its joy [Glück] it derives from becoming aware of calamity [Unglück]; all its beauty comes from rejecting the illusoriness of beauty."[47] But this interpretation, how well or not well it may work as a description of twelve-tone music, is not adequate when confronted with Glass's style of minimal music—let alone his self-positioning as being explicitly post-modernist. A detailed, scene-by-scene analysis of *Koyaanisqatsi* would surely be able to identify a lot of sequences where the film offers the recipient sources of positive emotions and occasions for being impressed by the modern technologies depicted here and the way of life based on them. From this point, it would likewise be easy to demonstrate how Glass's compositions can evoke a positive semantic space built around attributes like vitality, power, successful organization, and so on.

Of course this does not mean that Glass's music in general would not be able to carry, among other things, negative connotations, especially as a film score. In fact, it is open to a wide range of readings—as are *Koyaanisqatsi* as a whole and also *Requiem for Fossil Fuels*. The 1980s film and the composition by Odland and Auinger admittedly differ with respect to their use of media. While a tagline from one of the film's trailers claims: "Until now, you've never really seen the world you live in,"[48] "sonic thinkers"[49] Odland and Auinger probably would suggest that the problem is to be found somewhere else: in the concept, which proposes vision as a privileged means of perceiving the world.

Another aspect that has to be mentioned are the plans that Odland and Auinger have for future performances of their *Requiem.* They intend to stage the piece at venues that have in themselves strong connections to the prospects of a transforming world society and the environmental boundaries that it has to deal with. As a place where "we'd like to go," they first name the Cathedral Basilica of Saint Mark in Venice, which is a place opening up several associations: this particular church building is appreciated as a major scene of music history, being connected with the work of composer Giovanni Gabrieli and the emergence of Venetian polychoral style, often cited as a historical reference for contemporary composers dealing with the distribution of sound in space. Furthermore—and much more original—Odland and Auinger declare Saint Mark's to be a "Cathedral of Global Climate Change," referring to the fact that the church's floor is below sea level; because of that, the building is flooded regularly. A further sea-level rise caused by global warming would be a massive threat to the Venice Lagoon and its monuments.[50] Other buildings that Odland and Auinger propose for performances of *Requiem for Fossil Fuels* are interpreted by them in much the same way as symbolic "cathedrals," which concretize certain aspects of environmental problems. Among them are a Dubai shopping center, designated as a "cathedral of consumption," and a coalmine in the German Ruhr Valley, which they understand as a "cathedral of coal."[51] Finally, they also mention a performance of the *Requiem* in what they call a "cathedral of sustainability," leaving undetermined what kind of place this could be, although one can derive from the context that they imagine a location in an urban setting—neither in an unimpaired countryside nor in a primeval forest. Apparently, this kind of place does not exist yet, and I wonder if it could be one that still has to be invented completely from scratch. Maybe the most convincing choice would be some "cathedral of the fossil fuel age" which may eventually be appropriated for another, possibly "greener" purpose in the future, since the adaption and reuse of existing buildings, like it is proposed and demonstrated, for example, by the architect Muck Petzet,[52] is more and more recognized as an important strategy to "decarbonize" housing as an important part of our living environment. In this way, *Requiem for Fossil Fuels* with its characteristic standpoint could be united with another story about transition.

NOTES

1. At the moment, it remains difficult to trace an exact origin of the term. In the German context, the concept of decarbonization was recently, for example, employed by economist Ottmar Edenhofer, lead author of the Intergovernmental Panel on Climate Change's Fifth Assessment Report (2013) and advisor to the German government. He also appeared at a corresponding conference about decarbonization right before COP21 hosted by German Umweltbundesamt. See Workshop "Decarbonisation—100% Renewable Energy and more," accessed April 6, 2016, http://www.umweltbundesamt.de/en/decarbonisation-workshop.

2. The premiere of *Requiem for Fossil Fuels* took place in New York City's Judson Memorial Church on October 19, 2007; however, a first version in form of a permanent sound installation (without singers performing live) was on display at Sophienkirche, Berlin, in 2004.

3. Surely the most striking examples of this development are divestment campaigns (see Bernd Sommer and Harald Welzer, *Transformationsdesign. Wege in eine zukunftsfähige Moderne* (Munich: Oekom, 2014), 190–92), the ecologically motivated large-scale withdrawal of private capital from fossil fuel industries. Even the Rockefeller Family Fund renounced furthering *ExxonMobil*, claiming that "history moves on, as it must" (quoted in Rupert Neate,"Rockefeller Family Charity to Withdraw All Investments in Fossil Fuel Companies," *The Guardian*, March 23, 2016, accessed April 6, 2016, http://www.theguardian.com/environment/2016/mar/23/rockefeller-fund-divestment-fossil-fuel-companies-oil-coal-climate-change). Also remarkable are developments like a recently passed resolution of Germany's Bundesrat, which proposes "to ban the internal combustion engine starting in 2030" (Bertel Schmitt, "Germany's Bundesrat Resolves End of Internal Combustion Engine," *Forbes*, October 8, 2016, accessed October 17, 2016, http://www.forbes.com/sites/bertelschmitt/2016/10/08/germanys-bundesrat-resolves-end-of-internal-combustion-engine/#411907ed31d9).

4. An indication of sound studies' transition from a loose interdisciplinary field of research toward a discipline is the recent arrival of two handbooks, see Trevor Pinch and Karin Bijsterveld, eds., *The Oxford Handbook of Sound Studies* (Oxford: Oxford University Press, 2012). and Jonathan Sterne, *The Sound Studies Reader* (New York: Routledge, 2012).

5. Although his concepts and views have often been criticized, it is still impossible not to mention the work of R. Murray Schafer at this point, who summarized his early work in the monograph *The Tuning of the World*. Murray R. Schafer, *The Tuning of the World* (New York: Alfred Knopf, 1977).

6. Bruce Odland and Sam Auinger, "Reflections on the Sonic Commons," *Leonardo Music Journal 19* (2009): 63–68.

7. See Alain Corbin, *Village Bells: The Culture of the Senses in the Nineteenth-Century French Countryside* (New York: Columbia University Press, 1998).

8. See Matthew Gandy and B. J. Nilsen, eds., *The Acoustic City* (Berlin: Jovis, 2014).

9. Gascia Ouzounian and Sarah Lappin, "Editorial: Recomposing the City: New Directions in Urban Sound Art," *Journal of Sonic Studies* 11 (2016). Posted

January 16, 2016, accessed April 6, 2016. https://www.researchcatalogue.net/view/236505/236507.

10. Odland and Auinger, "Reflections on the Sonic Commons," 63.

11. Ibid.

12. Other major locations of recording were Rotterdam, Netherlands, and Berlin, Germany.

13. A literal translation of the German expression would be "car-conform city."

14. See Hans Berhard Reichow, *Die autogerechte Stadt* (Ravensburg: Otto Maier, 1959).

15. However, the spacious and "organic" distribution of roads proposed and idealized by Reichow and like-minded planners in the years around 1960, which is today widely regarded as specific for the zeitgeist of that period, was also an object of criticism. Lewis Mumford wrote as early as 1958: "In that tangled mass of highways, interchanges, and parking lots, the city would be nowhere: a mechanized nonentity ground under an endless procession of wheels" (Lewis Mumford, *The Highway and the City* (New York: Mentor Books, 1964), 255).

16. Koolhaus, *Delirious New York*, 10. I am sure that both Koolhaus' statement and my interpretation of it are in some way dependent on a transatlantic or European perspective on the city.

17. Ibid.

18. The vocal score used for the performances exists only as a manuscript and is not published. I would very sincerely like to thank Sam Auinger for allowing me to examine the score and providing me with additional material and information.

19. This aspect of synthesis may be one of the reasons why Odland and Auinger regard *Requiem for Fossil Fuels* as an epitomization of their life's work.

20. The most popular example is probably the chorus of the Christmas carol "Angels We Have Heard on High."

21. The combination of "instruments and/or vocals plus tape" is very common in art music compositions since the 1950s; it can even be said that it has stabilized itself as a genre of contemporary music on its own right.

22. The specific software used here by Odland and Auinger—and also by many other sound artists and electronic music performers—is *Ableton Live*.

23. However, there are exceptions to some of the overall characteristics mentioned in my description of the piece: In this case, time-stretched audio has been used in the *Dies Irae* section according to the composers.

24. As noted before (see note 23), there are some exceptions to the overall characteristics mentioned here.

25. While writing this paper, I discovered a similar effect as I prepared a meal for a lunch break: The tubular noodles I fried in a pan slightly filtered the click sounds emitted by the heated oil.

26. The physical preconditions require that the singers sing in a tuning derived from the natural overtone scale, differing from the commonly used standard equal temperament.

27. This abbreviation, representing a very loud sound, is hidden in the title of *Requiem for Fossil Fuels*, which is also designated as *Requiemfff* by Odland and

Auinger, probably hinting at the loudness of the fossil-fueled soundscapes their piece incorporates and the enormous energies causing them.

28. See, for example, Augoyard et al., *Sonic Experience*, 38–40.

29. For the German context see, for example, von Borries, *Klimakunstforschung*.

30. See the autobiographic introduction in Welzer, *Selbst denken*, 7–11.

31. Here, I am just reproducing a common attitude toward a large part of so-called eco art. Although I agree with some part of this criticism (compare the example of the polar bear below), this attitude is not congruent with my own perspective.

32. See Slovic, *Green Culture*, 82–84.

33. Ibid., 102.

34. Küng, *Das Christentum*, 67 (my translation).

35. Killingsworth and Palmer, *Green Culture*, 21.

36. In Christian belief the apocalypse is a particular moment in eschatology, again followed by an eternity.

37. See Schellnhuber, *Selbstverbrennung*.

38. Because of this, the designation "sound studies" has been adopted in the German discourse, nearly without discussion about a German alternative. The German equivalent ("Klangstudien") would obviously have a tendency toward "the sound of music" ("der Klang der Musik"), while pushing aside associations of unwanted noise, complex soundscapes, functional sounds, and so on.

39. It is also interesting to note that there is no plural form to the German "Lärm" (a possible translation of "noise," meaning unwanted noise or din). To me, this reveals a difference: In this verbal representation, there can be different noises, but there is just one "Lärm." To speak about different qualities of "Lärm" in German, one needs to use the compound "Störgeräusche" (disturbing noises).

40. It is exactly this spectrum of interpretation, for which John Cage, one of the most influential figures in avant-garde music's reevaluation of everyday sound, argued: "Wherever we are, what we hear is mostly noise. When we ignore it, it disturbs us. When we listen to it, we find it fascinating" (Cage, "Future of Music," 3).

41. See Cleophas and Bijsterveld, *Oxford Handbook of Sound Studies*.

42. This information can be found on the composers' website, where the piece can also be listened to; see http://www.o-a.info/rfff/listen.html, accessed on April 6, 2016.

43. Brandstrader, "The Music of the Gears."

44. To me, it is inevitable to mention *Helicopter String Quartet* (1992/93) by Karlheinz Stockhausen at this point: probably the major *musical* exponent of a technology-focused 1960s zeitgeist (see again Welzer, *Selbst denken*, 7–11), Stockhausen blocks out all negative connotations (especially environmental concerns, which often accompanied performances of the composition) of helicopters in favor of a celebration of the helicopter's power as a means of flying, overcoming distances and communication.

45. Carson, "Essence of Life."

46. I have to admit, as I personally appreciate the music of Philip Glass, that this is why interpretations of *Koyaanisqatsi*, which seem to overlook the possibility of technology-friendly readings of at least some parts of the film (in the spirit of Harald

Welzer´s before-mentioned ambivalent fascination with the 1960s technology and culture), have always been annoying to me.

47. Adorno, *Philosophie der neuen Musik*, 126 (my translation).

48. "Koyaanisqatsi," IMDB.

49. This is a characterization the two sound artists use for themselves, see Odland and Auinger, "Reflections on the Sonic Commons," 63.

50. Currently, expensive activities to protect this UNESCO World Heritage site are going on. At the 2014 meeting of UNESCO World Heritage Committee, the Italian delegation "reported that the system of mobile gates called MoSE (Experimental Electromechanical Module) to control high waters temporarily isolating the lagoon from the sea is under construction and will be operational by 2016" (UNESCO 2014).

51. Zeche Zollverein (Zollverein Industrial Complex) in the city of Essen is also listed as UNESCO world heritage. It is appreciated as "an important example of a European primary industry of great economic significance in the 19th and 20th centuries" (UNESCO 2001), whose quality as a material historic document can also be expected to enrich the more world society veers away from the paradigms of this epoch.

52. See Petzet and Heilmeyer, *Reduce, Reuse, Recycle.*

BIBLIOGRAPHY

Adorno, Theodor W. *Philosophie der neuen Musik*. Frankfurt am Main: Suhrkamp, 1975.

Augoyard, Jean François et al. *Sonic Experience: A Guide to Everyday Sounds.* Montreal: McGill-Queen's University Press, 2006.

Borries, Friedrich von. *Klimakunstforschung*. Berlin: Merve, 2011.

Brandstrader, J.R. "The Music of the Gears." *The New York Times*, October 28, 2007. Accessed April 6, 2016. http://nyti.ms/25Nt8fS.

Cage, John. "The Future of Music: Credo." In *Silence: Lectures and Writings*, 3–6. Middletown, CT: Wesleyan University Press, 1973.

Carson, Greg. "Essence of Life." *Koyaanisqatsi.* special ed. DVD. Directed by Godfrey Reggio. Santa Fe: Santa Fe Institute for Regional Education, 2012.

Cleophas, Eefje, and Karin Bijsterveld. "Selling Sound. Testing, Designing and Marketing Sound in the European Car Industry." In *The Oxford Handbook of Sound Studies*, edited by Trevor Pinch and Karin Bijsterveld, 102–24. Oxford: Oxford University Press, 2012.

Corbin, Alain. *Village Bells: The Culture of the Senses in the Nineteenth-Century French Countryside*. New York: Columbia University Press, 1998.

Gandy, Matthew, and B. J. Nilsen, eds. *The Acoustic City*. Berlin: Jovis, 2014.

Killingsworth, M. Jimmie, and Jaqueline S. Palmer. "Millenial Ecology. The Apocalyptic Narrative from Silent Spring to Global Warming." In *Green Culture. Environmental Rhetoric in Contemporary America*, edited by C. G. Herndl and S. C. Brown, 21–45. Madison: University of Wisconsin Press, 1996.

Koolhaus, Rem. *Delirious New York. A Retroactive Manifesto for Manhattan*. New York: Monacelli Press, 1994.
"Koyaanisqatsi." *IMDB.* Accessed April 6, 2016. http://www.imdb.com/title/tt0085809.
Küng, Hans. *Das Christentum. Wesen und Geschichte*. München: Piper, 2007.
Mumford, Lewis. *The Highway and the City*. New York: Mentor Books, 1964.
Neate, Rupert. "Rockefeller Family Charity to Withdraw All Investments in Fossil Fuel Companies." *The Guardian*, March 23, 2016. Accessed April 6, 2016. http://www.theguardian.com/environment/2016/mar/23/rockefeller-fund-divestment-fossil-fuel-companies-oil-coal-climate-change.
Odland, Bruce, and Sam Auinger. "Reflections on the Sonic Commons." *Leonardo Music Journal* 19 (2009), 63–68.
Ouzounian, Gascia, and Sarah Lappin. "Editorial: Recomposing the City: New Directions in Urban Sound Art." *Journal of Sonic Studies* 11 (2016). Posted January 16, 2016. Accessed April 6, 2016. https://www.researchcatalogue.net/view/236505/236507.
Petzet, Muck, and Florian Heilmeyer. *Reduce, Reuse, Recycle. Architecture as Resource. German Pavilion/13th International Architecture Exhibition La Biennale di Venezia 2012*. Ostfildern: Hatje Cantz, 2012.
Pinch, Trevor, and Karin Bijsterveld, eds. *The Oxford Handbook of Sound Studies*. Oxford: Oxford University Press, 2012.
Reichow, Hans Berhard. *Die autogerechte Stadt*. Ravensburg: Otto Maier, 1959.
Schafer, R. Murray. *The Tuning of the World*. New York: Alfred Knopf, 1977.
Schellnhuber, Hans Joachim. *Selbstverbrennung. Die fatale Dreiecksbeziehung zwischen Klima, Mensch und Kohlenstoff*. Berlin: C. Bertelsmann, 2015.
Slovic, Scott. "Epistemology and Politics in American Nature Writing." In *Green Culture. Environmental Rhetoric in Contemporary America*, edited by C. G. Hernd and S. C. Brown, 82–110. Madison: University of Wisconsin Press, 1996.
Schmitt, Bertel. "Germany's Bundesrat Resolves End of Internal Combustion Engine." *Forbes*, October 8, 2016. Accessed October 17, 2016. http://www.forbes.com/sites/bertelschmitt/2016/10/08/germanys-bundesrat-resolves-end-of-internal-combustion-engine/#411907ed31d9.
Sommer, Bernd, and Harald Welzer. *Transformationsdesign. Wege in eine zukunftsfähige Moderne*. Munich: Oekom, 2014.
Sterne, Jonathan. *The Sound Studies Reader*. New York: Routledge, 2012.
Umweltbundesamt. Workshop "Decarbonisation—100% Renewable Energy and more." Accessed April 6, 2016. http://www.umweltbundesamt.de/en/decarbonisation-workshop.
UNESCO. "Venice and its Lagoon. State of Conservation." Accessed April 6, 2016. http://whc.unesco.org/en/soc/2830.
Welzer, Harald. *Selbst denken. Eine Anleitung zum Widerstand*. Frankfurt am Main: S. Fischer, 2013.

Chapter 10

Urban Ecology

Envisioning and Aestheticizing Modernity in Samuel Beckett's Not I

Lisa FitzGerald

The growth of the cityscape has been at the core of modernity. The United Nations' Population Division predicts that there will be 9.7 billion human beings on our planet by the year 2050.[1] Most of this population will live in urban centers, and the growth of cities forms a central trope in many apocalyptic visions of the future. But why does the rise of urban space necessarily correlate with an unsustainable future? Surely the increasing concentration of the human population in denser areas counters the exploitation of land and natural resources that has shadowed humanity's expansion. Apocalyptic narratives predicting untenable urban growth have not always formed part of the history of the West. Greg Garrard has noted that the "prospect of an imminent *eschaton* or End of Time opened up for Judeo-Christianity in the two centuries on either side of the Christian Year Zero."[2] The significance of the apocalypse as a trope, and its connection to the unsustainable growth of the human population, is a discourse that needs to be deconstructed in narratives of the city.

Apocalyptic narratives signaling the end of humankind underscore the sense of emergency that environmental discourse finds itself in. Lawrence Buell has pointed out that the apocalypse is the "single most powerful master metaphor that the contemporary environmental imagination has at its disposal," and the burgeoning population has a lot to do with the growth of this type of narrative as a symbol of the unsustainable drive toward the future.[3] This is growth on an unprecedented scale. Dipesh Chakrabarty has aptly noted that what has emerged in contemporary discourse is the "challenge of having to think of human agency over multiple and incommensurable scales at once."[4] Anthropogenic climate change, mass extinction, and the proliferation of waste have all become part of humanity's overextension on earth, and cities are currently viewed as part of the problem. This essay examines how the urban city has moved from the utopian ideal to the apocalyptic nightmare. I will be reading Samuel Beckett's *Not I* and suggesting that the cityscape and its immensity of scale have stimulated more sustainable practices, and corresponding narratives, for the human population.

The emergence of the apocalyptic landscapes of the mind, the post-World War II wastelands of modernity, the polluted, the disenfranchised, and the aesthetics of apocalyptic fiction, is part of a wider, and more fraught, narrative of place. Although Samuel Beckett's texts became more self-generating in his later years and increasingly unhooked from concrete depictions of place, the material traces of place, in particular the post-World War II landscape, are evident in his work. These material traces of spatial experience, and the perceived unsustainable nature of the city, are what emerge in an ever-increasing toxicity of place. But is a sustainable city not an option? Are urban inhabitants more viable in their consumption than the rural population? The urban ecology that, I argue, is evident in narratives such as Beckett's *Not I* indicates corporeal submergence in the broader cityscape. These narratives can be read as green urban texts. This fusion between person and the environment has an energy that sustains rather than depletes. I will focus on the materiality of the body and the surrounding environment, examining how that interchange opposes any sense of the apocalyptic. James Knowlson has noted of Beckett's later work: "Life material remains. It is simply located at several removes below the surface."[5] I hope to excavate below that surface to examine the materiality that infuses Beckett's *Not I*, envisioning it as a synthesis between the body and its urban surroundings.

Early twentieth-century modernism has had its impact on social spaces and the emergence of the cityscape as a symbol—both negative and positive—of human advancement and progress. Lee Rozelle examines the interplay between modernist writers and urban environments, finding that "modernist representations of urban spaces" prefigure the "terror of our current crisis."[6] Their work anticipates contemporary narratives in, as Rozelle notes, "an awe-inspiring and terrifying (post)nature."[7] The suggestion is that modernist writers describe spaces that cannot be equated with the regenerative sustainability of rural sites; that the growth of this (post)natural urban space is a portentous symbol of excess. This is an inversion of the utopian ideals of urban living: the emergence of a space dedicated to the exchange of culture and enterprise; systemic networks built to foster high-density communal living. The growth of urban spaces lessens humanity's impact on the rural environment. Urban density directly correlates to more sustainable living as the human population moves upwards instead of outwards. Nevertheless, a burgeoning cityscape has become the visual iconography of a dystopian (if not apocalyptic) future.

So how can the negative framing of the city as unsustainable be countered? Modernity has created a condition that is antithetical to romanticized ideas of the urban environment. There is no nature; the condition of modernity is that both man and environment have become mechanized. How do we visualize urban centers (and urban networks) in contemporary culture? Does urban space impair the freedom we attribute to more rural sites? Is the alienating and ominous fragmentation of space in contemporary culture conducive to the performance of urban space and large-scale surveillance? In her seminal work, "A Cyborg Manifesto," Donna Haraway states that "we are all chimeras, theorized and fabricated hybrids of machine and organism; in short, we are all cyborgs."[8] Have we and the environment that surrounds us become automated and, if so, how should we represent that automation? I argue that it is in this modern urban cityscape that *Not I*'s protagonist, Mouth, resides, and that *Not I*'s portrayal of urban living is, essentially, a positive one. The location of the urban population in a relatively small geographic area creates a sustainable model for living, and *Not I* is representative of how the human body interacts within the urban networks of the city.

Samuel Beckett's interest in the material ran parallel to his personal configuration of the environment—fragmented, dismantled, and broken down. He wrote a letter to Mary Manning Howe on July 11, 1937, declaring an interest in a "ruptured writing, so that the void may protrude, like a hernia."[9] He wrote *Not I* in English in the spring of 1972 and produced a play designed (in his own words) to work on the "nerves of the audience, not its intellect."[10] It is a play that closes down the theatrical space where the focus is not on a figure or an actor onstage but on a body part: a mouth. The narrative is rushed, vaguely coherent and delivered in an anguished, prolonged torrent of words. The text is divided into four distinct movements and is repetitive and circular in structure, reiterating much of the same imagery throughout. Although dense and barely coherent in places, there is a narrative in the story that can be broken down into an introduction and four movements. This narrative is emblematic, I argue, of modernity, and the energy released in the course of the play is suggestive of a sustainable, ecological, and progressive model for living in urban space.

The play's breathless and aggressive syntax echoes a life lived at a prodigious pace and vulnerability in the face of the ever-increasing speed of modern urban life. She starts with the birth: "out … into this world … this world … tiny little thing … before its time."[11] She sees herself and narrates her life in the third person, arguably, a modern tick of storytelling evident in social media. She is unanchored from family, having had a traumatic upbringing, and was abandoned at birth: "parents unknown … unheard of … he having vanished … thin air … no sooner buttoned up his breeches."[12] She lives her life on the margins, not speaking for the most part, but occasionally erupting into rambling and verbose rants that she cannot control: "mouth on fire … stream of words … in her ear … practically in her ear … not catching the half … not the quarter … no idea what she's saying … imagine! … no idea what she's saying!..and can't stop … no stopping it … she who but a moment before … but a moment! ... could not make a sound."[13] The way the play is structured—the rhythm, the speed, and the urgency—is symptomatic of a pace of life that urban modernity offers. Jeremy Miller and Michael Schwartz have noted that: "To possess speed is to be modern; to control speed rather than be controlled by it is perhaps the most important form of contemporary power."[14] The notion that speed is linked to modernity is a well-established truism, but, in Mouth's case, the impact is destabilizing.

One of the most crucial points in *Not I*, and the reasoning behind the title, is Mouth's inability to accept a traumatic experience as her own. The title refers to a negation of the self, and she insists on speaking in the third person, "what? … no! … she … SHE!"[15] This denial illustrates Mouth's consistent evasion of personal awareness. She is unable to process information. The result is a circular, ceaseless performance from where neither the narrator nor the audience can escape. The destructive or apocalyptic drive that underscores *Not I* is, perhaps, her self-nullifying behavior, but there is confusion as to whether there is a self to negate. There is an ambiguity as to whether Mouth is a concrete body or not. Paul Lawley has called Mouth a "'no matter' an absence." He describes her as a "function rather than a being, a conduit through which pours the words."[16] Although Mouth has no bodily form, the voice emanating from the mouth presupposes a body. The voice is female, and we can also make certain assumptions from the narrative about her life, such as her abandonment as a child, her marginalization, and that her psychological state is fragile. Her psychological trauma is what drives her to seek dissolution, but what kind of space does the body dissolve into? What environment is the theater space representative of?

Various critics have argued that the stage space that surrounds Mouth is psychosomatic. Brian Gatten states that we are in Mouth's purgatorial world; that the catastrophic event that happens at Croker's Acres is her death, and she is trying to piece together her life's narrative: "everything we see on stage is a projection of Mouth's imagination, or else a metaphorical representation of her existential state in life."[17] The environment is depicted as a space that reflects Mouth's state of being. She and her surroundings are intertwined. Paul Lawley claims that what matters in *Not I* is the "magnetic" stage image: "It is unignorable, attracting the text to itself and thus insisting upon a present-tense dimension to the story. The text hovers in panic between a past other, of which and of whom it can safely talk, and a present self (hopelessly fragmented though that is), which it must continue to deny: Not I."[18] This stage image is the depiction of a modern life, lived at a phenomenal pace.

Her world as it is represented on stage is the ceaseless, surface white noise that is associated with living in the modern urban environment. However, although the psychological reasons behind Mouth's bodily fragmentation and dissolution are destructive, the processes themselves might not necessarily be the same. The dissolution of the body into the environment is, perhaps, an urban way of being, where the emphasis is not on the individual but on the collective. The stage environment in *Not I* is not, I argue, her purgatorial world or her psyche, but urban modernity. The flood of language that is emitted throughout the performance, the deliberately confined, dense theatrical space for both actor and audience, and the circular structure with its pattern of ceaseless repetition are all expressions of urban modernity.

If *Not I* relates to the modern landscape, then what kind of place does it depict? She embodies urban narratives and performs the spatial practice (to use Michel de Certeau's term) of navigating the city. Her bodily fragmentation, her anxieties, her endless circular narrative is not that of a corporeal body. We should not be looking to anthropomorphize the body part but should instead accept her as a fragment of living modernity. Mouth has become the modern body and, as such, has become technologized. She refers to her body as "the whole machine."[19] and talks about being disengaged from herself as a whole, saying "some flaw in her make-up ... incapable of deceit ... or the machine ... more likely the machine ... so disconnected ... never got the message."[20] Perhaps her narrative is not singular but multiple. Rather than separating her from her environment through embodiment, we should see her as indivisible from her surroundings in the way that an embodied person is not. The city is made up of thousands of individuals that in many ways cease to be individual and operate as one, in order to have a streamlined and cohesive community on a much greater scale than historically known. This is not an apocalyptic narrative prophesizing the endgame but an urban aesthetic. Mouth has become a fragmented part of urban modernity in the sense that the energy and forcefulness of her disjointed narrative creates a stage space where body and environment break down into one entity.

The key to *Not I* lies not only in the narrative gushing forth but in the tensions that exist between Mouth's denial of her past trauma (resulting in her circular entrapment) and her wish for dissolution. In many ways, Mouth's resistance to fragmentation is also a vestige of past ways of living. The modern urban body is one that interacts with its surroundings on a much more intuitive scale. The dehumanizing factors involved in bodily fragmentation certainly can be viewed as apocalyptic, but this breaking down of the body is also radical in its materiality. Mouth is not a body, but a piece of a body. The stage image is of the body fragmented. Mouth has been described as an orifice and her monologue described as a bodily purging, a physical act. For the audience, the blackness of the space around the mouth and the impression of the mouth as symbolic of all bodily orifices trigger, in the words of Anna McMullan, a "body … turned inside out, and continually recycled by the voice as waste material."[21] There is no presence and no physical body on stage that would form a traditional performance. What is being performed is the absence or the end of the individual and the beginning of the body becoming totally immersed in the urban environment. It is interesting to link this idea with Daniel Dennett's image of the self as constructed through the material we gather in our environment. The self creates a constant narrative, devised as we make sense of the world around us: "We (unlike professional human story tellers) do not consciously and deliberately figure out what narratives to tell and how to tell them; like spiderwebs, our tales are spun, but for the most part we don't spin them; they spin us. Our human consciousness and our narrative selfhood is their product, not their source."[22] The narrative that we weave to form our own core identity is a direct response from our environment and our sensory connection with it. Perhaps *Not I*'s Mouth is an attempt to overcome that divide between body and environment.

If we are a product of a narrative generated from our surroundings, then what space is surrounding the talking mouth in *Not I*? Her anguished monologue certainly reflects the uncertainties associated with urban living. In 1940, Walter Benjamin writes about encountering the city. He makes the observation that "fear, revulsion, and horror were the emotions which the big city crowd aroused in those who first observed it."[23] He describes moving through traffic, which: "involves the individual in a series of shocks and collisions. At dangerous crossings nervous impulses flow through him in rapid succession, like the energy from a battery. […] Thus technology has subjected the human sensorium to a complex kind of training."[24] This energy, the nervous impulses that Benjamin talks about, reflects Mouth's nervous energy. The experience of moving through the city is impossible given that Mouth is just that, a mouth, but, through her performance, that external urban world is brought inside, into the space of the auditorium.

Michel de Certeau writes that "every story is a travel story—a spatial practice."[25] In his chapter "Walking in the City," he argues that the urban space of the street emerges through the embodied person. In moving through the city, we make use of urban space that we cannot see objectively but feel through our senses, resulting in a more insightful way of engaging with the city, as opposed to attempts at map-making and the objectification of the urban experience. He uses New York as an example of these two modes of engaging with urban space. First, there are our attempts to negotiate the urban landscape through mapping or reading the space and turning it "into a text that lies before one's eyes."[26] This omnipresent viewpoint—where we stand outside the urban space as a voyeur—is challenged by engaging with the city through walking. This is the experience of imagining the city as a spatial practice. Mouth's spatial practice in *Not I* is one of moving through a distinctly urban space. I would argue that performative texts such as *Not I* show an awareness of this flow of energy between, and are conscious of, the body and its place in the environment. The concentrated networks within an urban environment, and its inhabitants, create a confluence of energies that are regenerative and sustainable.

Stacy Alaimo's concept of trans-corporeality (which emphasizes the porous body) allows us to pick up on Buell's networks of modernity but to move toward the notion of networks of materiality, where matter moves fluidly between human and nonhuman. Alaimo discusses the ethical and political implications of thinking about the space between the human and what she calls more-than-human nature. She points out that "imagining human corporeality as trans-corporeality, in which the human is always intermeshed with the more-than-human world, underlines the extent to which the substance of the human is ultimately inseparable from 'the environment.'"[27] I argue that, in much the same way, the impact of the networks of modernity has affected Mouth in *Not I*. Mouth has become the porous body with the technological tools of modernity flowing through her. She makes manifest the physiological impact of living in a modern landscape. This focus on the materiality of body is a process that allows us to narrate on different scales, overcoming the challenge of thinking along microscopic versus macroscopic modes of perception. Mouth could be deemed an aesthetic narrative of the trans-corporeal body, performing the interchanges between collapsing notions of the self and its environment. These interchanges result in entanglements where the body can never be separated from the wider environment. The porous body is steeped in the networks of materiality to which it is exposed. Matter thus is agential and illustrates or narrates the global inequalities that have become a feature of the contemporary environmental crisis.

Mouth illustrates the porosity of the body in *Not I*. The tropes of living in urban modernity—such as the energy, the technology, and the urban space itself—are central to Mouth's way of being. We negotiate urban space differently to rural space; as de Certeau notes, "The networks of these moving, intersecting writings compose a manifold story that has neither author nor spectator, shaped out of fragments of trajectories and alterations of spaces: in relation to representation, it remains daily and indefinitely other."[28] The fragmentation of the narratives of negotiated space is represented in *Not I*. Our bodies succumb to the urban space, and we become hive-like, communicating in new and distinctly modern ways. De Certeau's method for negotiating the city moves away from the voyeuristic experience of the urban city as text and into a more sincere experience of moving through urban space. This is a creative act understood and practiced by those who live "down below."[29] It is in the interacting with the environment at ground-level that the individual begins to break down.

But, to return to the idea of *Not I* as an apocalyptic narrative, the breaking down of the body into the landscape is not always seen a progressive and radical act. In *Rambling as Resistance*, Jason Kosnoski challenges de Certeau's idea of walking in the city by suggesting that, in our contemporary, modern urban city, the fragmentation of spaces and our inability to transcend them are destructive: "Individuals face more and more observation, control and discipline through both spatial and temporal strategies upon the body, yet the fragmentation of spaces, cities and regions ensure that individuals do not traverse the entirety of their environment with freedom or creativity."[30] Negotiating urban space is not always generative and can be suppressive, particularly to those expressing dissident tendencies. Which position does Mouth take? Is her engagement with her environment a creative act (embracing how the city and its intrinsic modernity have evolved), or is her anxiety a result of her oppressive surroundings rather than a past trauma? There is evidence in *Not I*, supporting Kosnoski's assertion, that urban space actually impairs our freedom. There is a repeated ray or beam of light—which she mistakes for the moon—that shines on Mouth and echoes the ubiquitous surveillance devices used in monitoring urban populations. She describes the beam of light as omnipresent, "always in the same spot."[31] Its presence is relentless and searching, "ferreting around … painless … so far … ha! … so far."[32] She initially thinks the beam of light is the moon but then changes her mind, "probably not … certainly not always the same spot … now bright … now shrouded … but always the same spot … as no moon could … no … no moon … just all part of the same wish to … torment."[33] This beam of light is unnatural, penetrating and threatening. In much the same way as large-scale surveillance is used in a modern context, the effect is heightened anxiety and tension.

The alienating and ominous fragmentation of space is certainly evident in *Not I*. And this enclosing of urban space is essentially an argument that returns the body to that of the viewer, framing rather than immersed in the environment. The contemporary urban landscape contains within its geographical spaces the concerns that are prevalent in much of Beckett's work: alienation, loneliness, paralysis. Yet the word landscape itself is problematic. Cultural Geographer John Wylie describes landscape as "a tension between proximity and distance, body and mind, sensuous immersion and detached observation. Is landscape the world we are living *in*, or a scene we are looking *at*, from afar?"[34] In much the same way that de Certeau distrusts the mapping or objective viewpoints of the city, the narrative of *Not I* is immersed totally within the city space as opposed to framing it impartially. Raymond Williams argues that "the very idea of landscape implies separation and observation."[35]

He describes a framed environment, to be looked at from a distance, and something (in the same way scenery in a theater might be) not to be interacted with. These definitions of landscape suggest an inability to be part of an environment that is being narrated, something that Mouth clearly challenges.

These tensions between the body and its environment, between being both immersed and outside the space, are reflective not just of the audience in the theatrical space but in Mouth's way of seeing herself. She does describe her environment—a "little mound in Croker's Acres," a "busy shopping centre" or the "godforsaken hole" where she was born—as something that stands outside of herself.[36] However, in his work, *Landscape and Memory,* Simon Schema notes that we are: "accustomed to separate nature and human perception into two realms, they are, in fact, indivisible. Before it can ever be repose for the senses, landscape is the work of the mind. Its scenery is built up as much from strata of memory as from layers of rock."[37] It could be argued, in accordance with other critiques of *Not I*, that the traumatic event that has instigated Mouth's denial of identity has left her unable to place herself in the world. She remains outside her environment, a voyeur; her experience of place is only a simulacrum of the true urban city. Mouth cannot truly know her environment, and that is at the core of her trauma. However, rather than seeing the landscape (from the traditional Western mode), as something to be framed or looked upon, I contend that Mouth is fully immersed in her environment. The environment in *Not I* does not represent the traditional landscape—following from Raymond Williams's notion of landscape as implying separation and observation—separated from the body or observer, but a landscape of modernity. The urban cityscape does not exist in opposition to the rural mode of living but as a part of a wider, emerging environmental ecology. The notion of the golden age of rural life emerges from the urban perspective. The theatrical space is bound by the physical confines of the urban place, but it gives us access to nature and the natural world through performance. Through the memories that she evokes in her monologue and her concrete engagement with the space of the stage, Mouth exists in the throes of modernity. Circular rather than linear, one where the body in space is not articulated with the entire body—but fragmented. From the outside in, Beckett's theatrical landscape internalizes the elements of modernity that form the urban landscape and performs that urbanism in *Not I*.

Not I is a play that embodies and narrates the complex processes at the heart of the modern, urban landscape. But can the city, as represented in literary texts and plays such as *Not I*, move beyond the apocalyptic vision of the future, and can there be a more sustainable and progressive outlook for urban modernity? Timothy Morton has stated that there is no longer such a thing as a natural landscape; that the condition of modernity is that both man and environment have become mechanized. That labor is hidden and aestheticized and has resulted in the creation of "Junkspace"—a phrase borrowed from architect Rem Koolhaas to denote our contemporary urban spaces. Morton argues that, "Capitalist thinking, and Capitalist machinery, actively 'disappear' the workers who operate it. [...] Humans 'man' the machines: more than ever, human beingness is now revealed as a product of mechanical processes."[38] Rem Koolhaus offers us this concept of the Junkspace as the ultimate wasteland of modernity, a place "so extensive that you rarely perceive limits, it promotes disorientation by any means (mirror, polish, echo)."[39] Mechanized and disenfranchised, there are no human forms in *Not I* but human parts. The result is a white noise, an automated art form that is symptomatic of a wider apocalyptic landscape.

What these impressions of modernity—the speed, fluidity, and shine—do have, is the ability to incite a fear of the city as a nexus for environmental crisis. Apocalyptic narratives have deemed the city as the site for a failed humanity. But as the majority of humanity now lives in urban areas, there is a need for new narratives that see the city not as a harbinger of environmental devastation but a citadel for sustainability. The impacts of the networks of modernity—auto, telephonic, and internet—have certainly affected visual culture. This contemporary aesthetic of the urban makes manifest the physiological impact of living in a modern landscape. *Not I* opens up a space that echoes this prevalent urbanism and in Mouth (its protagonist) creates a figure that embodies the energy, the interdependence, and the precariousness of urban living.

We must remember that *Not I* is a performed rather than a real representation of an urban space. Corina Martin-Jordache notes: "Beckett's cities seem to have gone a long way from the proud city-states of antiquity or the enlightened cities of the renaissance. They are the exhausted cities of the twentieth century modernity, post-*Waste Land* cities."[40] But what of Becket's plays? What of the theatrical space itself? Does it perform the same urbanism or urban space as the descriptions of the depleted cities as described by Martin-Jordache? Can the figures that populate Beckett's theatrical landscape internalize the elements of modernity that form part of the cityscape and *perform* that urbanism? And is that urbanism sustainable? I wanted to examine, in this essay, how place is registered in *Not I* and what kind of place it is. How does the theatrical space of *Not I* relate to the modern cityscape? How do spaces of interiority such as the theater itself or the thought process evoked in *Not I* describe or link to these exterior places? I have argued that the emergence of the apocalyptic landscapes of the mind in his work is part of a wider narrative of place, one that echoes society's deepening material and environmental anxieties. But I also want to point out that these are urban spaces that echo the circular, ceaseless energy as generated in the city.

Anxieties associated with modern living, of which urban centers form a large part, emerge with the attempts to encounter and narrate our own materiality and interaction with the space around us. I argue that *Not I* is an example of a truly modern creativity that represents the urban landscape. The interconnectedness of the body in the environment, being and language, the biological and the mechanical, is at the heart of modernist narratives of the city. The social history of the body and the inequalities perpetrated on it are evident in Mouth's attempts to interact in an urban environment. The fragmentation of the body in *Not I* goes some way to remedy this hierarchy of belonging in social space. Matter becoming agential—be it a body part, or humanity as a geological force—allows for narratives that supersede traditional representations of social communities, be they urban or rural.

Mouth's body is directly attuned to the modern urban space in accordance with Buell's definition of the environment as "a network or networks within which, on the one hand, humans are biotically imbricated (like it or not) and, on the other hand, nature figures are modified (like it or not) by techne."[41] Mouth is the alienated, fragmented body that is part-human and (in Buell's words) part-techne. This part-mechanization of man—the unnatural manner of speaking, the fragmented body—is inherent in the crisis of modernity, where there is a fear that the natural order has in some way broken down. Buell argues that the threat of toxicity which emerged in a post-industrial society is only part of the increased anxiety of modern living. But, rather than urban space being a central metaphor in heralding an endpoint of humanity (and there is no doubt that the anxieties evident in these narratives are valid), the city and its networks can be embraced as a site of true modernity.

This essay contends that the fragmentation evident in *Not I* is a breakup of the body in the environment. But this dissolution of the body into the environment does not have to be apocalyptic; the emergence of the endless networks found in the modern cityscape can also produce a vital and sustainable space for living. The representation of the city as an apocalyptic nightmare is essentially a fable and one that is waning; as Greg Garrard points out, "environmental apocalypticism has had to face the embarrassment of failed prophecy even as it has been unable to relinquish the trope altogether."[42] Rather than the cityscape envisioned by modernist architects such as Le Corbusier, what has transpired in Koolhaas's vision of the Junkspace is a partisan version of urban life: a "fuzzy empire of blur" that "fuses high and low, public and private, straight and bent, bloated and starved to offer a seamless patchwork of the permanently disjointed."[43] But this seamless patchwork is also a feature of ecological interdependence. The desire to preserve what is deemed the natural environment and to exclude the urban is detrimental to a future in which the city is evidently a large part.

The urban cityscape is the one that is recapitulated in *Not I*: the effect of modernization on the environment and consequently—because of our enmeshment with that environment—on the mind. Apocalyptic narratives have in many ways misinformed what it means to be human and living in a modern cityscape. The city is certainly fragmented, technological, material, and, as Samuel Beckett's *Not I* suggests, undeniably interconnected. This depiction of the urban city, however, is not necessarily as detrimental as the fantasy of the future previously has been. *Not I* is suggestive of an aesthetic that, while depicting the precariousness of urban living, also proposes an urban body that is capable of complete immersion. Mouth is a body that overcomes the strict binaries of individual and environment and represents the interplay between the two. This exchange—which, in many ways, depicts urban modernity—should surely mark a more progressive representation of the city than apocalyptic narratives of an unsustainable future.

NOTES

1. "2015 Revision of World Population Prospects," United Nations: Department of Economic and Social Affairs, accessed May 20, 2016, http://esa.un.org/unpd/wpp/.
2. Greg Garrard, *Ecocriticism* (London and New York: Routledge, 2004), 86.
3. Lawrence Buell, *The Environmental Imagination: Thoreau, Nature Writing and the Formation of American Culture* (London: Princeton University Press, 1995), 285.
4. Dipesh Chakrabarty, "Postcolonial Studies and the Challenge of Climate Change," *New Literary History* 43, no. 1 (2012): 1.
5. Knowlson, James, *Damned to Fame: The Life of Samuel Beckett* (London: Bloomsbury, 1996), xxi.
6. Lee Rozelle, "Ecocritical City: Modernist Reactions to Urban Environments in 'Miss Lonelyhearts' and 'Paterson,'" *Twentieth Century Literature* 48, no. 1 (2002): 100–115.
7. Ibid.
8. Donna Haraway, "A Cyborg Manifesto: Science, Technology, and Socialist Feminism in the Late Twentieth Century," in *Simians, Cyborgs and Women: The Reinvention of Nature* (New York: Routledge, 1991), 150.
9. Samuel Beckett, *The Letters of Samuel Beckett: Volume 1, 1929–1940*, eds. Martha Dow Fehsenfeld and Lois More Overbeck (Cambridge: Cambridge University Press, 2009), 521.
10. Quoted in James Knowlson and John Pilling, *Frescoes of the Skull: The Later Prose and Drama of Samuel Beckett* (London: J. Calder, 1979), 195.
11. Samuel Beckett, *The Complete Dramatic Works* (London and Boston: Faber and Faber, 1986), 376.
12. Ibid.
13. Ibid., 380.

14. Jeremy Miller and Michael Schwartz, "Introduction—Speed is a Vehicle," in *Speed—Visions of an Accelerated Age*, eds. Jeremy Miller and Michael Schwartz (London: Whitechapel Gallery, 1998), 17.

15. Beckett, *Collected Dramatic Works* 377, 379, 381, 382.

16. Paul Lawley, "Counterpoint, Absence and the Medium in Beckett's Not I," in *On Beckett: Essays and Criticism*, ed. S. E. Gontarski (London and New York: Anthem Press, 2012), 250.

17. Brian Gatten, "The Posthumous Worlds of Not I and Play," *Texas Studies in Literature and Language* 51, no. 1 (2009): 96.

18. Lawley, "Counterpoint, Absence and the Medium," 247.

19. Beckett, *CDW*, 380.

20. Ibid., 378.

21. Anna McMullan, "Performing Vision(s): Perspectives on Spectatorship in Beckett's Drama," in *Samuel Beckett: A Casebook*, eds. by Jennifer Jeffers and Kimball King (London and New York: Garland Press, 1998), 150.

22. Daniel C Dennett, *Consciousness Explained* (New York: Little, Brown and Company, 1991), 417–18.

23. Walter Benjamin, *Walter Benjamin: Selected Writings. Vol. 4, 1938–1940* (Cambridge, Mass. London: Harvard University Press, 2003), 327.

24. Ibid., 328.

25. Michel de Certeau, *The Practice of Everyday Life* (Berkeley: University of California Press, 1984), 115.

26. Ibid., 92.

27. Stacy Alaimo, *Bodily Natures: Science, Environment, and the Material Self* (Bloomington: Indiana University Press, 2010), 2.

28. de Certeau, *Practice of Everyday Life*, 93.

29. Ibid.

30. Jason Evan Kosnoski,"Rambling as Resistance: Frederic Law Olmsted, Michel de Certeau, and the Micropolitics of Walking in the City," *Situation: Project of the Radical Imagination* 3, no. 2 (2010): 127–28.

31. Beckett, *CDW*, 378.

32. Ibid., 381.

33. Ibid., 378.

34. John Wylie, *Landscape: Key Ideas in Geography* (London: Routledge, 2006), 1 (original emphasis).

35. Raymond Williams, *The Country and the City* (New York: Oxford University Press, 1973), 120.

36. Beckett, *CDW*, 380, 379, 376.

37. Simon Schema, *Landscape and Memory* (London: HarperCollins Publishers, 1995), 6–7.

38. Timothy Morton, *Ecology without Nature: Rethinking Environmental Aesthetics* (Cambridge, MA: Harvard University Press, 2007), 86–87.

39. Rem Koolhaas, "Junkspace," "Obsolescence," Special Issue, *October* 100 (2002): 175.

40. Corina Martin-Jordache, "Modernity, Urban Semiology and the Beckettian 'Cityscape'," *Journal of European Studies* 32, no. 127 (2002): 366.

41. Lawrence Buell, "Toxic Discourse," *Critical Inquiry* 24, no. 3 (1998): 657.

42. Greg Garrard, *Ecocriticism* (London and New York: Routledge, 2004), 100.

43. Koolhaas, "Junkspace," 176.

BIBLIOGRAPHY

Alaimo, Stacy. *Bodily Natures: Science, Environment, and the Material Self.* Bloomington: Indiana University Press, 2010.

Beckett, Samuel. *The Complete Dramatic Works*. London and Boston: Faber and Faber, 1986.

———. *The Letters of Samuel Beckett: Volume 1, 1929–1940*, edited by Martha Dow Fehsenfeld and Lois More Overbeck. Cambridge: Cambridge University Press, 2009.

Benjamin, Walter. *Walter Benjamin: Selected Writings. Vol. 4, 1938–1940.* Cambridge, Mass. London: Harvard University Press, 2003.

Buell, Lawrence. *The Environmental Imagination: Thoreau, Nature Writing and the Formation of American Culture*. London: Princeton University Press, 1995.

———. "Toxic Discourse." *Critical Inquiry* 24, no. 3 (1998): 639–65.

Chakrabarty, Dipesh. "Postcolonial Studies and the Challenge of Climate Change." *New Literary History* 43, no. 1 (2012): 1–18.

de Certeau, Michel. *The Practice of Everyday Life*. Berkeley: University of California Press, 1984.

Dennett, Daniel C. *Consciousness Explained.* New York: Little, Brown and Company, 1991.

Garrard, Greg. *Ecocriticism*. London and New York: Routledge, 2004.

Gatten, Brian. "The Posthumous Worlds of *Not I* and *Play*." *Texas Studies in Literature and Language* 51, no. 1 (2009): 94–101.

Haraway, Donna. "A Cyborg Manifesto: Science, Technology, and Socialist Feminism in the Late Twentieth Century." In *Simians, Cyborgs and Women: The Reinvention of Nature*, 149–81. New York: Routledge, 1991.

Knowlson, James. *Damned to Fame: The Life of Samuel Beckett*. London: Bloomsbury, 1996.

Knowlson, James, and John Pilling. *Frescoes of the Skull: The Later Prose and Drama of Samuel Beckett.* London: J. Calder, 1979.

Koolhaas, Rem. "Junkspace." "Obsolescence." Special Issue, *October* 100 (2002): 175–90.

Kosnoski, Jason Evan. "Rambling as Resistance: Frederic Law Olmsted, Michel de Certeau, and the Micropolitics of Walking in the City." *Situation: Project of the Radical Imagination* 3, no. 2 (2010): 115–44.

Lawley, Paul. "Counterpoint, Absence and the Medium in Beckett's *Not I*." In *On Beckett: Essays and Criticism,* edited by S. E. Gontarski, 245–52. London and New York: Anthem Press, 2012.

Martin-Jordache, Corina. "Modernity, Urban Semiology and the Beckettian 'Cityscape.'" *Journal of European Studies* 32, no. 127 (2002): 351–68.

McMullan, Anna. "Performing Vision(s): Perspectives on Spectatorship in Beckett's Drama." In *Samuel Beckett: A Casebook*, edited by Jennifer Jeffers and Kimball King, 133–58. London and New York: Garland Press, 1998.

Miller, Jeremy, and Michael Schwartz. "Introduction—Speed is a Vehicle." In *Speed—Visions of an Accelerated Age*, edited by Jeremy Miller and Michael Schwartz, 16–23. London: Whitechapel Gallery, 1998.

Morton, Timothy. *Ecology without Nature: Rethinking Environmental Aesthetics.* Cambridge, MA: Harvard University Press, 2007.

Rozelle, Lee. "Ecocritical City: Modernist Reactions to Urban Environments in 'Miss Lonelyhearts' and 'Paterson'." *Twentieth Century Literature* 48, no. 1 (2002): 100–15.

Schema, Simon. *Landscape and Memory*. London: HarperCollins Publishers, 1995.

United Nations, "2015 Revision of World Population Prospects," *United Nations: Department of Economic and Social Affairs*. Accessed May 20, 2016. http://esa.un.org/unpd/wpp/.

Williams, Raymond. *The Country and the City*. New York: Oxford University Press, 1973.

Wylie, John. *Landscape: Key Ideas in Geography*. London: Routledge, 2006.

Chapter 11

Sustainable Urban Development and Literature

A Central-European View

Katarzyna Szalewska*

Most urban concepts that are already well entrenched in Western thought and practice of urban studies are adapted with a delay in Poland. Furthermore, in the Polish environment, the known terms gain a different set of interpretations related to other legal, cultural, and economic conditions but above all due to the difficult history of Central-European cities. Only after 1989, did it become possible in Polish academic thought to analyze cultural dimensions of experience planning and the people's rights to the cities because obviously only socialist government had this right before. However, the process of urban residents regaining the city and cultural practices have occurred quickly since the 1990s, and the losses have been made up at lightning speed. Interest in the subject resulted in the development of urban studies and a great increase in urban literature, especially novels, whose main hero has become a city. These novels reinterpret terms known and rooted in Western thought, adapting them to Polish conditions, often leading to the discussion of the specifics of Polish urban practices.

Among these concepts discussed too late in Poland is New Urbanism. The concept is very ambiguous and is often combined with different ideas on aesthetics and pragmatics of urban space, focused mainly on the continuation and reinterpretation of Leon Krier's thought. What unites different statements about the city formulated from this perspective is an antimodernistic approach. In the words of one of the mainstream theorists of this New Urbanism, Douglas Kelbaugh: "What is new about the New Urbanism is its totality. It attempts to promote a sort of unified design theory of an entire region—form the small scale (building block, street) through the intermediate scale (corridor, neighborhood, district) to the large scale (regional infrastructure and ecology)."[1]

---

* Translated by Joanna Zahorska

New Urbanism has become in these deliberations a metaphor of postmodernist reflection of the city as a whole, taking into account the principles of sustainable development. Thus it serves as a starting point to show the Polish road to a "unified design theory," which is not defined by a total political or architectural foundation but scattered in informal actions. Since it is impossible to create a coherent narrative of space marked by trauma (and such is Polish urban space), literature has taken on the task of recalling and working through the trauma, discussing spatial scars, errors, and discontinuities, which in the future will result (and has already resulted) in real actions in urban space.

New urbanism is an ambiguous concept with a rich history, already overgrown with numerous interpretations, discussions, and criticism in Anglo-Saxon literature. The reception of this architectural-social reflection on the Polish ground, from the perspective representative of the urbanistic situation in the post-Soviet states of Central and Eastern Europe, is thoroughly different. As we can read in the *Encyclopaedia of Human Geography*,

> New urbanism is a collective term for an array of planning and design practices originally developed to respond to dissatisfaction with contemporary urbanization processes, especially those linked to urban sprawl. This dissatisfaction included the separation of land uses (a precept of traditional zoning), the domination of daily life by the automobile, and the loss of a sense of community. [. . .] It is, however, not without its critics, who have seen new urbanism as a veneer over the organization of space of capitalism that replicates race and class divisions within contemporary society, even as it embeds them in a utopian vision of communities of the future.[2]

In Polish conditions, the utopian vision of the future must inevitably clash with the spatial traumas of war and Communism, as well as with the shameful traces of wild capitalism and the unlearned lesson of the postmodernism of the 1990s.

Thus, if new urbanism[3] postulates, among other things, a return to the traditional layout of cities with a centrally situated historic center (old town, square, market square) and adjacent residential commercial spaces arranged in quarters, then, in Polish conditions, it must mean a return to trauma. During World War II (including as a result of the Warsaw Uprising), Warsaw and other main Polish cities were almost completely destroyed. Postwar ruins were soon replaced with socialist factories and housing estates of slab block buildings capable of accommodating the people mass-migrating to cities and, in the long run, the postwar demographic bulge. Therefore, if new urbanism means a change in proportions between motor traffic and pedestrian traffic in favor of the latter, then, in the context of Polish cities, one should bear in mind that renovated modern ring roads are often advances that have not appeared until recent decades, even the last several years; therefore, it is hard to write about the burdensomeness of transitive spaces that are, in fact, usually accepted by residents as a symbol of modern times and a substantial convenience. Lastly, if new urbanism postulates tackling alienation in urban crowds by restoring a sense of community, then, in Polish conditions, this task will be exceptionally difficult due to the population being completely mixed as a result of divisions imposed by the agreements of Yalta and Potsdam, which are particularly evident in cities in the so-called Recovered Territories (a term used in the period of the People's Republic of Poland referring to the territory of the former Free City of Danzig, as well as western and northern parts of contemporary Poland, which have been granted to Poland according to the resolutions of the Potsdam Conference), such as Gdańsk, Szczecin, and Wrocław.

There are two reservations concerning the above remarks. First of all, the period between 1945 and 1989 brought to Poland industrial construction, industrialization of urban spaces (and thus exclusion of these spaces from everyday spatial practices of residents),[4] and housing estates of slab block buildings, but also interesting urban solutions commented on in Western Europe. As Filip Springer notes in *Wanna z kolumnadą*, prior to World War II, the Polish spatial planning system used to be one of the best on the continent, oftentimes adopted by other countries, for example, Germany. Importantly, after the war, the spatial planning system was forced to give way to a central planning system; however, as Springer remarks, people responsible for urban planning remained active, in some cases with great success, for example, the author of threshold analysis Bolesław Malisz, who dedicated his work to limitations in urban development (such as economic, social, or landscape conditions). According to Springer, Malisz's algorithms facilitate choosing the relevant development path for authorities and hence the method is commonly applied worldwide but not in Poland.[5]

A similarly ambiguous evaluation should be given to spatial complexes developed after 1989 in a free market situation since they stand for, on the one hand, an openness to new ideas, including new urbanism, and, on the other, a dictate of economics, which oftentimes meant a primacy of developers' interests over aesthetics. The subject of contaminated spaces of Polish cities and the lack of their sustainable development is (slowly) becoming of interest to literature. A considerable contribution in this regard ought to be attributed to reporters. Reportage is probably the fastest growing field in modern Polish literature. One of the critics of urban orders and advocates of sustainable development is the aforementioned Springer (significantly, an author born in 1982, thus representing a new generation of residents and urban analysts who grew up after the breakthrough of 1989), who dedicated most of his books (e.g., *Źle urodzone. Reportaże o architekturze PRL-u*[6]; *Wanna z kolumnadą. Reportaże o polskiej przestrzeni*, and *13 pięter*[7]) to

Polish urban space in its various aspects: the presence of buildings from the time of the People's Republic of Poland in the space of a modern city; the problems of housing shortage and property market; and, lastly, urbanism with regard to its relation to aesthetics. All these issues connote a question about sustainable development, a slogan that comprises economic, aesthetic, and social dilemmas, and which, in Polish conditions, is most promptly posed in literature. And it is literature that will be analyzed in this essay, which comprises interpretations of contemporary Polish stories (primarily from the twentieth and the twenty-first centuries) from the perspective of the concept of sustainable development.

The aim of this essay is not to present the history of modern urban thought or discuss principles of New Urbanism and the idea of sustainable development but to show how Polish urban space is a kind of laboratory where various modern concepts of spatial practices are growing and being adapted and reinterpreted. The second part of this essay is devoted to Polish literature of the twentieth and twenty-first centuries, which has become a mirror reflecting the cultural, political, and economic processes affecting the functioning and perception of urban space. Literature is an important voice, often having a secondary impact on the actual operation carried out in real space (one example being the popularity, also touristic, of the city of Wałbrzych—thus far absent in the Polish topographic imagination—after the publication of the widely read contemporary novels of writer Joanna Bator). The last part of this essay extends the literary considerations to a cultural analysis of the selected cities; the example of Gdańsk shows the process of growing ideas and practices of urban planning in Poland.

THE IDEA OF SUSTAINABLE DEVELOPMENT AND CONTEMPORARY POLISH LITERATURE

In her novel *Cwaniary*, Sylwia Chutnik makes a map of Warsaw an attribute of tourists, wealth, and the center of the capital of Poland that smells of luxury and hence of spaces that are inaccessible to protagonists excluded from the topographical order symbolized by cartography. This is depicted in Chutnik's novel from the perspective of a protagonist named Halina, who runs into a group of tourists when crossing the courtyard of the Royal Castle in Warsaw. The group is described with disgust as a plague of pests, penetrating the entire city with maps crumpled in their pockets, interested only in all-inclusive programs and the costly smell of hotels.[8] The formation of modern Polish literature is marked by critical cartography,[9] for which a map does not constitute an unquestionable representation of the spatial order anymore but becomes an instrument of power instead. Hence, authors of the latest literary depictions of Polish cities join the discussion not only of the traditional conventions of a textual description and the creation of an urban *effet de réel* but also of actual spatial practices, urban planning concepts, and local spatial politics. In Polish literature, urbanistic issues have undergone various stages strictly subjected to economic and social factors. Following a period that might be called "heroic urbanism," which occurred in the second half of the nineteenth century and the beginning of the twentieth century (including works like *The Doll* by Bolesław Prus and *The Promised Land* by the Nobel laureate Władysław Reymont) and was related to the progressive industrialization and urbanization of the country, as well as the wartime gap, a time of "two cities" took place in the Socialist era of 1945–1989. Here, I draw my approach from the Polish poet Adam Zagajewski, whose autobiographical essay *Dwa miasta* talks about a dualistic experiencing of space: the visible, present space (in Zagajewski's work, this is represented by industrialized Gliwice, one of the cities inhabited by people displaced after the war) and the past space that is recalled (in the story, this role is played by Lviv, which was lost in 1945) at the same time. The duality observed in Zagajewski's work is characteristic of the entire range of urbanistic issues in Polish literature of 1945–1989, which either presents a city as an oppressive space of the "small realism"[10] or escapes to the memory and the dream of Lviv or Vilnius mythologized in memory and post-memory.[11] It was not until the works written after the year 1989 by a generation of authors who have not experienced displacement but were somewhat thrown into the present and the modernity of Polish cities that a critical reflection on the city here and now was introduced—not on the city that was lost or recovered but the existing city, the one that is seen, experienced, and used.

The female protagonists of Sylwia Chutnik's *Cwaniary*, like in many other novels of the last few decades, take strolls in the largest Polish cities, and this *flânerie* becomes, at the same time, an act of questioning the order of the center and a proposal to recall the urban entities that were disregarded by the city in the course of its development. Such an insurrectionist discourse can become: a spatial story of a woman, like in Joanna Bator's[12] prose; the speech of an inhabitant of a suburban housing estate of slab block buildings, as in the works of Dorota Masłowska[13]; the monologue of a participant in gay rights demonstrations, as in Michał Witkowski's[14] view on Wrocław in *Lubiewo*; or the discussion held in a market that serves as an urban *agon,* as in *Kieszonkowy atlas kobiet* by Chutnik.[15] Nowadays, writing the city by denying oppositions significant to this spatial practice (center/suburbia; interior/exterior; female/male; private/public) alters the topography of the novel, which triggers changes in the narrative point of view, focalization, language, construction of the protagonist, and type of the novel, drawing them closer to a contemporary anecdote than the grand narrative of past times.

The difference between the modern-day criticism of the city and the traditional topos of *polis* in literature is revealed by a comparison of the monumental Polish novel *The Promised Land* by Władysław Reymont and the recently famous *Piaskowa Góra* by Bator. It is a precarious comparison, as these books are separated by a temporal abyss, different conventions, and *intentio operis*. Still, they share the construction of the main protagonist and the role taken on by the city, which is, in both cases, an industrial center. Reymont begins his canonical novel, which is highly important to Polish literature and was published in the years 1897–1898, with a vision of Łódź, a factory town,[16] as an urbanistic monster:

> He took a cab and went off straight down Piotrovska Street [the main street of Łódź]. [...] He looked around him—on the dingy town, wet with rain, muddy on the side-walks, swarming with humanity; on the uncountable chimneys [...]; on the coal-wagons in their hundreds, forming an immense chain of which one extremity was entering some factory or other; on the multifarious conveyances rushing in every direction; at the multitudinous offices and warehouses, those full of people, these piled with goods; and on the strain and stress of life that seethed around him. On all these things he gazed, in something like despair. He felt his own powerlessness: felt that in no long time to come he should drop out of that enormous whirlpool—out of the huge machine called Lodz—like a bit of used-up matter which the monster needed no more.[17]

The above fragment, which is typical of some trends in modernist essay writing, has an anti-urbanistic poetics that presents the city, particularly an industrial one, as a monster, a post-apocalyptic inferno beyond human scale. Reymont strives to show a wide panoramic view on the society without hiding his fascination with urban modernity emerging at that time, in which the spectacle of daily life takes place and everything is possible. For that very reason, the main protagonist, Borowiecki, like Balzacian Rastignac, challenges the city—*pars pro toto*—a world of socioeconomic arrangements. In both of these novels, in the spirit of the nineteenth century, urban space is shown as a reflection of human aspirations, relationships, and struggles, as a set of entities that together form one organism (even if incapacitated by illness) with a clearly distinguished heart—in the case of a city, a representative city center, with both Borowiecki's and Rastignac's aim being to gain control over it.

In Bator's novel published in 2009, titled *Piaskowa Góra*, Wałbrzych, a tomb-town that encloses corpses of closed mines, becomes a palimpsest that bears traces of former days within, with no perspectives on the future. This is exemplified by Babel, a giant slab block building, home to one of the protagonists, Dominika. The silhouette of Babel seems vivid and well-defined, yet its foundations are embedded in soil that is soggy with the water from flooded mine shaft galleries that once used to lie beneath the city. In Bator's description, the hill of Babel is rotten and sunken into the ground.[18] The poetic image of the industrial town of Wałbrzych, bereft of people with its apocalyptic-nostalgic tone, is seemingly congruent with the description of Łódź. However, a substantial difference lies in the narrative point of view. Reymont's omniscient narrator looks upon the city of manufacturing plants and opportunities for migrants filled with swarming crowds. Łódź is an apocalyptic monster, but, at the same time, the protagonists try to tame it by looking into the future. The figural narrator of Bator's novel looks to the past, because, to him, the Wałbrzych of today or tomorrow does not exist. The characters of the novel are either frozen in the past, stuck in its traumas and nostalgia for bygone days, or they leave Wałbrzych. The urban landscape is void of people, the only thing left being a housing estate of slab block buildings full of orphans of the former mining industry. Here, the mining town is an apocalyptic monster (a depiction of a city as Babylon, a monster, a space where apocalypse occurs, which is common in anti-urbanistic literature). Nevertheless, in the postmodern era, it is not a "disaster" that constitutes a challenge to be faced by the protagonist (as in Reymont's view) but an apocalypse that has already taken place—a *fait accompli*. A path that leads from Reymont to Bator is a path that leads from the city of the future to the city of the past and post-memory at the very same time. Urbanistic heroism, which is characteristic of the literature of the rising modernity (a belief that a city can be "taken over," that it may serve for the purpose of upward mobility and personal success), is transformed in postmodernity into critical urbanism (a criticism of a city as a space that is hostile to an individual and prevailing economic and social relations).

(URBAN) CASE STUDY: GDAŃSK

The existence of a city, not as a geographic-administrative construct, but as a social-cultural category, is preconditioned by the cultural practices cultivated in its landscape. Michel de Certeau interprets them as actions taken by entities that operate in the urban space, thus identifying these practices with stories. Stories, on the other hand, can be told by an institution (which is evidently inspired by Lefebvre's analyses[19])—municipal authorities, urbanists, and individual users, that is, inhabitants who perform quotidian, private, spatial practices. In Gdańsk, one of the largest cities in Poland, an absence of the latter can be observed. However, the concept of sustainable development provides for a coexistence of top-down institutional acts of spatial processing, developing, and planning with needs, visions, and daily practices. What act as adhesive that binds these two collective urban entities—or, as de Certeau would call them, two urban stories—are cultural texts. Literature, art, and happenings—they all introduce urban space into Lacanian symbolic order by developing imaginary space in the city, which may become real and, above all, has the strength that unites individual visions and urban narratives into one single story with numerous narrators yet a coherently depicted world. Stories conveyed by means of literature lie, yet this lie is essential for the city to function not only as a geographic construct but also as a symbolic space. Institutional and political efforts to create a coherent urban rhetoric—to use de Certeau's imagery once again—are noticeable in numerous undertakings. And so, post-modernist tenement houses implanted in Gdańsk on the River Motława, which runs through the city, are supposed to create an effect of old times and strengthen the Hanseatic narrative of Gdańsk as a formerly rich European center. On the other hand, the initiative to found the Museum of the Second World War and the European Solidarity Centre is an attempt to strengthen Gdańsk as a place of memory of important periods in twentieth-century history that puts to use the city's past. Without judging the examples of constructing de Certeau's rhetoric of a stroll, mentioned above only as ideas, it should be noted that they do not translate into everyday spatial practices of inhabitants. This is partially due to the very topographic conditions. First of all, Tricity[20]—cut off by the sea on one side and by moraine hills and a landscape park on the other—due to geographic conditions, develops not concentrically but in a linear way, with a line-shaped main street and considerable distances separating subsequent points. This is a carriageway unfavorable for hiking, which, for de Certeau, is a prerequisite for creating a narrative. The center of Tricity becomes, to use analyses of Polish philosopher Tadeusz Sławek, nothing else but a system of intersections that takes on the role of the city center in order to achieve as much clarity as possible. At present, the metaphorical meaning of the center as an intersection of sky and earth, body and soul, thought and sensation—to use Sławek's imagery—"grows" out of literalness. It is a city center meaning a system of intersections for improved flow of people and vehicles. Sławek notes that the city "gets lost" in this kind of center; it presents a newcomer with a language of grammatical and semantic rules increasingly different from the ones that he or she is used to.[21]

Second of all, Gdańsk has virtually no city center with the old meaning. It is not a Medieval snail-town, as I would call cities such as Paris, Siena, and many others of Western Europe untouched by the cataclysm of war, where a distinct, historic *axis mundi* determines everyday strolls, and where subsequent boroughs grow concentrically around this center according to the chronology of the urban bibliography. By my terminologically inept suggestion, Gdańsk is a rhizome-city, surpassing snail-towns with its abundance of centers that are formed in individual quarters. Like every big city, Gdańsk has many centers, but, in contrast to snail-towns, the particularly tourist-attracting, historic Old Town, promoted as an element of the image of the city center of Gdańsk, is not characteristic of such towns; for topographic reasons, it serves only as a tourist attraction and a location of institutionalized events. Therefore, as a rhizome-city, Gdańsk has many centers, yet not one of them serves as the said *axis mundi,* the center of the (urban) world, which plays significant symbolic and foundational roles, providing a landmark that stabilizes the urban order.[22] Here, I would yet again refer to the analyses of Sławek, who views a city center as a "celebration of the common and the international," even when consisting of highly different elements; suburbs will always preserve their "minority" character—there, I am a person with the status of an emigrant.[23] Centers in cities deprived of the concept of sustainable development are dying, subjected to the processes of dislocation and change of their purpose. When transposed onto the area of stories in the understanding of de Certeau, Gdańsk has many centers, thus many competing macro-narratives, few of which have the chance to weave themselves into a passerby's private micro-narrative. This results in either discursive cacophony or silence.

An urban text in a well-known compilation of Vladimir Toporov[24] is a specific super-narrative, a sum of stories that overgrew a city, that feed on the city as a text. Although this Russian researcher in semiotics analyzed Petersburg as a specific story-city with topoi that recur in its description, this concept has become a term applied to most cities that have their own cultural representations. An urban text is a polyphonic code of writings, voices, myths, urban legends, rumors, anecdotes, the history of the city, and the created literature on the text. The urban text finds room for things absent in a "text-city" regarded as a sphere of the visible—for the myth, the subjective "feelings" of the city, and the literature. Here, another paradox comes into play, as the official urban text of Gdańsk formed under the institutionalized tourism and cultural policy is even excessively rich. In this text, Gdańsk is simultaneously the amber capital and the city of historical time periods: the beginning of World War II and of the Solidarity uprising; a center of the culture of Kashubia,[25] Kociewo,[26] and Vilnius,[27] a multicultural port city owing to its history and its closeness to the sea; a city of freedom; a former German and Polish city, at the very same time, due to its role in the history of these nations. I have already mentioned the lack of translation of a municipal macro-narrative into micro-stories of the residents in the context of culturality. Likewise, stories about the Jewish Gdańsk, the German Gdańsk, and the myth of the Free City[28] are absent—or, rather, present only in literary practices and not in spatial practices. For the younger generation, this also involves stories about Solidarity and the shipyard as the place of Gdańsk's *genius loci*, or even about the port open to the sea, considering that there is hardly any recognizable and original maritime art, and, in the urbanistic context, Gdańsk has turned away from the sea, which yet again was determined by both topography and the socialist industrial legacy.

Surprisingly, the excessively rich urban text has translated in Gdańsk into a much-loved work of literature that exerts considerable influence on the creation of this Gdańsk text, which is, however, one-dimensional, maintained in the idiom formed by Günter Grass, and persistently holding on to a static approach of the city as a palimpsest. Whereas, according to Rybicka, the metaphor of the palimpsest—often employed in accentuating the overlying urban spaces, varying in both their cultural and temporal aspect—seems "insufficient" in the contemporary context, as it omits the issue of conflicting areas between the layers. In Rybicka's words, a palimpsest "describes ontologies without describing their mutual relations," which makes it a static, metaphorical model. Thus the city becomes a reservoir of memory. In the present-day circumstances, Rybicka views the concept of urban memory landscape more as a "field of conflict" where various memory discourses collide.[29] In Gdańsk, this conflict manifests itself with growing intensity, revealing discrepancies not only between the official story and the spatial practices of residents, but also between individual constituents of the urban super-narrative. Gdańsk is not a place of memory since there is hardly any memory in Gdańsk due to its resettlement character[30] and the dominant formation of settled migrants, which are not yet well-rooted. Recently, memory is being created, implemented, and reconciled, which constitutes a fundamental difference. Hence, Gdańsk is not a place of memory, but a field of acting memorial-narrative forces and conflicts of interpretation; its text is not a palimpsest to be learned, but a work-in-progress.

In this regard, progress in this work is possible only in the circumstances of a sustainable development of the city, where individual narratives—economic, political, cultural, everyday narratives—will form a common depicted world of the city. For the time being, this is not the case. In the novel *Mercedes-Benzi* by Paweł Huelle, one of the most well-known Gdańsk authors, we are presented with a depiction of Lviv and Vilnius saved from the torment of Soviet occupation, compared by the author to a patient recovering from a disease, regaining "strength and former beauty after a bout of typhoid, whereas Gdafisk and Wroclaw needed to have limbs amputated, heart, kidney and liver transplants, and false teeth fitted." Huelle further explores this imagery by claiming that the spine of these cities has been severely damaged and, though treated, prevents them from living life to the fullest.[31] The organicist metaphor allows us to see the city as a living being whose growth and development require individual systems to cooperate. Meanwhile, in Huelle's vision, Gdańsk, like many other cities in this part of Europe, reminds us of a patient during convalescence. Dying of wartime wounds, it was brought to life by incompetent surgeons of the Communist period, subjected to a lifting procedure as a result of early capitalist enthusiasm, and now—in a half-dead, hybrid form—it seeks its own identity.

This hybridism is a consequence, above all, of implants being situated within the urban body; in the Central- and East-European landscape, this refers to housing estates of block buildings,[32] yet not only those made of the so-called "giant slabs" (prefabricated concrete panels), an architectural design that dominated in Poland from the 1950s to the 1980s, but also the modern ones modeled on Western-European apartment buildings. Housing estates, prevalent in the Central-European urban landscape, in the case of Polish cities bereft of "centers"—centers of culture, meetings, and common spatial practices of residents—are not only evidence of the expansion of transitive spaces but also a spatial mark of amnesia, of building on a trauma concealed beneath the earth. When considering the roles of memory and oblivion in constructing an urban space in his monograph *Urban Memory: History and Amnesia in the Modern City*, Adrian Forty indicates the symbolic meaning of concrete:

> concrete has [...] been so generally associated with the erasure and obliteration of memory. Concrete makes everywhere the same. It cuts people off from their past, from nature, from each other. This sense that concrete is a symptom, if not a cause, of alienation has been around since the early 1960s, if not longer. Take, for example, J-L. Godard's 1966 film Two or Three Things I Know about Her, in which lingering shots of concrete being poured on the construction of the Paris périphérique are accompanied on the sound track by musings on the city's loss of meaningfulness, its demise as a communicative medium. Or another French example, Henri Lefebvre's 1960 essay Notes on The New Town, where, despite the absolute clarity and legibility of the functions of the concrete buildings, the result is an utterly impoverished environment: "Here I cannot read the centuries, not time, nor the past, nor what is possible." [...] All this is nicely summed up by graffiti sprayed onto a multi-storey car park in Marburg in Germany in 1992–3—"Beton ist Koma," "Concrete is Coma. The Urban Bedroom Becomes a Place of the City's Coma"[33]

This issue is problematized in a similar way by the Gdańsk writer Barbara Piórkowska in *Utkanki*, a collection of feminist readings of urban space, in which she writes about one of the latest housing estates of slab block buildings in Gdańsk, namely, Ujeścisko. She clearly differentiates between the part of the housing estate that gives birth and raises its offspring in the Upper World and the one that remains in close connection with the dead, sidelined in the Lower World.[34] There are thus two cities: the contemporary one, focused on everyday reality, and the one that was denied, subjected to urbanist censorship, and pushed into the collective urban subconscious. The lack of sustainable development can be also expressed in the language of psychoanalysis. A dialogue between the upper and the lower city, two levels of urban space, is impossible due to slab block buildings. "Therefore the question can be put—what in fact do these blocks block out?," asks Andrzej Leder, conducting a psychoanalytical reading of Warsaw's architecture:

> If it were not for them a void would fill the metropolis—estranged, open hills of rubble with the odd isolated surviving edifices. [...] After the hit, the ruined capital nonetheless was revived. Thus a mass of grey housing fills the massive space of loss, creating some lacunae for life—even if this life is closed in appallingly repetitive and claustrophobic stairwells and kitchens. [...] These apartment blocks witness "that the remaining psychical functions were extensively paralyzed or reduced." Many of these buildings [...] are produced by the anal desires of the party chief Gomułka. Dwarf-like, they appear as if someone had laboriously spat them out, though at the same time, thanks to the repetition compulsion, they are able to fill the void of trauma with their unchangeable aspect. They exist thanks to the obsessive obligation to repeat, tied to the loss of libidinal energy and simultaneously, to a great extent, consuming the remains of it. The very same stairwells, row after row of windows, balconies in vertical stripe and satellite antennae in their thousands.[35]

Gdańsk has just started to tackle the trauma mentioned by Leder, and only as a consequence of working through it and relieving the grief after the past cultural and geographic divisions will it be possible to introduce sustainable development into the city. Lastly, Gdańsk, like many other cities of this part of Europe, has just begun to form its own generational places—that is, to use Aleida Assmann's typology of places as memory media—points in space that determine generational continuity.[36] This continuity, which is materialistic in an urbanistic, demographic, and narrative sense, has just begun to form, yet optimism is inspired by bottom-top initiatives undertaken more and more often at the borough level in Gdańsk, leading to the creation and the strengthening of centers, their stories and practices of passersby performed within their framework. (Many of these initiatives are a result of the generation of Gdańsk-born people making their voice heard, for whom the city constitutes a significant location in their biographies, and who break free from the burden of post-memory.) Moreover, Gdańsk has numerous locations that are remains of some cultures, still testifying to a loss, a break in the continuity between the past and the present, as well as separate places of trauma that block the possibility of positive interpretation. These types of space are still waiting for their stories, which must be born only by means of a spatial therapy and discursive reworking, including the literary. Otherwise, following Assmann's train of thought, they will become not archives but their negatives—a "garbage" of memory. It seems that, for the time being, the cultural landscape of Gdańsk is dominated by the last type specified by Assmann—ghost places,[37] which serve as arenas for imaginary games and the return of denied contents, still awaiting their mutual reconciliation.

NOTES

1. Jill Grant, *Planning the Good Community: New Urbanism in Theory and Practice* (London–New York: Routledge, 2006), 45.

2. "New Urbanism," in *Encyclopaedia of Human Geography*, ed. B. Warf (London and New Delhi: Thousand Oaks, 2006), 334.

3. For more information about new urbanism see Grant, *Planning the Good Community*; Aaron Passell, *Building the New Urbanism: Places, Professions, and Profits in the American Metropolitan Landscape* (London–New York: Routledge, 2013); Ilse Helbrecht and Peter Dirksmeier, eds., *New Urbanism: Life, Work, and Space in the New Downtown* (Surrey–Berlington: Ashgate Publishing Company, 2012).

4. I understand the notion of spatial practices in Michel de Certeau's meaning as a cultural activity in space undertaken by passersby who thus create their own story about the city. According to de Certeau, "The act of walking is to the urban system what the speech act is to language [...]" and "it has a triple 'enunciative' function: it is a process of appropriation of the topographical system on the part of the pedestrian [...]; it is a spatial acting-out of the place [...]; and it implies relations among differentiated positions [of pedestrians] [...]." Such an approach allows one to draw lines that run from the real movement in space to discursive movement, to see in specific literary representations of spatial experiences a representation of actual spatial practices, by which I mean each and every cultural act performed by an entity in space. For de Certeau, walking is an individual act of speech, *parole*, taking place within the framework of the system, *langue*, which forms the city as a planned, organized, and institutionally managed topographic order. Obviously, he is interested mostly in the spatial stories that are created as a result of a derogation from the scheme, avoidance of well-known routes, urban plans and prohibitions, by means of which, therefore, the place is not only executed but also constructed. Michel de Certeau, *The Practice of Everyday Life*, trans. S. Rendall (Berkeley: University of California Press, 1984), 97–98.

5. Filip Springer, *Wanna z kolumnadą. Reportaże o polskiej przestrzeni* (Wołowiec: Czarne, 2013).

6. Published in 2011 in Cracow.

7. Published in 2015 in Wołowiec.

8. Sylwia Chutnik, *Cwaniary*. Illust. M. Zabłocka (Warsaw: Świat Książki, 2012), 45.

9. Critical cartography, a concept first formulated in the 1980s by John B. Harley, defines a map not as an imitation of the reality that employs conventional symbols, but as its socially and culturally determined representation of reality or even—in its radical trend—as an implement for creating spatial order, an element of Foucauldian discourse of knowledge/power. See Denis Wood, *The Power of Maps* (New York: Guildford Press, 1992).

10. This term is used for a trend in Polish prose distinctive of the literature of the 1960s, which is characterized by realism, a tendency to describe the reality of ordinary people's daily lives. Moreover, this term has pejorative connotations as it is associated with opportunism, moral minimalism, and a dislike of change.

11. The author of the term "post-memory," Marianne Hirsch, defines this phenomenon as a hereditary trauma that manifests itself in narratives of the second generation, thus among the descendants of a generation that experienced a kind of collective trauma and then passed it on to their offspring in stories and conversations. Family stories become overloaded with emotions and, mediated in pictures and narratives, they become noticeable in texts created by the second generation, which—though having no experience of trauma—preserves a specific implanted "memory." See Marianne Hirsch, "Family Pictures: Maus, Mourning, and Post-Memory," *Discourse: Journal for Theoretical Studies in Media and Culture* 15, no. 2 (Winter 1992–93): 3–29.

12. Joanna Bator, *Piaskowa Góra* (Warsaw: W.A.B., 2012); Joanna Bator, *Ciemno, prawie noc* (Warsaw: W.A.B., 2012).

13. Dorota Masłowska, *Wojna polsko-ruska pod flagą biało-czerwoną* (Warsaw: Świat Książki, 2003).

14. Michał Witkowski, *Lubiewo* (Cracow: Wydawnictwo Korporacja Ha!art, 2006).

15. Sylwia Chutnik, *Kieszonkowy atlas kobiet* (Cracow: Korporacja Ha!art, 2009).

16. Łódź (Lodz) is the third most populous city in Poland. In 1820, based on a document dated 18th September, signed by prince viceroy Józef Zajączek, the government of the Kingdom of Poland incorporated Łódź into a cluster of industrial settlements in the industrial region of Kalisz and Mazovia and made it into a cloth-making and weaving center. Since then, the image of the city as an industrial center has strengthened. In 2015, some areas of the city were recognized as "historic monuments" under the collective name *Łódź—wielokulturowy krajobraz miasta przemysłowego*.

17. Władysław Reymont, *The Promised Land Vol. 1*, transl. M. H. Dziewicki (New York: Alfred A. Knopf, 1927), 178.

18. Bator, *Piaskowa Góra*, 438.

19. See Henri Lefebvre, *The Production of Space*, transl. D. Nicholson-Smith (Malden–Oxford–Victoria: Blackwell, 1991).

20. This term is used in reference to Polish agglomeration formed by three cities: Gdańsk, Gdynia, and Sopot.

21. Tadeusz Sławek, "Akro/nekro/polis: wyobrażenia miejskiej przestrzeni," in *Pisanie miasta. Czytanie miasta*, ed. A. Zeidler-Janiszewska (Poznań: Wydawnictwo Fundacji Humaniora, 1997), 33.

22. According to Sławek, a city center confirms the unchanging character of cardinal directions that serve the purpose of spatial orientation of the city, as well as bring harmony to the city-world relation. By going to the city center, a person "renews" and "approves of" the law and stability of geographic coordinates. Ibid., 16–17.

23. Ibid., 17.

24. See Vladimir Toporow, *Miasto i mit*, trans. B. Żyłko (Gdańsk: Słowo/Obraz Terytoria, 2000).

25. Kashubia—a cultural region in northern Poland near Gdańsk, thereby located in Pomerania province, inhabited by Kashubians (native Pomeranians) using a regional language—an ethnolect called Kashubian.

26. Kociewie is a cultural region in the vicinity of Gdańsk inhabited by *Kociewiacy* of an ethnic identity weaker than that of Kashubians, and is still differentiated as a historically and culturally different part of northern Poland.

27. Traces of Vilnius in Gdańsk are remains of obligatory population displacements from their homeland (the so-called Eastern Borderlands, i.e., lands to the east lost as a result of the Yalta Conference) and settling them in the so-called Recovered Territories after World War II.

28. The Free City of Danzig (*Wolne Miasto Gdańsk* in Polish, *Freie Stadt Danzig* in German) is an autonomous city-state that existed in the interwar period under the protection of the League of Nations. It was formed as a result of the defeat of the German Empire in World War I.

29. Elżbieta Rybicka, "Pamięć i miasto. Palimpsest vs. pole walki," *Teksty Drugie* 5 (2011): 210.

30. From among the residents of the Free City who remained in Gdańsk after World War II, the vast majority of the German population (ab. 120–130 hundred people in mid-1945) was transferred to Germany, whereas after the Free City was taken by the Third Reich, the Polish population was forced to sign a *Volksliste*, obliged to undergo a verification procedure organized by the new authorities. From mid-1945, Gdańsk was a destination of numerous transports of forcibly displaced people of the Eastern Borderlands. Cf. Norman Davies, *Vanished Kingdoms: The History of Half-Forgotten Europe* (London: Allen Lane, 2011); Peter Oliver Loew,. *Danzig*: *Biographie einer Stadt* (Munich: Verlag C. H. Beck, 2011).

31. Paweł Huelle, *Mercedes-Benz* (London: Serpent's Tail, 2005).

32. A giant slab is a constructional element used in the building of, most of all, slab block residential buildings. Its characteristic feature, which influences also the aesthetics of residential complexes built in this way, is modularity, since it is a prefabricated element applicable in multiplication. Giant slab buildings were raised in Poland from the 1950s to the 1990s (the peak of its popularity being the 1970s), transforming urban landscapes into deserts of concrete facades (though today renovated and aestheticized), which have become a symbol of the proverbial grayness of the time of the People's Republic of Poland, understood in a political, social, economic, and aesthetic sense. Miron Białoszewski sought a poetic form for expressing spatial experiences of living in such a huge slab, and, longing for the brick, created an "epic poem" of the *mrówkowiec*, as he used to call a slab block building, in the form of numerous short compositions (especially, *Chamowie* and *Odczepić się*).

33. Adrian Forty, "Concrete and Memory," in *Urban Memory: History and Amnesia in the Modern City*, ed. M. Crinson (London: Routledge, 2005), 75.

34. Piórkowska, *Utkanki*, 17. Interestingly, a very similar image can be found in *Listy z dolnego miasta* by Krzysztof Gedroyć, who depicts Białystok, a city exceptionally distant to Gdańsk in terms of location, mythology, and historical circumstances.

According to Gedroyć, the Uptown serves the role of a "cradle of time," while in the Downtown, the "time-child" finds no continuance with the past, the present, and the future, as well as—as Gedroyć marks—those kinds of existence that remain there unnamed for the time being. Gedroyć, *Listy z dolnego miasta*, 17.

35. Leder, "Psychoanalysis of a Cityscape."

36. Assmann, "Przestrzenie pamięci. Formy i przemiany pamięci kulturowej."

37. In her work, Assmann lists among the media of externalized memory also places that transform into places of memory due to events of high importance (either religious, historical, or biographical). Assmann argues that memory can be stored within places even in the times of mass oblivion, as a memory is equally activated by a given place, and a place is activated by a given memory. She further explores this issue to conclude that, with all possibilities for conveying messages being blocked, ghost places are formed to serve as arenas for uninhibited imaginary games or for the returning meanings that have been earlier suppressed. Ibid.

BIBLIOGRAPHY

Assmann, Aleida. "Przestrzenie pamięci. Formy i przemiany pamięci kulturowej." Translated by P. Przybyła. In *Pamięć zbiorowa i kulturowa. Współczesna perspektywa niemiecka*, edited by M. Saryusz-Wolska. Cracow: Universitas, 2009.

Bator, Joanna. *Ciemno, prawie noc*. Warsaw: W.A.B., 2012.

———. *Piaskowa Góra*. Warsaw: W.A.B., 2012.

Chutnik, Sylwia. *Cwaniary*. Illustrated by M. Zabłocka. Warsaw: Świat Książki, 2012.

———. *Kieszonkowy atlas kobiet*. Cracow: Korporacja Ha!art, 2009.

Davies, Norman. *Vanished Kingdoms: The History of Half-Forgotten Europe*. London: Allen Lane, 2011.

de Certeau, Michel. *The Practice of Everyday Life*. Translated by S. Rendall. Berkeley: University of California Press, 1984.

Forty, Adrian. "Concrete and Memory." In *Urban Memory: History and Amnesia in the Modern City*. Edited by M. Crinson. London: Routledge, 2005.

Gedroyć, Krzysztof. *Listy z dolnego miasta*. Kielce: Charaktery, 2003.

Grant, Jill. *Planning the Good Community: New Urbanism in Theory and Practice*. London–New York: Routledge, 2006.

Helbrecht, Ilse, and Peter Dirksmeier, eds. *New Urbanism: Life, Work, and Space in the New Downtown*. Surrey–Berlington: Ashgate Publishing Company, 2012.

Hirsch, Marianne. "Family Pictures: Maus, Mourning, and Post-Memory." *Discourse: Journal for Theoretical Studies in Media and Culture* 15, no. 2 (Winter 1992–93): 3–29.

Huelle, Paweł. *Mercedes-Benz*. London: Serpent's Tail, 2005.

Leder, Andrzej. "Psychoanalysis of a Cityscape: A Case of the Post Traumatic Stress Disorder: The City of Warsaw." In *Psychoanalytic Reflections on a Changing World*, edited by H. Brunning. London: Karnac, 2012.

Lefebvre, Henri. *The Production of Space*. Translated by D. Nicholson-Smith. Malden–Oxford–Victoria: Blackwell, 1991.

Loew, Peter Oliver. *Danzig: Biographie einer Stadt*. Munich: Verlag C. H. Beck, 2011.

Masłowska, Dorota. *Wojna polsko-ruska pod flagą biało-czerwoną*. Warsaw: Świat Książki, 2003.

"New Urbanism." In *Encyclopaedia of Human Geography*, edited by B. Warf. London and New Delhi: Thousand Oaks, 2006.

Passell, Aaron. *Building the New Urbanism: Places, Professions, and Profits in the American Metropolitan Landscape*. London–New York: Routledge, 2013.

Piórkowska, Barbara. *Utkanki*. Illustrated by S. Lipnicki. Gdańsk: Instytut Kultury Miejskiej, 2014.

Reymont, Władysław. *The Promised Land*. Translated by M. H. Dziewicki. Vol. 1. New York: Alfred A. Knopf, 1927.

Rybicka, Elżbieta. "Pamięć i miasto. Palimpsest vs. pole walki." *Teksty Drugie* 5 (2011): 201-211.

Sławek, Tadeusz. "Akro/nekro/polis: wyobrażenia miejskiej przestrzeni." In *Pisanie miasta. Czytanie miasta*, edited by A. Zeidler-Janiszewska. Poznań: Wydawnictwo Fundacji Humaniora, 1997.

Springer, Filip. *Wanna z kolumnadą. Reportaże o polskiej przestrzeni*. Wołowiec: Czarne, 2013.

Toporow, Vladimir. *Miasto i mit*. Translation and foreword, collected by B. Żyłko. Gdańsk: Słowo/Obraz Terytoria, 2000.

Witkowski, Michał. *Lubiewo*. Cracow: Wydawnictwo Korporacja Ha!art, 2006.

Wood, Denis. *The Power of Maps*. New York: Guildford Press, 1992.

Chapter 12

Exploring the American City's Environment

Friction between Fluctuating Urban Sustainability, Modern Myths of Space, and Memory Places in the Works of Paul Auster

Mehdi Kochbati

The activity of reminiscence is a collective endeavor initiated in socialized places built by people. According to Pierre Nora, there are places where "memory is selectively embodied and which, by the will of men [sic] and the work of centuries, remains as the most vivid of symbols."[1] These memorials or memory places are part of a sustainable and a long-term common sense and are placed in a dynamic of ancient and millennial continuities. Indeed, the space of the American city seems to be not only a framework for Paul Auster's "memory sustainability as intrinsic value"[2] but also a place of memory. Although the perception of "sustainable development or sustainability has been used in different fields at various scales,"[3] this chapter looks at the concept of sustainability in relation to postmodern American literature and particularly Auster's representation of memorial sustainability. The latter is embodied in memorial writing in the specific case of the American city and the vast territories of the west. In this framework, Lidia Varbanova defines the city in accordance with the notion of dynamism and sustainability:

> The city [...] is the space which gives us a notion of "belongingness"—a space of mixtures of [...] diverse people [...] [and] community life [...]. This is a dynamic concept, focusing on creativity [...] and culture as main indicators for having a [...] city, sustainable [...]. The important aspect is also the "act of sharing"—of local cultural history and memory, of cultural and artistic resources and achievements, of public spaces.[4]

In Auster's memory writing, urban space, a symbol of this "notion of belongingness," triggers memorial sustainability. The city as an artifact often conditions the existential and memorial path of Auster's protagonist, branding the first outlines of his quest for sustainability. This urban space, as an "act of sharing" and a point of intersecting lines between multiple existences of the self, is a promising and propitious space for providential scheming and experiences of "otherness" in "public spaces." Although the city is portrayed as an oppressive and engulfing environment, it serves as a matrix for a sustainable mnemonic action prompted by the mental, memorial, and geographical wanderings of the protagonist. Nevertheless, the city represents a space folded in on itself and is gradually transformed into a place of loss and violent dispossession.

It is important to emphasize here that the three volumes of *The New York Trilogy* retrace the journey of three men in their respective quest for sustainable memorial benchmarks. They all engage in a long process of physical and mental dis-identification in order to achieve a sustainable identity and a memorial rebirth. It is in this bewildering metropolis, often likened to a maze of signs, that the writer Quinn in *City of Glass*, detective Blue in *Ghosts*, and the anonymous narrator in *The Locked Room* are swallowed by the urban environment while leading an investigation with anti-detective tones to discover the secret of the traces left by a missing character. They are soaked into this modern urban environment in order to merge completely with it. New York becomes their only guide and imposes in return a logic of material dispossession taken to the extreme. In *The New York Trilogy*, the city, whose description is often concealed by the memorial narrative, highlights an underlying lack of any memorial "rhetoric of sustainability."[5] These journeys in the urban environment compose an "aesthetic of sustainability,"[6] form part of a "visual rhetoric"[7] of memory, and convey a message of sustainability. In *City of Glass*, the first volume of *The New York Trilogy*,[8] New York City hinders any visual rhetoric of sustainability. Indeed, the city is the location of non-space ("nowhere")[9] and the loss of the subject both in the urban space and in his own subjectivity, preventing any memorial sustainability. Urban space stems rather from a "nowhere" that induces the dislocation of a free and harmonious space. The description of urban space is assimilated to emptiness and non-representation[10] which contradicts any notion of sustainability.[11] However, in *City of Glass*, it should be stressed that the gradual disappearance of the urban environment hinders any sustainability to the landscape[12] of the city. Retreating into the apartment of Peter Stillman and renouncing his urban wandering, Quinn articulates his own disappearance.[13] His vanishing existence overlaps with that of the city. The changing and shifting city duplicates the space of the memorial sustainable writing (in the red notebook) subjected to the extinction. The latter matches with the evaporation of the memorial writing's performer. Although Quinn considers the city as a physically and mentally prison-like universe, he gradually sinks into Manhattan and allows himself to be carried away by this hostile universe accompanying his desire for a sustainable memorial and identity quest. Ignoring his existence, he tracks down the steps of his double (Stillman) up to the borders of the city.

Nonetheless, Quinn gradually surrenders to the alienation produced by urban space until achieving the ultimate annihilation. This obliteration overlaps with his settlement in a minimalist room. The latter is unfurnished but allows him to watch over Stillman's apartment. The room's space engulfs him progressively by taking possession of his body. Here, urban space is deemed to be a force that gradually destroys the self by absorbing it. Moreover, the impulse of inscribing memorial sustainable writing in the archival materials (notebooks) is an underlying theme in the three volumes of *The New York Trilogy*. The novel highlights the necessity of testimony to fight against the evanescence of both the urban environment and the collective memory in the city. Based on the fact that "[t]he preservation of the city collective memory is [...] a relevant issue in a sustainable city,"[14] Quinn focuses on the irresistible desire to write the evanescent memory of New York City. He records all the actions, movements, and wanderings of Stillman through the city in a chronological linearity. Right after the interview with Stillman's son and Virginia Stillman, Quinn is invaded by the overwhelming desire to acquire a red notebook.[15] In *City of Glass*, the writing in the notebook is an act of memorial sustainability revealing the intimate equation between writing, mnemonic action, and struggle against oblivion. Just as the verb "[to] remember"[16] (twenty-seven times) that defers death in *The Invention of Solitude*, the registration of the gradually vanishing memory in the notebook rejects death in *City of Glass*.

The depletion of the notebook is accompanied by the progressive vanishing of the urban environment. However, the notebook is gradually approaching its imminent ending and Quinn is facing the anxiety associated with the impossibility of writing his memorial quest. He has a premonition: the end of the notebook is inseparable from that of memorial sustainability along with his own existence "to face the end of the red notebook with courage."[17] Confined in his room, this premonition does not prevent him from creating a new mode of storytelling. Quinn's observations, related to Stillman and the fading city, turn into a memorial and existential meditation concerning the exhaustion of the sustainable urban space, of his room (part of the urban space) and the notebook. In this respect, it is important to tighten up the definition of sustainability according to its "paradoxical" lack of attainment and historicity:

> Sustainability can never be achieved, for it defines itself by its own pathways. Each stage is a reflection of its predecessor and a prognosis for the next. [...] [T]he notion of sustainability is both paradoxical and beguilingly ambiguous.

> Like so many contradictory concepts, such as fairness, justice or equity, sustainability eludes a common definition. Every individual [...] will perceive sustainability in the context of its history, its aspirations, its power relations, and its response to opportunities to reveal its local identities.[18]

Similarly to the tales of Scheherazade[19] (integrated in *The Invention of Solitude*), Quinn adheres to the principle of the sustainability of memory and history, which is intrinsically associated with that of memorial writing: "as long as you go on speaking, you will not die."[20] Quinn is attempting to postpone death through sustainable memorial writing. The shrinking of his writing (he keeps writing smaller and smaller) comes to preserve the pages of the red notebook. When writing faces the end of the white page of the notebook, the phantom of death settles up, and Quinn vanishes forever from the narrative. However, Quinn's notebook ends with an optimism which calls into question the progression of memory writing and raises the question of its sustainability in the future: "The last sentence of the red notebook reads: 'What will happen where there are no more pages in the red notebook?'"[21] It should be underlined here that at the heart of the geographical and memorial wanderings of Auster's characters (Daniel Quinn, Anna Blume, A. Fogg, the anonymous narrator, Fanshawe, Peter Aaron), the notebook underlines the sustainability of memory of the "journey" in the west and/or in the city.

Despite the abundance of mobile trajectories related to the journeys of Auster's characters, the city of New York remains the favorite place for Auster's memorial sustainability. In spite of the constant commitment of Auster's protagonists to new geographical wanderings, they return to the city at the end of their itinerary. Although *Mr. Vertigo* is located in St. Louis and *The Music of Chance* starts in Boston, the city of New York remains unquestionably the privileged urban space for the mnemonic action. *City of Glass*, *Ghosts*, *Moon Palace*, and *Leviathan* also have the city of New York as a backdrop. In this context, New York remains the privileged framework for mnemonic action as illustrated in *The New York Trilogy*. Like the room, the matrix of memory writing, the city of New York seems to be the geographical matrix of the mnemonic action and memorial sustainability. It is the room at 6, Varick Street[22] in Brooklyn, New York, where *The Invention of Solitude* emerges. It is important to emphasize here that sustainability stresses the notion of a "historic city"[23] according to the concept of urban sustainability. In this context, on the level of the construction of individual and collective sustainable memory, New York as a "historic city" is the first frame of reference for Auster's novels since this place is closely associated with his own past. It is this city where he grew up and which is now the environment in which he evolves. Brooklyn, in particular, is a favorite part of New York for Auster's sustainable memorial writing.

Far from being linear, the exploration of time in Auster's sustainable memory writing breaks away from a traditional chronological approach. As such, the sustainable memorial writing is often proof of a rupture with the time dimension, highlighting "an unbroken sequence of embodied experiences from the past and into the future."[24] Moreover, the memorial-narrative sustainability is constantly blurred and creates confusion by a succession of random events which introduce a multiple chronology. Actually, the self is facing a world deprived of temporality where the chronological flow of the "past, present and future"[25] overlap and are imbricated inside each other. In this context, Linda Hutcheon deals with memory and its relationship to sustainable time from the perspective of a postmodern vision that denounces the illusory nature of temporality. According to Hutcheon, postmodernism does not negate the existence of the past. It nevertheless edges our knowledge and interception of the events of the past bursting into the present. These events can only be intercepted on the basis of their tracks, their texts, the construction of these facts and their meaning.[26] Actually, postmodernism emphasizes the "real" and, in particular, "the past real"[27] that can only be detected through signs. In *City of Glass*, the confinement of Quinn in this enclosed place and his pursuit of the Stillmans, despite the demise of the narrative, separates him from any notion of project, future, and sustainability, highlighting the conflict of "competing forces [...] between physical existence and memory."[28] Indeed, in *City of Glass*, the city remains a space of confinement and is often equated with a dead-end place, with no way out, except for an existence combined with the past. The absence of any projection into the future brands a disruption with all temporal sustainable continuity that would drive Quinn's trajectory forward. The existence of Quinn is separated from all sustainable temporality and combined with containment in the enclosed space of the room of his New York apartment. In an interview with Gerard de Cortanze, Auster declared himself to be inspired by an American reality of which the epicenter remains the city of New York: "We live somewhere and this place constitutes in itself an essential world to each individual. An artist, a writer, anyone responds to the environment where he lives. In my books, I respond to the reality surrounding me, an American reality."[29] The places of his past structure the major part of Auster's sustainable memorial writing. Characters are also imbued with the American experience specific to the writer. In this context, it is useful to remember that Auster's wandering character ends up most often in New York, the place of sustainable memory in Auster's works. Gaston Bachelard defines the place of memory, or more precisely the concept of "topophilia,"[30] as the product of the association of images with the memory spaces invested and committed by the mind of humankind. According to Bachelard, individual memory of places is not a medium to revive the past. Instead, it falls, rather, within the space where the sustainable memory is anchored. Space would be a sustainable memory that highlights the "materiality of a place,"[31] which is associated with a system of places (*loci*) and images that shape a pathway. This can take the form of a particular route, a didactic or initiation ritual which appeals for reminiscence. In the framework of Auster's memorial sustainability, the city combines the image with the memory of urban space in which reminiscence is established. The city is part of the evolution of Auster's character figuring out his falling movement and ascent.

In Auster's writing, the city fills the heart of the memorial narrative. The titles of the first works already place emphasis on the theme of space, its variations and components. *City of Glass*, *The Locked Room*, *Moon Palace*, and *In the Country of Last Things* underline this theme. Each volume refers to a vanishing, fading, and changing urban space hindering any sustainable memory of the city. In *The New York Trilogy*, there is a dialectic relationship between the construction of a sustainable memory of the city (through wanderings and journeys) and the vanishing of any sustainable benchmarks. This gradual fading of the city's landmarks hinders any sustainable and material anchoring in space. The description of the city is often neglected by the memorial narrative that rarely depicts it. The ubiquitous images of an urban setting contribute to reminiscence rather than a meticulous depiction of the city. The displacement movement of the visual experience (non-description of the places) in favor of the mnemonic action remains the product of the disruption of the linguistic contract between signifier and signified in the urban space and the collapse of the precarious ground on which this mode of representation stands.

SUSTAINABLE MEMORIAL CITY VS. VANISHING OF SUSTAINABLE BENCHMARKS

The representation of urban space both produces memorial sustainability and an identity crisis of the self. In *The New York Trilogy*, the dialectic relationship between sustainability and identity is illustrated through the urban space. The latter is the place where the self is lost and the "locus" where new subjectivities are recaptured and recollected. The space of the city affects the self's identity: it fades gradually and splits till reaching a multiplicity of fragments. In *City of Glass*, the endless space of the city creates an irrefutable movement which is very difficult to control. In this urban space, the self is soaked by the changing universe of the city according to a particular spatial configuration highlighting a movement of "turbulence and switching trends."[32] Each Auster character faces the progressive loss of identity until vanishing completely. Towards the end of *City of Glass*, Quinn becomes "other" by immersion into the adverse and oppressive environment of the city. Driven by an "invisible hand of the modern nothingness,"[33] Quinn strolls through the lost spaces of the urban environment. By a duplicity effect and taking on the appearance of a private detective, he becomes the fleeing man whom he is trying to track (Stillman's father). The failure of his quest and his abortive wanderings marginalize and exclude him from the city. The latter is distinguished by a bewildering reality in which space suffocates the self. To dodge the oppressive space of the city that swallows him, Quinn is forced to invent a new sustainable reality responsible for warding off the alienation exerted on him by the urban environment. It is the quest for Stillman's traces in the streets of New York that leads Quinn to map out a new sustainable linguistic reality through the route of "The Tower of Babel."[34] It is important to underline that the abundance of signs that Auster's character must grasp and intercept imparts the city with a particular linguistic sustainability. The latter is performed through the description of mental and memorial itineraries of characters. In the case of *City of Glass*, the "Tower of Babel"[35] incises the memorial and identity quest of Quinn in his journeys through the city, referring to sustainable symbolic meanings which are aroused and empowered by these expeditions. In this volume, there is an attempt to dominate space by a specific language configuration belonging to memorial sustainability. This urban configuration is highlighted by the cartography of the "TOWER OF BABEL."[36] The concept of "reterritorialization"[37] or the process of forming a sustainable territory is very important to underline here. Quinn's urban wanderings trigger a mental perambulation and draw a symbolic path expressed with the sign of "T.O.W.E.R.O.F.B.A.B.E.L."[38] This "reterritorialization" embodied by this trajectory is equated with a reconstruction of a memorial sustainability for New York City. This reconstruction is located at a halfway point between an individual memory linked to Quinn's own childhood and an American collective memory related to the first urban constructions of the city of New York. This memorial sustainability of the city conducted through its historic monuments reactivates (in the city) the ancient myth of the united language in "The Tower of Babel."[39] Similarly, the city in *Moon Palace* is branded by its arbitrary and fleeting nature, hindering any memorial sustainability.

During his inner journey, burdened with rebuilding his identity and memorial path, M. S. Fogg leads an erratic existence. Fogg's trajectory is not linear but interspersed with losses and excesses and marked by an extreme dislocation of landmarks. The impossibility of setting a spatial, sustainable reference, real or metaphorical, forces the self to get rid of any concrete inheritance, as shown in Fogg's experience of austerity, splitting up from his books. Having lost his job, he is forced to gradually sell even his furniture to survive in a logic of material asceticism opposed to any sustainable existence and memory. It is relevant to note here that, in *Moon Palace*, the identity path is placed under the sign of nomadism and constant movement, strengthening the cohesion of the self and the world: "Now I was being plunged into a world of particulars."[40] Fogg follows the same identity path as Thomas Effing (his grandfather). Duplicity, or repetition of the same, emerges here as a constant theme related to memorial sustainability: "It was [...] like following the route of an explorer from long ago."[41] Effing, like Fogg, shares the experience of the retreat into a cave—the first in the vastness of the western desert territories, the second in a recent past, in a cave in Central Park. Similarly, in *In the Country of Last Things*, the books of the great library are consumed by fire. The latter is the only functional and sustainable structure of the city. Anna Blume takes refuge at "The Woburn House," the locus of her memory writing. The city in the novel represents an evil force and is anything but a protective space for the characters. Instead, it draws a model of oppression and alienation of the self immersed in a ghostly city where human beings disappear at the same speed as things. Space rejects Blume, who finds herself in a jungle-city described as a source of suffocation and marked by its absence of sustainability through the whitening and vanishing of its geographical landmarks. The outcome of the attempt by Anna and her companions to flee the city through its battlements and walls separating the desert and the sea from the city remains unknown. The changing reality of the city raises questions that will remain unanswered and stresses the informational gap of the memorial narrative.

SUSTAINABILITY IN URBAN SPACE AND REACTIVATION OF AMERICAN MYTHS OF THE WEST

In *Moon Palace*, Fogg's urban wanderings around the streets of New York strengthen his meditations and his desire for a sustainable memorial quest. These wanderings make possible an accurate introspection through which Fogg discovers a filiation with Solomon Barber, allowing him to rebuild a sustainable individual memory. Fogg leaves Chicago to begin his initiatory and memorial journey in New York and work with Effing (his grandfather). The New York area also provides the framework for Fogg's tribulations. Here, the urban space accelerates the fall of the character who does not express any desire to act on his fate. Instead, he lets himself be gradually swallowed up by the city. After finding refuge in his room, believing he can do so to prevent his imminent fall, Fogg is quickly evicted from his apartment due to his lack of material means. He turns into a homeless man, wandering the streets of New York. With only a backpack and a few personal effects, Fogg's erratic existence following expulsion from his apartment leads him to the south of the city, where he finds refuge in the artificial jungle of Central Park.[42] Fogg's refuge in the city that engulfed him in the beginning turns into a place of rebirth (Central Park), paving the way for a second existence.[43] The park area is opposed to the confinement and absence of sustainability that represents the space of the city as such.

The city of New York as reflected in *Moon Palace* remains an undefined place, devoid of any specific geographical reference.[44] Although the urban jungle of Central Park is at the heart of the city, it breaks with the vanishing and prison-like environment outside of it: "I found myself [...] absently running my hand along the stone wall that divides the park from the street. I looked over the wall, saw the immense, uninhabited park, and realized that nothing better was going to present itself to me at that hour."[45] Central Park offers an opening that allows Fogg to escape the suffocating atmosphere of the urban environment. The New York artificial jungle, "a man-made natural world,"[46] is a source of sustainability and peace since it protects the hero (Fogg) from the street's violence. The peaceful haven of the New York jungle provides him with the freedom to start his memorial quest. He transforms the park into a propitious observation and meditation space for mental and memorial evasion. The park offers the possibility of geographical wanderings that extend into infinity. This in turn triggers mental and memorial wanderings allowing the character (Fogg) to retreat from the world in a deep state of loneliness.[47] The "homelessness" of Fogg, first wandering the streets of New York before striding through the soothing space of Central Park, offers the possibility of a creative solitude: the separation from the world is undertaken in order to undergo a rebirth. Despite the symbolism of the "artificial nature" of the park, the latter remains the source of Fogg's memorial quest in the urban environment: he endorses a deep exploration of his interiority and consequently reminiscence. In the heart of the city of New York, the park is closely associated with the notion of refuge and sustainability. It is a return to the womb as suggested by the image of the matrix cave located in the heart of the New York urban jungle.

Furthermore, the symbolism of the park refers to nature and thus summons the American Transcendentalist literature of Henry David Thoreau, posing in *Moon Palace* the premises of a reconstruction of a sustainable, American collective memory. The symbolism of the park provides New York City with a mythical dimension which recalls the American collective memory of open spaces. It is important here to define collective memory in relation to sustainable "National Narratives":

> Man [sic] [...] needs stories to interpret the world around him [sic] and to forge his [sic] individual and/or group identity. National Narratives are cognitive instruments that grasp together information into a coherent whole; they organize

> texts, episodes of the past, symbols, and practices, which make up Collective Memory, into a story. Collective Memory (as a combination of texts, episodes of the past, symbols, and practices).[48]

This collective memory is that of an American Transcendentalist literature and, in particular, poetics related to wild nature sublimated by Transcendentalist writers like Herman Melville and Thoreau. These are appealing for a return to elementary life and build a "transcendental space associated to nature."[49] Thoreau refers to nature as a moral force. In *Walden*, the appeal to nature is a pattern to purify and to bring forth an excellence from the internal nature of the individual.[50] The existence of Thoreau in *Walden* is governed by nature, which acts as a purifying and liberating force to the individual. Nature as the subject of writing prompts the exploration of the subject through a pure language. Thoreau attempts to retrieve this heroic and sustainable nature in opposition to American Puritanism, which demolished the spirit of nature through trade expansion. In *Moon Palace*, the city is branded by evaporation, desertion, and withdrawal of landmarks, hindering any memorial sustainability.

The transition from the city to the vastness of the west in *Moon Palace* allows settling a "sustainability of memory,"[51] which is barely permitted in the city. The west conquers the space of Auster's memorial writing as it motivates the reactivation of the American myths of freedom and the Frontier conquest, installing memorial sustainability. An individual memory is gradually reconstructed in the west, with Fogg's discovery of the real identity of his father (Barber). The sustainable collective memory is built in turn through the memorial narrative of Effing, which recalls the founding myths of the immensity of the American territory and revives ancient myths, such as the "Gold Rush" and the "Open Road,"[52] ubiquitous in novels by North American authors (Jack Kerouac and Don DeLillo) accompanying the crossing of the desert.[53] As such, a whole body of imaginary travel literature (Robinson Crusoe, Cabeza de Vaca, Thomas Harriot, etc.) regarding expeditions into unknown territories is explicitly integrated into the memorial narrative of Effing. These memorial narratives (of Effing), reinstated and restored retrospectively in his New York apartment, make Fogg discover a whole facet of an American collective memory: the west, the desert, and the first people of America. The enclosed space of the apartment, a symbol of interiority, opens as Effing's memorial-narrative progresses: "In the park, I did not have to carry around this burden of self-consciousness. It gave me a threshold, a boundary, a way to distinguish between the inside and the outside. If the streets forced me to see myself as others saw me, the park gave me a chance to return to my inner life."[54] His memorial narratives prompt an abundance of images belonging to desert landscapes, stimulating his imagination. Effing's narratives are considered to be an initiation into sustainable memorial journeys and itineraries, transcending the walls of his New York apartment, crossing the desert, and entering into American founding myths.

If the path of Fogg duplicates that of Effing (his grandfather), then the "man-made" nature seems to reproduce the experience of the return to origins accomplished by Effing during his desert crossing of the west. Indeed, Central Park takes on a mythical dimension, which recalls the conquest narrative of the Frontier articulated by Effing, highlighting the similarity between the vastness of the desert of the west and urban space: "There were eight hundred and forty acres to roam in."[55] The geographical wanderings in the park assign a mythical dimension to the city of New York. New York cries out, by the immensity of its explored territory, the American collective memory: "I liked wandering back and forth among these different sectors, for it allowed me to imagine that I was traveling over great distances, even as I remained within the boundaries of my miniature world."[56] In addition, Fogg's geographical wandering in the New York jungle strengthens his concept of freedom: his decision to live in the park stems from a desire for introspection, a resistance to the laws of the city, and a willingness to fight against the conformity of the urban space. He refuses to melt away into the masses and prefers to withdraw into voluntary exile in the "natural" area of the park. The solipsistic experience in the park is heavy and recalls the suffocating room in *The Invention of Solitude*. The transition from the room to the park in the case of *Moon Palace* highlights a new facet of loneliness: that of solitude lived in the middle of the crowd, which endorses an autarchic existence, introversion, and dislocation of the self.

Nevertheless, as in the memorial narrative of the desert exploration conveyed by Effing, Central Park is not immune to night terrors and betrays the sustainability of the place. The park also exerts its power of repression and confinement. As the desert area of the west is for Effing, the park is a space for alienation and estrangement for Fogg, producing a mirror effect between the two men's experiences. Just like the advent of death in the cave occupied by Effing when lost in the desert, the park space threatens Fogg's life: ubiquitous death hovers over the urban maze, hindering any sustainability. This similarity between the two erratic existences (of Effing and Fogg) is duplicated in their memorial and intergenerational quest. The adventure in the New York jungle creates a dark image of the city. The area of the park, originally described as a sustainable haven of peace and "inner colloquies,"[57] turns into a ghostly place. Fogg's terrifying misadventures[58] experienced in the darkness of the night brand a breakdown of nature's sustainability originally described as in resonance with the American Transcendentalist texts. The reactivation of the vagrant myth in the vastness of nature is opposed to the artificially natural space of the park that engulfs self-revelation in its idyllic form, showing that its hidden side is made of nightmarish realities: "The rain continued until dawn [...] exploding with monumental bursts—screeching battalions of cats and dogs, pure wrath tumbling from the clouds."[59] Fogg's freedom in the space of Central Park ("you were free to do what you liked")[60] is a fictitious one. The illusion of freedom lived in the urban jungle is denounced by the absence of any sustainability.[61] The city is described at the beginning as a monster, a "biblical behemoth"[62] swallowing the self, proving that freedom and the city are two antagonistic entities: "Sitting in repose, with his bald behemoth's head jutting from the folds of his massive neck, there was a legendary quality about him, a thing that struck me as both obscene and tragic."[63] Being part of the city, sustainability and freedom are elsewhere than Central Park, this nature created by humans. This artificial nature is a reflection of the hostility of the urban environment surrounding the park. However, it is only by meeting Effing that Fogg reactivates the myth of freedom through Effing's memorial narrative. The latter appeals to the American past—the conquest of huge open spaces—and overcomes the absence of sustainability of the urban environment. Memorial sustainability and freedom[64] are only reactivated by an expansive movement compelled both by the realm of imagination and mnemonic action. This revives the traces of a past linked to a symbolic, subsidiary memorial and generational transmission (Effing is revealed to be Fogg's grandfather).

The New York Trilogy, which hinders any sustainability of identity, emphasizes a space of loss of the individual. The geography of space is often confused. Characters, like Quinn, disappear from the unfamiliar area of the city as they disappear from the memorial narrative. Anchoring in the space of the city would prevent a character's mobility and would inflict a stagnation of both route and destiny contrary to the project of memory writing based on geographical and memorial wanderings. Although it is often represented as a source of alienation of the self, urban space triggers Auster's sustainable memorial writing and prompts the construction of an identity and an initiatory journey identifiable as a sustainable quest for an individual and collective memory in *The New York Trilogy* and *Moon Palace*.

NOTES

1. "Celebrations, emblems, monuments and commemorations eulogies, dictionaries and museums constitute these memory symbols" (my translation). "Fêtes, emblèmes, monuments et commémorations, éloges, dictionnaires et musées constituent ces symboles de la mémoire," in Pierre Nora, ed., *La République*, Vol. 1 of *Les Lieux de la mémoire* (Paris: Gallimard, 1984), VII.

2. "Like all economically and ideologically fungible commodities, [...] memory, especially collective memory, and [...] even the most private memories, find their significance through modes of collective validation. Memory has rarely been permitted to find its level of maximal sustainability as intrinsic value." Djelal Kadir, *Memos from the Besieged City: Lifelines for Cultural Sustainability* (Stanford: Stanford University Press, 2011), 77.

3. Shu-Yi Wang, "Tradition, Memory and the Culture of Place: Continuity and Change in the Ancient City of Pingyao, China" (PhD diss., University of Colorado at Denver, 2008), 43.

4. Lidia Varbanova, "Our Creative Cities Online," *in The Creative City: Crossing Visions and New Realities in the Region*, ed. Nada Švob-Đokić. Culturelink Joint Publications Series, no. 11 (Zagreb: Institute for International Relations, 2007), 9.

5. "[T]he 1960s rhetoric of modernisation has been replaced by one of economic and environmental sustainability." Pamela J. Stewart and Andrew Strathern, eds., *Memory and History: Anthropological Perspectives* (London: Pluto Press, 2003), 160.

6. "The symbolic and aesthetic manifestations of the image reinforce identification with 'authentic place' and celebrate discernible difference between places." Terry J. Williamson, Antony Radford, and Helen Bennetts. *Understanding Sustainable Architecture* (London and New York: Taylor & Francis, 2003), 30.

7. According to Foss, visual rhetoric is "a product individuals create as they use visual symbols for the purpose of communication." Sonja K. Foss, "Framing the Study of Visual Rhetoric: Toward a Transformation of Rhetorical Theory," in *Defining Visual Rhetorics*. eds.Charles A. Hill and Marguerite Helmers (Mahwah, New Jersey: Lawrence Erlbaum, 2004), 143.

8. *The New York Trilogy* is a travel journal. The theme of wandering is strengthened by implicit and explicit references to fictional travelogues and accounts. These are implied by the presence of *Moby Dick* (Auster, *City of Glass*, 63), *Arthur Gordon Pym* (Auster, *City of Glass*, 85), *Robinson Crusoe* (Auster, *City of Glass*, 40), *Don Quixote* (Auster, *City of Glass*, 118), *History of the World* by Walter Raleigh (Auster, *The Locked Room*, 325), *The Journeys of Cabeza de Vaca* (Auster, *The Locked Room*, 325), *Marco Polo's Travels* (Auster, *City of Glass*, 6), *The Arctic Adventure* of Peter Freuchen (LR. 300), *Neverland* by Fanshawe (Auster, *The Locked Room*, 272) and many others. Paul Auster, *City of Glass* (Los Angeles: Sun & Moon Press, 1985); Paul Auster, *Ghosts* (Los Angeles: Sun & Moon Press, 1986), 1987. Paul Auster, *The Locked Room* (Los Angeles: Sun & Moon Press, 1986).

9. Auster, *City of Glass*, 4.

10. "By wandering aimlessly, all places became equal, and it no longer mattered where he was. On his best walks, he was able to feel that he was nowhere. New York was the nowhere he had built around himself, and he realized that he had no intention of ever leaving it again." Ibid.

11. The filmic work of Paul Auster reflects also the concept of memorial sustainability in the city. The memorial sustainability of the urban landscape is illustrated in the two films, *Smoke* and *Blue in the Face*. Smoke transcribes the protagonist's fluctuations between collapse and ascension, in a space limited to Auggie Wren's tobacco office in Brooklyn. All the protagonists are lost in the city and trying to

escape the urban nowhere. Nevertheless, the two films establish some form of memorial sustainability by capturing the scenes of the visual landscapes of New York. Similarly, Brooklyn Boogie is a hymn to the modern urban space of New York and Brooklyn. *Smoke*, dir. Paul Auster, Wayne Wang, and Wang Wong (Santa Monica, CA: Miramax, 1995); *Blue in the Face*, dir. Paul Auster, Wayne Wang, and Harvey Wang (Santa Monica, CA: Miramax, 1995).

12. "Landscapes are not static entities and require constant effort to maintain. [. . .] [M]any iconic places have been created by people through cultural practices [. . .] and urban development. [. . .] There are a set of semiotics of place—varying with landscape, culture and memory—which condition our language, norms and management actions." Graham Harris, "Created Landscapes and Our Changing Sense of Place," in *Seeking Sustainability in an Age of Complexity* (Cambridge: Cambridge University Press, 2007), 101.

13. "This period of growing darkness coincided with the dwindling of pages in the red notebook. [. . .] Quinn no longer had any interest in himself. [. . .] He remembered the moment of his birth and how he had been pulled gently from his mother's womb. He remembered the infinite kindness of the world and all the people he had ever loved. [. . .] Nevertheless, he tried to face the end of the red notebook with courage. He wondered if he had it in him to write without a pen, if he could learn to speak instead, filling the darkness with his voice, speaking the words into the air, into the walls, into the city, even if the light never came back again." Auster, *City of Glass*, 156–57.

14. Walter Leal Filho, ed., *International Experiences on Sustainability* (Frankfurt am Main: Peter Lang, 2002), 228.

15. "Now that he had embarked on the Stillman case, he felt that a new notebook was in order. [. . .] For reasons that were never made clear to him, he felt an irresistible urge for a particular red notebook at the bottom." Auster, *City of Glass*, 46.

16. Paul Auster, *The Invention of Solitude* (New York: Sun Press, 1982), 168–69.

17. Auster, *City of Glass*, 156.

18. Tim O'Riordan, *Globalism, Localism and Identity: New Perspectives on the Transition of Sustainability* (Stockholm: Routledge, 2002), xix–xx.

19. Auster, *Invention of Solitude*, 153.

20. Ibid., 149.

21. Auster, *City of Glass*, 157.

22. Auster, *Invention of Solitude*, 136.

23. "A historic city is similar to a natural matrix holding significant history and cultural legacy, including tangible and intangible culture. [. . .] One can draw a parallel between naturally sustainable development—which requires the input of economists and sociologists, as well as environmentalists—and culturally sustainable development." Wang, *Tradition, Memory and the Culture of Place*, 43.

24. "Sustainability [. . .] is in some respects a literary construct, the outgrowth of particular ways of conceiving time; therefore, the different registers of time [. . .] both produce and are reinscribed by descriptions of sustainability as an ethics. [. . .] [S]ustainability evokes a succession of individual lifetimes—an unbroken sequence of embodied experiences from the past and into the future that presupposes societies and cultures developing against the backdrop of the timeless present of an abiding

Nature." Bruce Clarke and Manuela Rossini, eds., *The Routledge Companion to Literature and Science* (London and New York: Routledge, 2010), 65.

25. "Both people and fictional characters respond to their respective sensory worlds within the temporal realms of the past, present and future." Steven E. Alford, "Chance in Contemporary Narrative: The example of Paul Auster," *Lit: Literature Interpretation Theory,* 11, no. 1 (2000): 73–74.

26. "How can we know the past today? [...] How can we come to know the past real? Postmodernism does not deny that it existed; it merely questions how we can know real past events today, except through their traces, their texts, the facts we construct and to which we grant meaning." Linda Hutcheon, *A Poetics of Postmodernism: History, Theory, Fiction* (New York: Routledge, 1988), 225.

27. "Postmodern art merely foregrounds the fact that we can know the real, especially the past real, only through signs, and that is not the same as wholesale substitution. The postmodern still operates, in other words, in the realm of representation, not of simulation, even if it constantly questions the rules of that realm." Ibid., 230.

28. "[S]ustainability has not been established out of ideal components but from the balanced equilibrium of opposing forces. [...] [T]he approximation, [...] simultaneity and balance between all seemingly competing forces have contributed to the city's urban sustainability. What gives the city a degree of consistency and sustainability is the smooth fluctuation between incompatible themes. [...] [T]he city has an inherited paradox between myth and reality, between a physical existence and a memory." S. S. Zubir and C. A. Brebbia, eds., *The Sustainable City VIII: Urban Regeneration and Sustainability* (Southampton, UK: WIT Press, 2014), 618.

29. "On habite quelque part et ce lieu constitue en soi un monde essentiel à chaque individu. Un artiste, un écrivain quiconque répondent à l'environnement où ils habitent. Dans mes livres, je réponds à la réalité qui m'entoure: une réalité américaine." Gérard de Cortanze, *La Solitude du Labyrinthe: Interviews with Paul Auster* (Arles: Actes Sud, 1997), 91. (my translation).

30. Gaston Bachelard, *La Poétique de l'espace* (Paris: Quadrige/PUF, 2004 [1961]).

31. "The principle of sustainability [...] must take into consideration the vitality as well as the materiality of a place. [...] [C]ultural sustainability can be interpreted as the balance between different cultural models, identities, [...] cultural heritage and modern culture-production [...] in the relationship between 'urban memory' [...] and in a postmodern resemanticisation of heritage." Elizabeth Auclair and Graham Fairclough, eds., *Theory and Practice in Heritage and Sustainability: Between Past and Future* (Abingdon and New York: Routledge, 2015), 69.

32. Yves-Charles Grandjeat, "Le hazard et et la nécessité dans l'œuvre de Paul Auster," in *L'oeuvre de Paul Auster: Approches et lectures plurielles*, ed. Annick Duperray (Arles: Actes Sud/Université de Provence, 1995), 154.

33. "He has walked the main city avenue [...] as if pushed by the invisible hand of modern nothingness, he spontaneously turns into the man he is trailing." Nicolas Caleffi, "Paul Auster's Urban Nothingness," *The Italian Review of Social Studies Alfa Zeta* 54 (April 1996): 3.

34. Auster, *City of Glass*, 53.

35. Ibid.

36. "[…] the Tower of Babel stands as the last image before the beginning of the world." Ibid.

37. "The process of territory formation, deterritorialisation and reterritorialisation are inextricably entangled in any given social field. […] Deterritorialisation is always a complex process involving at least a deterritorialising element and a territory which is being left behind or reconstituted. […] Deterritorialisation is always bound up with correlative process of reteritorialisation, which does not mean returning to the original territory but rather the ways in which deterritorialised elements recombine and enter into new relations." *The Deleuze. Dictionary*, ed. Adrian Parr (New York: Columbia UP, 2005), 70–71.

38. Auster, *City of Glass*, 53.

39. "[…] the myth of the tower of Babel […] [has] a double inversion: in the first place the value of earth and heaven were reversed, and in the second place the myth of the separation of the peoples was taken over by the notion of peoples.[…] This symbol is a striking example of the mechanism of collective memory." Yuri Lotman, *Universe of the Mind: A Semiotic Theory of Culture* (Bloomington: Indiana University Press, 1990), 111.

40. Paul Auster, *Moon Palace* (New York: Viking, 1989), 121.

41. Ibid., 22.

42. Ibid., 55.

43. Saved by his two friends, Kitty and Zimmer, he finds shelter under the roof of Thomas Effing who hired him as a secretary. Fogg reads the blind man parts of universal literature and writes his obituary.

44. "This was New York, but it had nothing to do with the New York I had always known. It was devoid of associations, a place that could have been anywhere." Ibid., 56.

45. Auster, *Moon Palace*, 55.

46. Ibid., 62.

47. "It became a sanctuary for me, a refuge of inwardness against the grinding demands of the streets. […] the park offered me the possibility of solitude, of separating myself from the rest of the world." Ibid., 56.

48. Paata J. Kervalishvili and Sousana A. Michailidis, eds., *Philosophy and Synergy of Information: Sustainability and Security* (Amsterdam and Washington, D.C.: IOS Press, 2012), 12.

49. "In Thoreau's life in Walden, nature has worked as a purifier and a simplifier of the restless nineteenth century into the simple truth of freedom. […] [B]y living near nature and writing about it, Thoreau intends Walden to contain a true language. […] The goal for which Thoreau enters nature is to recover this chanticleer-like heroic nature." EunHyoung Kim, "Individualism and the Sectional Crisis: Emerson, Thoreau, Hawthorne, Melville, and Their Responses to Slavery and Racism." (PhD diss., State University of New York at Buffalo, 2008),100.

50. "[…] Thoreau writes startlingly, "Nature is hard to overcome, but she must be overcome" (Thoreau, *Walden*, 221). […] The moral paradox on which *Walden* rests, or perhaps totters, is its appeal to *external* nature as a vehicle for cleansing and

sublimating *internal* nature." Robert Milder, *Hawthorne's Habitations: A Literary Life* (New York: Oxford University Press, 2013), 63.

51. "Remembrance can be made to endure so that the values it embodies are sustained over time. I call this the 'sustainability of memory problem.' Monuments, archives, museums, testimonials, and artifacts of various kinds, each created to memorialize some element of the past, are often transient in terms of their memorial effects." Jeffrey M. Blustein, *Forgiveness and Remembrance: Remembering Wrongdoing in Personal and Public Life* (New York: Oxford University Press, 2014), 11.

52. "The trope of dropping out and taking to the open road is a further point of continuity [...] in the Beat generation. Since a literary model of aimless, bohemian wanderings already exists [...] fantasies of escape, mocking the naive belief that Kerouac held a secret alternative to the bruising, dull realities of everyday life [...]. In Don DeLillo's Americana (1971), David Bell—who, like John Johnson, works in the entertainment industry—attempts a journey 'into the depths of America, wilderness dream of all poets'." Andrew Tate, *Douglas Coupland* (Manchester: Manchester University Press, 2008), 124.

53. It is relevant in this regard to stress that *The Music of Chance* revives for its part the American myth of the "Open Road" and for which other cities like Los Angeles or California revive a bygone American dream. In *The Music of Chance*, the geographical wanderings of Nashe and Pozzi through the territory of the United States contribute to shaping a sustainable memory through their existential and memorial journeys and thereby reactivating the myth of freedom. The sustainable memorial writing of Nashe is imbued with his wanderings around the American territory, traveling by car along the highways of the United States. Paul Auster, *The Music of Chance* (New York: Viking, 1990).

54. Auster, *Moon Palace*, 57.

55. Ibid., 56.

56. Ibid., 63.

57. "My head turned with bookish theories, battling voices, elaborate inner colloquies." Ibid., 61.

58. "One night, a gang of kids chased me across Sheep's Meadow. [...] Those were close calls, but the most terrifying moment came on a cloudy night toward the end. [...] [A] shadowy figure rose up and pointed [...] a gun in my direction." Ibid., 64–65.

59. Ibid., 66.

60. Ibid., 57.

61. Zimmer and Kitty Wu put an end to Fogg's wanderings by sheltering him.

62. Auster, *Moon Palace*, 235.

63. Ibid.

64. "More frequently than Thoreau, Melville tempered his view of the western frontier, the terrain of the future, with skepticism about such fictive promises of American liberty. [...] In fact, the American west frequently united Edenic possibility and demoniac desire, its mythology, created largely in advance of its history, provided a simultaneously geographical and psychological stage on which Americans dreams of liberty, prosperity, and world leadership could be enacted." Eric J. Sundquist,

Empire and Slavery in American Literature: 1820–1865 (Cambridge: Cambridge University Press, 1995), 19.

BIBLIOGRAPHY

Alford, Steven E., "Chance in Contemporary Narrative: The Example of Paul Auster." *Lit: Literature Interpretation Theory,* 11, no. 1 (2000): 59–82.

Auclair, Elizabeth, and Graham Fairclough, eds. *Theory and Practice in Heritage and Sustainability: Between Past and Future*. Abingdon and New York: Routledge, 2015.

Auster, Paul. *City of Glass*. Los Angeles: Sun & Moon Press, 1985. New York: Penguin Books, 1987.

———. *Ghosts*. Los Angeles: Sun & Moon Press, 1986. New York: Penguin Books, 1987.

———. *In the Country of Last Things*. New York: Viking, 1987. New York: Penguin Books, 1990.

———. *The Invention of Solitude* (1982). New York: Sun Press, 1982. New York: Penguin Books, 1988. London: Faber and Faber, 1988. ["*Portrait of an Invisible Man*" and "*The Book of Memory*"].

———. *Leviathan*. New York: Viking, 1992. London: Faber and Faber, 1992.

———. *The Locked Room*. Los Angeles: Sun & Moon Press, 1986. New York: Penguin Books, 1988.

———. *Moon Palace*. New York: Viking, 1989. London: Faber and Faber, 1990.

———. *Mr. Vertigo*. New York: Viking, 1994. London: Faber and Faber, 1994. New York: Penguin Books, 1995.

———. *The Music of Chance*. New York: Viking, 1990. London: Faber and Faber, 1991.

———. *The New York Trilogy* (*City of Glass* 1985; *Ghosts* 1986; *The Locked Room* 1986). London: Faber and Faber, 1988. New York: Penguin Books, 1990.

Bachelard, Gaston. *La Poétique de l'espace*. Paris: Quadrige/PUF, 2004 [1961].

Blue in the Face. Directed by Paul Auster, Wayne Wang, and Harvey Wang. Santa Monica, CA: Miramax, 1995.

Blustein, Jeffrey M. *Forgiveness and Remembrance: Remembering Wrongdoing in Personal and Public Life*. New York: Oxford University Press, 2014.

Caleffi, Nicolas. "Paul Auster's Urban Nothingness," first appeared as "Paul Auster. L'inessenza urbana." *The Italian Review of Social Studies Alfa Zeta* 54 (April 1996): 18–23.

Clarke, Bruce, and Manuela Rossini, eds. *The Routledge Companion to Literature and Science*. London and New York: Routledge, 2010.

de Cortanze, Gérard. *La Solitude du Labyrinthe: Interviews with Paul Auster*. Arles: Actes Sud, 1997.

The Deleuze Dictionary. Edited by Adrian Parr. New York: Columbia UP, 2005.

Foss, Sonja K. "Framing the Study of Visual Rhetoric: Toward a Transformation of Rhetorical Theory." In *Defining Visual Rhetorics*. Edited by Charles A. Hill and Marguerite Helmers. Mahwah, New Jersey: Lawrence Erlbaum, 2004.

Grandjeat, Yves-Charles. "Le hazard et et la nécessité dans l'œuvre de Paul Auster." In *L'oeuvre de Paul Auster: Approches et lectures plurielles*, edited by Annick Duperray, 153–63. Arles: Actes Sud/Université de Provence, 1995.

Harris, Graham. "Created Landscapes and Our Changing Sense of Place." In *Seeking Sustainability in an Age of Complexity*, 101–14. Cambridge: Cambridge University Press, 2007.

Hutcheon, Linda. *A Poetics of Postmodernism: History, Theory, Fiction.* New York: Routledge, 1988.

Kadir, Djelal. *Memos from the Besieged City: Lifelines for Cultural Sustainability.* Stanford: Stanford University Press, 2011.

Kervalishvili, Paata J., and Sousana A. Michailidis, eds. *Philosophy and Synergy of Information: Sustainability and Security.* Amsterdam and Washington, D.C.: IOS Press, 2012.

Kim, EunHyoung. "Individualism and the Sectional Crisis: Emerson, Thoreau, Hawthorne, Melville, and Their Responses to Slavery and Racism." PhD diss., State University of New York at Buffalo, 2008.

Leal Filho, Walter, ed. *International Experiences on Sustainability.* Frankfurt am Main: Peter Lang, 2002.

Lotman,Yuri. *Universe of the Mind: A Semiotic Theory of Culture.* Bloomington: Indiana University Press, 1990.

Milder, Robert. *Hawthorne's Habitations: A Literary Life*. New York: Oxford University Press, 2013.

Nora, Pierre, ed. *La République*. Vol. 1 of *Les Lieux de la mémoire*. Paris: Gallimard, 1984.

O'Riordan, Tim. *Globalism, Localism and Identity: New Perspectives on the Transition of Sustainability.* Stockholm: Routledge, 2002.

Smoke. Directed by Paul Auster, Wayne Wang, and Wang Wong. Santa Monica, CA: Miramax, 1995.

Stewart, Pamela J., and Andrew Strathern, eds. *Memory and History: Anthropological Perspectives*. London: Pluto Press, 2003.

Sundquist, Eric J. *Empire and Slavery in American Literature: 1820–1865.* Cambridge: Cambridge University Press, 1995.

Tate, Andrew. *Douglas Coupland.* Manchester: Manchester University Press, 2008.

Thoreau, Henry David. *Walden; or, Life in the Woods.* New York: Dover Publications, 1995 [1854].

Varbanova, Lidia. "Our Creative Cities Online." In *The Creative City: Crossing Visions and New Realities in the Region*, edited by Nada Švob-Đokić, 9–18. Culturelink Joint Publications Series, no. 11. Zagreb: Institute for International Relations, 2007.

Wang, Shu-Yi. "Tradition, Memory and the Culture of Place: Continuity and Change in the Ancient City of Pingyao, China." PhD diss., University of Colorado at Denver, 2008.

Williamson, Terry J., Antony Radford, and Helen Bennetts. *Understanding Sustainable Architecture*. London and New York: Taylor & Francis, 2003.

Zubir, S.S., and C.A. Brebbia, eds. *The Sustainable City VIII: Urban Regeneration* and Sustainability. Southampton, UK: WIT Press, 2014.

Chapter 13

Trash That Connects

The Everyday Rhizome of Cédric Klapisch's Paris

Caitlin Yocco-Locascio*

Trash is one of the most mundane elements of everyday life. It surrounds us, collected in bins, moving from table to can, can to curb, and then curb to landfill. As the world's population continues to grow, trash will be exponentially produced. Because populations consistently migrate toward cities, urban spaces are, and will increasingly need to be, the epicenters of innovation in the reduction and sustainable practices of waste. In this way, the city is the leader in trash—in creating the problem and in solving it. In many ways, cities are becoming more desirable to live in because they offer a more sustainable alternative to suburban or rural life. They offer smaller spaces that require fewer utilities and have access to public transportation. Cédric Klapisch studies the intricacies of urban space in his 2008 film, *Paris*, in which trash is shown in a number of iterations through a complex cast of characters living in France's capital city. Throughout the film, Klapisch repeatedly shows how these characters are inextricably linked, in multiple ways, through trash. Not only does trash connect the people it touches, but it also inadvertently implicates them in a uniquely rhizomatic urban space.

Klapisch's film proposes a different view of living in the city—an urban rhizome—a particular structure that resists hierarchy or social distinctions. In order to understand the role that trash plays in the film, this chapter will first explain the concept of the rhizome and then closely examine key scenes in *Paris* in which trash demonstrates a new approach to connection. Trash as a connector can be organized into three categories: material trash, the body as trash, and film as trash. The goal is ultimately to illustrate the assertion that, in *Paris*, Klapisch makes a social statement that the city is a space wherein social hierarchy exists, but that networks such as waste removal and renewal function without regard to

* This chapter was written under the direction of Vlad Dima.

these class distinctions. They instead create a rhizome that offers a different way of conceptualizing life in the city and the roles that various individuals play in it. *Paris* shows that the city is a space sustained by all walks of life, its players woven together by the discarded pieces of their material existence. These three categories of trash appear throughout the film and overlap to create a constellation of examples that mirrors *Paris*'s complex cast of characters.

The narrative of *Paris* revolves around Pierre, a professional dancer confined to his apartment while he waits for a heart transplant, and the network of people living in the streets below. He is often visited by his sister, Elise, a social worker, and voyeuristically observes a pretty girl across the street, a student who has an affair with her professor, Roland. Elise works with a Cameroonian family to fill out paperwork so their brother, Benoît, can immigrate as well. Simultaneously, the film also visits Cameroon, where we see Benoît in the process of leaving his country, at the luxury resort where he works and meets a group of Parisian fashion models. This summary could continue to describe the web of connections that grows outward and folds back in on itself as the cast of characters mix and mingle, but the intricacies of the characters' interconnection are already clear. Klapisch reveals a complicated network that resembles Gilles Deleuze and Félix Guattari's rhizome, a social organization that spreads organically in every direction in disregard to any social hierarchical system.

The following pages will show that the rhizomatic construction of the city undermines preexisting social hierarchies in Parisian society and offers the spectator a new perspective on the connectivity of everyday life. Material trash circulates throughout the city and beyond, generally following a trajectory of descent but connecting people of all social status along the way, constantly reminding us of trash's role in urban sustainability. It is a product to be managed, but often leaves an unwanted trail of waste behind. The body as trash shows how even a human's most precious possession, the body, can be viewed as organic material to consume, reuse, and connect. Finally, trashy cinema shows the spectators how the cinematic apparatus itself reflects many of the important concepts that Klapisch promotes in his film, such as recycling, turning film itself into a sustainable medium. Trash in all of these forms comes together to re-materialize into a rhizomatic vision of the city. While recycling has always been viewed as a sustainable practice, these new conceptions of trash and its renewal expand the sustaining nature of the city, showing that it is not only paper or glass that regenerates but also the inhabitants of the city themselves and even art that showcases urban spaces.

THE RHIZOME

The concept of the rhizome opens Deleuze and Guattari's *A Thousand Plateaus*. In this introduction, the authors offer new ways of looking at the world and of organizing life. The rhizomatic model is based on "[p]rinciples of connection and heterogeneity: any point of a rhizome can be connected to anything other, and must be."[1] This new configuration opposes the arborescent model based on hierarchies generally consisting of an origin, the taproot, which gives way to a structured body from which a hierarchy of organized branches form. On the contrary, a rhizome is "entirely oriented toward an experimentation in contact with the real"[2] in that it is not a structure organized by abstract ideas such as social status but exists only as material elements existing in a random but supremely interconnected constellation. This way, any interaction with or within a rhizome involves many different choices and a myriad of possible outcomes. A rhizome always has multiple entryways, and if at any point a segment is broken or cut off, more rhizomatic branches will grow new lines in all possible directions.[3] This concept allows for freedom and movement. Characters are not bound to one position in the city or one position on the social ladder. A person's place in the collective is instead determined by interaction, experience, and connection. In fact, a person's place can only be thought of in terms of being one part of a larger whole. Much like the rhizome, the city offers an infinite number of potential paths that are only determined as they are taken. The characters of *Paris* explore and design these routes as they move through the streets and buildings. The same can be said for trash as it circulates and regenerates along predetermined or spontaneous paths. The spectators see both as Klapisch contrasts established routes of municipal waste removal with impromptu cases of recycling like organ donation and filler film. These unplanned routes of recycling are a sharp contrast to the rigid social structures that are inevitably a part of Parisians' lives.

The metro, waste removal, and the bureaucracy of immigration organize the people and space of Paris. These structured networks dictate hierarchical relationships—professor/student, citizen/immigrant, and so on—but trash circulates, and recycling appears in the most unusual of situations to show Paris as an organic, and what I call egalitarian, rhizome. Though the rhizomatic presentation of a community of people in Paris does not eradicate these hierarchies, it does undermine them to show an alternative perspective that brings with it serious implications concerning social structures. An essentialist perspective on social hierarchy would separate the characters into discrete social classes, but the film underlines the overlap that naturally occurs in everyday life. Throughout his filmography, Klapisch's forte is his subtle political activism. Klapisch's current style is inscribed in an evolution of politically engaged filmmaking that has been ongoing in France since the 1980s. Like the *cinema du look* movement of the 1980s, Klapisch's oeuvre often features marginalized characters. In the 1990s, French filmmakers began to turn to the suburbs of Paris for inspiration. The *cinéma de banlieue* focused on working-class populations and the violence they endured on the outskirts of major metropolitan areas. *Paris* also showcases working-class individuals, particularly street vendors. While his film does not move to the Parisian suburbs, it does move to Cameroon to draw attention to an outside space that is culturally and politically connected to the city of Paris. Klapisch is presently often characterized as a director of *cinéma social*, an aesthetic and thematic movement that highlights the everyday life of working-class individuals in a realistic way. The goal of the *cinéma social* movement[4] is to move its spectators to reflect on issues underlined on screen. Klapisch again takes up film as a tool to visually promote his message about the value of all social classes in the Parisian metropolis. He begins with material trash, a banal and universal substance that touches all walks of life.

MATERIAL TRASH

In a key sequence of scenes revolving around trash, Klapisch highlights the network that exists around us and facilitates our everyday life. The city and its habitants need networks to sustain their everyday practices of consuming, connecting, and moving. Trash moves throughout the city but does not necessarily end in a landfill,[5] thus showing more flexibility in its trajectory and displaying more rhizomatic characteristics than other urban networks. The spectators also see mail connect people through objects that can move beyond the limits of a demarcated city or state and the metro as it transports people and objects to geographical points of interest. While there are certainly hints of bureaucracy and hierarchy involved in the networks indicated in this sequence, the way that networks play out on an everyday level does not recognize hierarchies. The most concrete example of nonhierarchical connection through networks comes when the Cameroonian man is able to send a postcard to his brother a continent away despite the political and geographical obstacles. Paris, a microcosm of culture and wealth, symbolically contained on the postcard, moves to Africa, a space of former colonization that rests much lower on the Western-conceived social hierarchy. This missive sends the hope of a better life to an eager immigrant, though its clichéd image of Notre Dame in black and white presents a mediated simulacrum that omits a large part of reality. With the help of the metro, the Cameroonian immigrant can move freely through the space of Paris, a city that is becoming more and more exclusive to the rich.[6] Finally, the camera follows the postcard destined for Cameroon and leaves Paris for Africa, breaking through the confines of the city. It would be impossible to ignore the social or political implications of a Cameroonian immigrant's presence in Paris, but Klapisch's film, through its insistence on connection over all else, implores its audience to consider the equality that exists underneath societal expectations and conventions. The city often furnishes the connections between its inhabitants, acting as a sustainable environment for human interaction.

Importantly, the sequence, with all its implications about immigration, history, and the future, begins at a family's dinner table as the remains of the meal are scraped into the trash. The spectators see a white family, of a certain privilege to live within the city limits of Paris, who takes out the trash that is managed by an African immigrant of a previously colonized territory. The sequence draws attention to social structures that exist but also underlines the connections that cross these barriers. The scene begins as Pierre and Elise have just finished eating, and, while standing in front of the green-tiled backsplash, Elise scrapes the remnants of their meal into a garbage can lined with a green bag. The movie cuts to a man, a Cameroonian immigrant, walking on the street level, as he brings green garbage bins out to the curb to be emptied. The camera pans right to follow the man to the curb and then back into the building that he exited. Just as he slips into the building, Elise exits with the same green trash bag from the previous shot and deposits it into the bin that the man just brought to the curb. The garbage cans line the green railing of the metro station entrance. Later in the film, it is revealed that the man is a client of Elise's who hopes to bring his brother to Paris from Cameroon, but during this sequence they are simply people interacting with the same objects. The audience can assume that there is some social difference given the color of the two characters' skin, but their relationship at this point is mediated completely by the trash that moves between them. They are further connected thanks to Pierre who holds a position of power to survey from above. Just as Elise lifts the lid of the bin, the shot ends and cuts to an aerial view observing the action unfolding below. After a few seconds, the shot cuts to a medium shot of Pierre, confirming it was a similar, yet slightly different angle that the spectators were given. When examined closely, Pierre's height above the street is higher than that of the high-angle shots that the spectators are shown of the street-level interaction. Klapisch is offering his spectators a chance to actively participate in the scene. Instead of positioning them at Pierre's level, he lowers them and creates a personalized perspective, an Other[7] that the spectators appropriate as their own. In this way, the spectators become active participants of the scene through cinematic suture.[8] They become part of the rhizome growing through the passing of trash from one character to another. Pierre looks down to the street where he sees the dark hole of the entrance to the metro, suggesting another level of underground space created by the network of tunnels. This time, the Cameroonian man has returned with another garbage bin to add to the line-up. He is frustrated that Elise has carelessly placed her garbage in the bin, an early sign that Elise is, on some level, an unaware and privileged Parisian, and he is an overworked immigrant designated to clean up messes. The man fixes the problem and this part of the sequence, alternating between Pierre's point of view and street-level shots, ends. The next part of the sequence descends to the street.

Klapisch begins by framing the Cameroonian man in a medium shot. The latter walks toward the camera, and, as he approaches, it pans left to follow his movement. As the frame widens, a map of Paris and a Parisian mailbox become visible. All this is located in front of a staircase that leads down to the metro entrance. In one shot, Klapisch plays with the layers and spaces of Paris that he manipulates on screen. The diegetic[9] world includes the street level of Paris but also hints at the extradiegetic metro. The map of Paris reduces the implied space of the city to a two-dimensional but much more expansive graphic representation. The yellow mailbox is an explicit example of the simultaneous reference to the local and the global as one slot is designated to deliver within *Paris—Banlieue*—and the other to deliver to *Autres departments/Etranger*.[10] In a close-up, the camera focuses on the postcard that the man is preparing to send. "Paris" is written at the top, above the iconic Cathédrale de Notre Dame, yet another two-dimensional symbolic representation of the space in which the action is developing. As the man kisses the postcard, he slips it into the slot to leave the city as though he at once holds all of Paris in his hand and now is sending it beyond its own limits. While the scope of the camera and the size of the objects change in this sequence, they continue to refer to the same geographic space. Klapisch begins his process with long shots that contain evocative street scenes of Paris. He then moves to medium shots to focus on the Cameroonian man, the map, the metro entrance, and the mailbox. Paris is still represented in these images and objects, but the space of the city is reduced to the size of a mailbox, a postcard, or a single monument. Klapisch continues to restrict the size of the shot as he forces the space of Paris into increasingly smaller objects. This allows for the Cameroonian man to contain the city and send it off to his brother in Africa. The shot then cuts to follow the man around the green railing and remains at street level as he descends the stairs into the dark, expansive space of the metro. The next shot returns to Pierre's aerial point of view, as if it does not dare yet to descend into the bowels of the city.

Green and yellow have been present throughout the entire sequence, but the last shot is particularly striking. As Pierre looks down, he sees the yellow mailbox, the yellow "M" of the metro sign, a yellow taxi, and the yellow vests of the garbage men who arrive on their truck. All this yellow is contrasted by green. Klapisch places a green garbage truck with men in green garbage suits, who pick up green garbage cans next to the green railing of the metro entrance. With only a few exceptions, this shot is almost entirely devoid of any other noticeable color. The colors draw attention to some of the most banal objects of everyday life: the metro, the postal system, and the trash. Green and yellow visually represent life, renewal, and regeneration, referencing sustainable forces in nature. Unlike the taxis or private cars of the street level, the metro groups people together on their trajectories and, at times, forces interaction. The postal system encourages connection and communication between friends, family, businesses, and clients. It is a facilitator of conversation that also opens to the outside world. The waste removal and recycling networks are particularly interesting because they concern the discarded pieces of life. The daily remnants continue to circulate after their original value has been lost. Recycling gives new life to waste through renewal and regeneration so that resources can be used even past their expiration date, making it a sustainable practice. What these systems have in common is that they are networks that connect the entire population of Paris no matter what social standing or class the individuals hold. By placing trash in a scene that clearly highlights the rhizomatic nature of metropolitan networks such as the metro and postal systems, trash is elevated to rhizome status. It becomes another example of connectivity in urban space, regardless of whether city dwellers are aware of it or not.

One exception to this rule of blindness may be Pierre, who is in a spatially, though not necessarily socially, privileged position elevated above the city to observe the small things that most others take for granted. For the majority of the film he is isolated to the vantage point of his balcony, but his privilege is also a lonely curse. The next cut transitions us to a medium shot of Pierre, who looks down at the street, then wanders back into his apartment. As he leaves, the Eiffel Tower is revealed. Since Pierre's heart condition was diagnosed, he has remained largely immobile above the city. He is positioned so he can observe but has limited interaction with others. In this way, he has become the human representation of the Eiffel Tower in the constellation of characters in the film. His name suggests that he is the rock,[11] a first cell around which Klapisch grows the rest of his rhizomatic film. Narratively speaking, the film begins and ends with Pierre, and, in the meantime, occasionally returns to him. Since the rhizome resists hierarchical organization, Pierre does not hold a more meaningful position in the structure but rather is given a specific position to observe the connections as they grow outward.[12] In his observations from above the city, most of his relations form through his viewing experience, much like that of the spectator. Both are invited to recognize beauty in everything that goes on below, even in the collecting of trash or in the color of the metro sign. The next scenes in this sequence finally descend into the metro.

Dressed in a yellow jacket and fitting nicely into the aesthetic of this scene, the Cameroonian man penetrates the underground system of the metro. The first shot is a close-up of his face as he is riding the train. In the background, the lower section of the Eiffel Tower is visible, double-framed by the window of the train and the camera, anchoring him concretely in Paris as he travels through the space shown previously on the map. Klapisch employs a sort of thematic montage, a repetition of the iconic monument in two consecutive shots, to continually hint toward Pierre, the film's human metaphor for the Eiffel Tower. The last shot of this sequence is a medium shot from the exterior of the metro as it exits the underground tunnel. The camera starts angled to the left, the direction from which the metro issues, and pans right while tilting slightly upwards to follow the motion of the train as it continues on this network through the city. The movement of the train mirrors the sequence's trajectory that starts within Parisian city limits but will end in Africa. The final scenes of this sequence move to Cameroon, where Benoît receives the postcard from Paris.

The connection between France and Africa is steeped in cultural, social, and historical tension. In the introduction to his book on trash, Kenneth Harrow observes that, "if it is the jetsam of the material world, what's left over when the rich have eaten, then trash must define not only the scraps but the eaters of the scraps as well."[13] The previously analyzed cinematic sequence illustrates this double reference. Elise, a white Parisian of assumed Gallic decent, finishes a meal with her brother and scrapes the remains into the trash. The camera follows the "jetsam of the material world" as it travels from Elise to the African immigrant of lower social class, whose job it is to manage this waste, ending the sequence in Africa itself. Though trash is no longer physically present on screen, in this scene, Klapisch moves from high to low, beginning in the walk-up Parisian apartment, down to street level, underground in the metro, and through the postal service to Cameroon. In terms of movement, trash generally has a trajectory that, "when measured in terms of loss and lack of value […] has to be downward,"[14] thus evoking a descent from Europe to Africa. Klapisch illustrates this last step on the trek from Paris to Cameroon in a jump cut to a scene clearly outside the confines of France's capital. In a medium shot, the camera follows the path of a woman in an African village carrying the postcard from the previous shots. She gives the card to Benoît. As he turns the card over, Klapisch inserts a close-up of the document, this time focusing on the written message. Even without the picture of a Parisian monument, "Paris" is clearly written at the top. Klapisch essentially creates the same shot twice: two Cameroonian men hold Paris in their hands though the time and place of each is radically different. These parallel shots illustrate a long-standing tension that exists between France and previously French-colonized nations of Africa, such as Cameroon, Senegal, and Algeria. It is based on a pyramid of value where waste continually moves downward from the riches of the West toward the poverty of the developing countries below. The two areas share a common history of colonization, a history that is soaked in power dynamics and social hierarchy. Klapisch underlines these facts to remind his spectators of this ever-present reality.

Benoît's eventual movement from Cameroon to France seemingly subverts the downward movement of trash from the riches of Europe to the poverty of Africa. Despite this apparent reversal, the spectators anticipate his struggle as he attempts to insert himself into Parisian society. The last shot of this character during the closing minutes of the film shows him standing on a bridge over the Seine, holding the postcard of Notre Dame up to the actual cathedral in comparison. Benoît, the hopeful Cameroonian, is representative of "Africans as illegal immigrants striving to get to Europe's shores where they will either drown in the attempt, wash up with the flotsam and jetsam, or wind up selling sunglasses and balloons in the street; hopes trashed; victims of power in a globalized world system."[15] The spectators have already seen an example of the immigrant's reality in Paris through Benoît's brother. Though Klapisch is careful to include all walks of life in this film, his portrayal of the Cameroonian couple only amounts to a caricature of the African immigrant family. The wife comes across as loud and demanding and is not given subtitles as though the content of what she expresses has no value. Though the brother appears multiple times in the film, the spectators never learn anything substantial about him. He is simply the example of an African immigrant struggling in Paris. Nevertheless, Klapisch's inclusion of Benoît and his brother in the rhizome shows their value and position in the community as a whole.

In a similar way, African cinema harnesses a global status as *les déchets humains*[16] in order to "provide the *damnés de la terre* with the power to stimulate change."[17] Recognizing the fact that the outside world views Africa as a global landfill and its populations as human waste, African directors appropriate these misconceptions to suggest that "[t]here is more here than loss; more than decay and a descent. In fact, the descent is here only because of the belief that the depiction of it will serve to enable a change, an ascent, to become possible."[18] Klapisch's representations of African immigrants in *Paris* do exactly that. By showing Benoît's story—from working in a luxury resort in Cameroon while living himself in poverty, then immigrating on a makeshift raft with other Cameroonians, to standing in awe of the historic monuments in Paris—Klapisch shows a simultaneous descent and ascent. The detailed analysis of the trash scene shows the descent of material waste from Europe to Africa, but Benoît's immigration to France responds with an example of ascent. Instead of showcasing the typical narrative of the struggling immigrant, the director inserts Benoît's story in a panoply of diverse experiences, all equally placed in connection to each other, validating each character's experience as equally meaningful to the others. It is left to the spectators to determine whether they read Benoît's trajectory as arriving at a predetermined rung on the ladder of social hierarchy or as a hopeful display of equal value in the rhizome of city dwelling.

THE BODY AS TRASH

Paris also presents many examples of the body as trash, a concept that will focus on Pierre and women in the film. On a literal level, just as sustainable initiatives in a city must consider post-consumer waste and recycling, the human body is part of those activities as it produces its own by-products and must be properly cared for after death. The concept of the body as trash will also look at a figurative interpretation, as women's body are sometimes viewed as an object to be used and discarded. To begin, Klapisch offers Pierre to his audience, a preliminary example of the body as trash, and uses him as a *fil conducteur* around which he structures the rest of his story. The first sequence of the film deems the body organic material that has its own expiration date. The audience sees an ultrasound, a second screen that gives a summary of our main character and tells us where he is. They see his name, the date, and time alongside two recordings of his heart with his heartbeat visible underneath. Thanks to the GE Healthcare logo branded in plain sight and the doctor's voice-over, we can easily determine that Pierre is in a medical appointment. As the valves of his heart open and close, the doctor explains these functions, checking off a list of healthy processes. In the meantime, the camera backs up to show Pierre's back between the ultrasound screen and the spectators. He, too, is watching his heart. Next, the camera provides another shot of Pierre though he is obscured by a blurry figure that is most likely the doctor. When the camera finally shows us the doctor's face, the first full shot of a human, the doctor hesitates, becoming aware of a problem. It is not until after the doctor discovers an irregularity that Klapisch shows Pierre's face. In doing so, Pierre is from the start defined by his sickness. His body is identified as defective and, throughout the rest of the film, as an example of bodily trash, slowly marching toward its impending expiration date. All of this happens during the first five minutes of the film. For the next two hours, Pierre is biding his time hoping for a transplant that will correct the fatal flaw. While he is waiting, there are many other examples of the body as trash, most of them concerning the women in Klapisch's cast of characters.

Over the course of the film, a group of Parisian fashionistas bounces around, first appearing in a luxury resort in Cameroon, then at a fashion show in the capital city. After a *défilé de mode*, the women come across the group of market vendors who represent a different demographic, but one that is very much present in the film: the working class. The market is a crucial service that sustains the city by providing essentials goods for consumption. In the few shots that show the men working their jobs, it is clear they understand the importance of their role and take pride in the quality of the items they sell. The often-macho vendors immediately intrigue the flirtatious women who have stumbled into their territory. The men invite them into their warehouse to woo them with exotic fruits, vegetables, flowers, and meats, hinting at a larger system of importing goods from around the globe. While the Parisians are lucky to have access to the products, it is a reminder that cities do necessitate unsustainable practices such as the energy expended to bring foreign items to the city. The men and women pair off as they venture into different realms of merchandise. The next sequence of interest introduces "Miss Bidoche,"[19] who declares her love for meat, which explains her nickname. She and a street vendor make their way into the meat locker, seducing each other among the hanging animal carcasses on their way to the markets. At first glance "Miss Bidoche," dressed in a fur coat and refusing to wear a sanitary white jacket, finds her place among the meat, waiting to be devoured by her predator, the street vendor. He pursues her until he makes his kill, removing her fur coat, "skinning" her like the other pieces of meat,[20] and locking her in an embrace. But this is not exactly what is at play in the scene. "Miss Bidoche" takes control of the situation and reverses the normal predator–prey gender roles to wield the power over her target,[21] the vendor. "Miss Bidoche" accepts and embodies a carnal view of herself that is often perpetuated by society.

Another woman in *Paris* embodies trash in a very different way. During the last third of the film, the spectators witness the unfortunate death of Caroline, a female market vendor. Throughout the film, she is shown as a woman working in a group of men but fitting in as "one of the guys." Nevertheless, she is occasionally the target of sexual advances and, in one particular scene, sexually harassed to the point of breaking. In the scene, the vendors are at a restaurant socializing. At one point, Caroline gets up, and one of the male vendors follows her. He grabs her and playfully forces her into a wheelbarrow position. This is uncomfortable and degrading to Caroline. She is pressured into a compromising sexual position while wearing a dress in front of an audience. Not only is this an example of the objectification of women, a way of devaluing them, but it also resonates because a wheelbarrow is often used to carry and transport waste. Perhaps because of this incident, Caroline makes arrangements to switch jobs, presumably for one in a less aggressive environment. In the following scene, she determines the details, and her new employer asks her when she will begin. In reply, she puts on her helmet and swings her leg over her motorcycle, stating, "Je me sens libre comme l'air," an equivalent phrase to "I am free as a bird." We see Caroline ride through the streets in a series of traveling shots until a rough cut to a car in her path surprises us (and Caroline). She collides with the car and flies through the air, enacting her previous comment about being as free as a bird. Immediately following this shot, Klapisch inserts an example of what I want to call "film as trash," which I will analyze more thoroughly in the next section. In the context of the sequence, this film as trash signals Caroline's death and that her expiration date has arrived. Later in the film, Jean, one of the street vendors, takes Caroline's ashes to the top of the Tour Montparnasse and lets them go in the wind. This final example of Caroline's ashes floating above the city suggests a sort of reconnection, that even in her death she is free from the body that could no longer sustain her. Though it can be argued that death is more avoidable for those with the resources to afford stability and good healthcare, death is an inevitable truth that will eventually serve all people regardless of social hierarchy. Shortly following Caroline's death, Pierre gets a call that a transplant is available. Though we cannot know for certain, the sequence of shots leads the audience to believe the heart belonged to Caroline. Since healthcare is universal in France, the transplant is performed without regard to social class and with the sole purpose of sustaining life.[22] Giving and receiving transplants is a sustainable practice of human life that does not look at social standing; it depends on the quality of the organ and its potential for prolonging the vitality of the body.

After opening the film with the discovery of Pierre's sickness, Klapisch returns to Pierre to end the story. In the closing scenes of the film, Pierre is given hope that his own damaged heart will become waste in exchange for a new one, a fascinating example of the sustainable practice of human organs made recycled material. Just as obsolete or defective electronics can be scrapped for their working parts and refurbished so as to create and consume less, the human body is examined so the working organs can be put to use in others. The sequence begins with a shot of Pierre in bed surrounded by black sheets and broken light, as though confined in a cell. His walk-up apartment has become a prison during his sickness. He answers a phone call, his face mostly in the dark, reminiscent of the opening scene of the film. The spectator gathers through Pierre's responses that the hospital is calling with the news that a transplant is available. Throughout Klapisch's filmography, he insists on the interconnectedness of inhabitants of urban spaces. Sharing and recycling vital organs certainly reinforces the same idea. In this way, the recycling of the heart from Caroline to Pierre is a life-saving example of the sustainability of the city.

Once Pierre begins his journey to the hospital, the movie cuts to another character, Philippe, who holds his newborn son for the first time and telephones his brother, Roland, to share the news. Contrasts of life and death fill the last minutes of the film. This is one way in which Klapisch begins to conclude his film without suggesting that the interconnectedness of the city will end alongside it. He asks his spectators to imagine the narrative that continues on even after the credits begin to roll. When Pierre enters the car, the driver informs him of a traffic blockage and asks him which route he should take. The driver complains about it but comments that there is a small chance that it will clear. The following shots alternate between close-ups of Pierre and traveling shots of monuments and people on the street. In the close-ups, Pierre observes from a hopeful perspective how lucky the Parisians are to be alive, even if they do complain too much.

Now Pierre is lying back in the car and his point of view shots are angled upward, showing the spectator the sky, the opposite perspective from when he was acting as the Eiffel Tower looking down on Paris below, which signals that he has rejoined the web of connection that he so longingly observed during his isolation. The very last shot of the film is a panorama of the Parisian landscape. Pierre is not at all discernible in the hustle and bustle, but he is happiest that way, a small point lost in the busyness of the city. The final encompassing view of the city leads the spectator to take one last look at the connection between Pierre and the city of Paris. This capital city is the center of France, very surely, and is a "vital organ" on the global scale. The space and traffic of Paris are analogous to Pierre's heart and its function. The network of roads, freeways, and boulevards disseminates people and cars through the city toward other regions just as Pierre's heart pumps blood and circulates it to the rest of his body. When the driver notes the traffic jam, he draws an interesting parallel to Pierre's heart condition. Just as the *embouteillage* might pass, so may Pierre's operation repair any defect and help his blood to circulate through his veins. The sustainable practice of organ donation is what gives Pierre this hopeful outlook.

FILM AS TRASH

The last variation of trash concerns the cinematography, a meta-reflection on the sustainability of film itself. There are many examples of "filler" shots, shots that are essentially useless, or film as trash, that, when considered in the larger context of the film, function as meaningful examples of superfluous material. In the age of recording on actual film stock, trashy film would be the scraps that the director picked up off the cutting room floor. Now, as the industry almost exclusively uses digital recording and editing, film as trash comes from the electronic trashcan of editing software. Three main examples will provide a basis for this iteration of trash. The opening sequence of *Paris* blends together people and place to set the film on its way. Then the scene that directly follows Caroline's death shows a more poetic example of film as trash and will be discussed at length. Finally, the closing sequence deconstructs the opening sequence's equation to point again to people and place as separate entities. All the examples underline the practices of waste and recycling within the metropolitan area and within the film itself.

The opening sequence of *Paris* uses a different strategy than the traditional opening of a film. Usually directors begin with the general and progressively zoom in to the specific. For *Paris*, this would mean to show broad panoramas of the city before leading to more focused attention on specific characters. Instead, Klapisch continually oscillates between the general and the specific. His first shot, a panorama of the Parisian skyline from a landing on the Eiffel Tower, shows the place of the film. This shot then pans to the left to show Roland, an important character. Following are shots of buildings in Paris from a low-angle, simulating the perspective of a passenger in a car. A jump cut reveals a man dancing in a cabaret-style show. Though we are unaware at the time, he is the main character of the film, and this shot is from his past. The next shot shows Sacré-Coeur, an iconic landmark of Paris, and then an apartment building. At first, this mise-en-scène appears to focus only on place, but one of the characters opens a balcony door to appear on screen. His minor role is reinforced by his small size on screen and his very brief appearance. A traveling close-up of his girlfriend materializes next. She plays a much bigger role, thus the camera shows more interest, giving her more time and a larger representation. She dominates the screen, and there are no identifying characteristics of the city in this shot. Klapisch continues this vacillation between place and people for the duration of the opening credits. It is as though he is breaking down his upcoming film into its parts—people and place—the raw material collected during filmmaking. Once the narrative begins, they will come together and formulate space, which is the interaction of people in a place, as Michel de Certeau asserts. De Certeau writes: "In short, *space is a practiced place.* Thus the street geometrically defined by urban planning is transformed into a space by walkers."[23] Like this formula of people and place, cinema combines its own unique set of ingredients to create and sustain a narrative through cinematographic means. Film uses moving images and sound to produce a particular storytelling machine. In *Paris*, the shots and narratives of Parisians add a dimension of life and connection to the urban space, much like de Certeau suggests. Klapisch uses scraps, his film as trash, to illustrate the importance of the city, its dwellers, and how they come together, recycled and appropriated to make a film. Between the opening and closing sequences of his work Klapisch inserts a very poetic example of film as trash to reflect on the city dwellers' mortality.

Klapisch places Caroline's death two-thirds of the way through his film. Immediately following her death, a scene, as mentioned above, in which she runs into another car while on her motorcycle and catapults through the air, the director pauses the narrative. Instead of following her death with the chaos of emergency care, he inserts a beautiful example of film as trash. The first shot is of two trash bags blowing in the wind,[24] a powerfully evocative illustration given Caroline's reference to feeling like air, and the fact that her body just "expired" and turned to trash. Shots of the street, the sidewalk, and buildings ensue, all of which are handheld and shake ever so slightly, continuing the trope of wind. These calm images pay homage to Caroline. They are mostly devoid of people to imitate the absence of life and reduce De Certeau's equation of space back to mere place. These are bits of shots that we could imagine the director chose to fill the space that Caroline's death just left. They are essentially throwaways that find new meaning in the context of the loss of life.

The final example of film as trash, the closing sequence of the film, focuses on Pierre's journey back into the city with the hopes of a new lease on life. As Pierre moves to the street level and reintegrates himself with the rest of the city, Klapisch reverses the normal establishing shot sequence of the opening of a film. He begins with Pierre, the specific, and zooms out to the general city of Paris, the opposite movement of the traditional opening. From this last, concluding shot, Klapisch returns to the same panorama that he used during the opening credits. This recycled piece of film revitalizes the trope of sustainability, one that was already shown through municipal waste removal and vital organs, and attaches it to the film. In fact, if the spectators pay close attention, they will notice that the images used during the opening and closing credits come from the film itself; they are recycled. This decision on the director's part also reflects the rhizomatic nature of the film on a cinematographic level. Instead of subscribing to the normative organization of a film—beginning, development, end—Klapisch chooses to duplicate scenes and undermine this linear structure. The beginning is also the end, which is also the middle. This repetition of previously used footage becomes very clear during the end credits, in which Klapisch deconstructs de Certeau's equation, subject + place = space, to show its unique importance to film. The characters and their lives add a vibrant aspect of life and interconnectedness to France's capital. His recycled panorama, appearing in both the opening and closing credits, underlines the place of the film, Paris. Next, close-ups of individual characters emphasize the subjects, the many characters that played major and minor roles during the film. Most of these specific shots are recycled from previous moments, so the attentive spectator realizes the sustainable status of film itself. Since these recycled shots portray not only the place of Paris but also the people who inhabit the city, Klapisch positions his characters as equally important to the monuments and the iconic views; they are also what makes Paris such a vibrantly intertwined and interdependent city.

CONCLUSION

With a growing global population that will inevitably produce more waste, it is important to consider sustainable alternatives for the future. In his film, Klapisch offers the spectators three other ways to conceptualize trash and recycling. He suggests that urban living is a unique organization that is not as dependent on social hierarchies as it seems. Rather, the community that grows in the city generates a rhizome, which feeds off forms of connection on all levels of interaction. Trash reiterates this idea in multiple ways. As everyday activities, waste production and sustainable practices permeate many facets of life. From municipal waste removal to the body and to film itself, human existence is dependent on waste, renewal, and regeneration. Trash circulates throughout the urban space, illustrating power dynamics and wealth distribution. Regardless of these connotations, trash is also egalitarian in that it touches all classes. The body illuminates different ways in which it generates and expires, acting as trash itself. The body can also be seen in terms of varying value and is sometimes a thing to be used or discarded depending on several factors. Finally, Klapisch uses his own cinematic apparatus to return to sustainable practices. He inserts and repurposes bits of film to reflect on important questions of life, death, and connection, as well as on the renewable nature of cinematography. These three iterations of trash ultimately underline the sustainability of urban space. The communities of people within it connect, share, and discard in a multitude of meaningful ways to create a rhizome of connection that continues to extend beyond the limits of the city.

NOTES

1. Gilles Deleuze and Félix Guattari, *A Thousand Plateaus*, trans. Brian Massumi (Minneapolis: University of Minnesota Press, 1987), 7.
2. Ibid., 12.
3. Ibid., 9.
4. For more one these cinematic movements and Klapisch's relationship to them, see Elizabeth Ezra, "Cats in the 'Hood: The Unspeakable Truth about *Chacun cherche son chat* (Klapisch, 1996)," *French Cinema in the 1990s: Continuity and Difference*, ed. Phil Powrie (Oxford: Oxford University Press, 1999), 211–22; Sue Harris, "The *Cinéma du Look*," *European Cinema*, ed. Elizabeth Ezra (Oxford: Oxford University Press, 2004), 219–32; Myrto Konstantarakos, "Which Mapping of the City? *La Haine* (Kassovitz, 1995) and the *cinéma de banlieue*," *French Cinema in the 1990s: Continuity and Difference*, ed. Phil Powrie (Oxford: Oxford University Press, 1999), 160–71; Phillipine Le Bret, "La part de gâteau indigeste de Cédric Klapisch," *La Péniche*, April 3, 2011, accessed December 14, 2014, http://lapeniche.net/la-part-de-gateau-indigeste-de-cedric-klapisch/; Audrey Mariette, "Commitment through the Work of Art, Commitment through the Name: The Case of French Directors of 'Social Films' in the 1990s-2000s," *International Journal of Politics, Culture, and Society* 23, no. 2 (2010): 157–73; Rémy Pellissier, "Interview de Cédric Klapisch: le 7e art en péril?," *Le Figaro*, May 14, 2008, accessed December 13, 2014, http://

evene.lefigaro.fr/cinema/actualite/interview-cedric- klapisch-paris-cannes-visions-sociales-1371.php.

5. For a more thorough example of the spontaneous trajectories of waste see Agnès Varda's film: *The Gleaners and I*, DVD, dir. Agnès Varda (New York: Zeitgeist Films, 2000).

6. For more information on the legislation against "ghettos for the rich," see Ian Brossat, "3 milliards pour le logement, 3 raisons de s'en réjouir," *Le Huffington Post*, December 17, 2014, accessed September 28, 2015, http://www.huffingtonpost.fr/ian-brossat/3-milliards-pour-le-logement-3-raisons-de-sen-rejouir_b_6336106.html. Although his piece shows the city's fight against this movement, the necessity of these types of protections points to a problem of global metropolitan areas becoming increasingly expensive and prohibitive for working-class and middle-class individuals.

7. Stephen Heath, "Notes on Suture," *Screen* 18 (1977/78): 56.

8. For more on cinematic suture, see Heath, "Notes on Suture," 48–56 and Kaja Silverman, *The Subject of Semiotics* (Oxford: Oxford University Press, 1983).

9. In employing the term diegetic, I am referring to the space visible on screen as well as any sounds emitting from sources included in the frame. Extradiegetic space is created by sounds or objects that reference imagined space beyond the limits of the frame. For example, Klapisch films many of his scenes in the streets of Paris. In any given shot there is the space on screen, the diegesis, but Klapisch can also extend that space by including an ambulance siren. Even though that vehicle does not appear on screen the spectators understand that it exists outside of the field of the camera on another street in Paris. When spectators see the metro entrance on screen they imagine the underground system it is referencing, the different lines, stations, and tunnels. In both cases the ambulance in the streets of Paris or the underground space of the metro are extradiegetic, not literally visible on screen but referenced so that the space beyond the screen extends in the imagination of the spectators. Non-diegetic sound is music in the soundtrack that does not refer to anything on screen or to any offscreen space, object, or person. Klapisch employs this type of non-diegetic sound multiple times during the film.

10. These labels create a binary opposition: Paris and its suburbs oppose other regions in France and all international destinations.

11. In French, "pierre" also means "rock."

12. While the situation in which Pierre rests above the city as a metaphorical Eiffel Tower may appear as a panopticon, a system of surveillance in which one person holds the privileged position to observe everything going on around him while those being observed do not know when they are being watched and thus forced to behave, this is not exactly the case. Pierre does hold a position wherein he can surveil without being detected, but those being watched do not know that Pierre is looking down from above. The Eiffel Tower may be a better example of a panopticon because Parisians and tourists alike are very aware of its presence above the city. In reality Pierre is a less imposing voyeur than the Eiffel Tower, even though he does embody it metaphorically.

13. Kenneth W. Harrow, *Trash: African Cinema From Below* (Bloomington: Indiana University Press, 2013), 1.

14. Ibid., 2.

15. Ibid.

16. Human trash; human waste.

17. Harrow, *Trash*, 2.

18. Ibid.

19. "Bidoche" is a familiar or slang expression that translates to "meat."

20. The scene with "Miss Bidoche" is reminiscent of Lady Gaga's controversial meat dress she wore to the 2010 Video Music Awards. See Maeve Keirans, "See What Lady Gaga's Meat Dress Looks Like Now—5 Years Later," MTV News, August 28, 2015, accessed November 5, 2015, http://www.mtv.com/news/2254239/lady-gaga-meat-dress-vmas/.

21. Kenneth Harrow explores the dynamics of portraying trashy women in African film. While there are similar questions of gender dynamics at play, the trashy women in the context of African film are also steeped in exoticism due to the history of colonialism, its fall, and the residual effects. Harrow's chapter and Klapisch's film provide examples of women harnessing the role of the trashy woman to reverse power dynamics and rebel against normative social structures. See Harrow's chapter "Trashy Women: *Karmen Gei*, l'Oiseau rebelle" in *Trash: African Cinema from Below* (126–29) for more on "trashy" women in African film.

22. Klapisch chooses to present a scenario in which the process of organ donation is legal and occurs within the confines of citizenship. *Paris* does not represent illegal examples of organ buying and selling through the black market, those of which principally involve illegal citizens who are making the choice out of desperation. For an example of this clandestine organ donation and its social implications, see the film *Dirty Pretty Things*, dir. Stephen Frears, DVD (Santa Monica, CA: Miramax, 2002).

23. Michel de Certeau, *The Practice of Everyday Life*, trans. Steven Rendall (Berkeley: University of California Press, 1980), 11 (original emphasis).

24. These trash bags on metal rings also evoke the lives of the victims of terrorist attacks in Paris in the 1990s. After multiple instances of bombs being placed in trashcans, the city decided to remove the bins for the freestanding bags to avoid further attacks. See "Trash Cans Removed after Paris Bombing" in *The Spokesman-Review* for more information. The attacks in January 2015 on *Charlie Hebdo*, and more recently on November 13, 2015, remind us that the city unfortunately remains a target for terrorism. "Trash Cans Removed after Paris Bombing," *The Spokesman-Review*, August 19, 1995, accessed October 29, 2015, http://www.spokesman.com/stories/1995/aug/19/trash-cans-removed-after-paris-bombing.

BIBLIOGRAPHY

Brossat, Ian. "3 milliards pour le logement, 3 raisons de s'en réjouir." *Le Huffington Post*, December 17, 2014. Accessed September 28, 2015. http://www.

huffingtonpost.fr/ian-brossat/3-milliards-pour-le-logement-3-raisons-de-sen-rejouir_b_6336106.html.

de Certeau, Michel. *The Practice of Everyday Life*. Translated by Steven Rendall. Berkeley: University of California Press, 1980.

Deleuze, Gilles and Félix Guattari. *A Thousand Plateaus*. Translated by Brian Massumi. Minneapolis: University of Minnesota Press, 1987.

Dirty Pretty Things. Directed by Stephen Frears. DVD. Santa Monica, CA: Miramax, 2002.

Ezra, Elizabeth. "Cats in the 'Hood: The Unspeakable Truth about *Chacun cherche son chat* (Klapisch, 1996)." *French Cinema in the 1990s: Continuity and Difference*, edited by Phil Powrie, 211–22. Oxford: Oxford University Press, 1999.

The Gleaners and I. DVD. Directed by Agnès Varda. New York: Zeitgeist Films, 2000.

Harris, Sue. "The *Cinéma du Look*." *European Cinema*, edited by Elizabeth Ezra, 219–32. Oxford: Oxford University Press, 2004.

Harrow, Kenneth W. *Trash: African Cinema From Below*. Bloomington: Indiana University Press, 2013.

Heath, Stephen. "Notes on Suture." *Screen* 18 (1977/78): 48–76.

Keirans, Maeve. "See What Lady Gaga's Meat Dress Looks Like Now—5 Years Later." *MTV News*, August 28, 2015. Accessed November 5, 2015. http://www.mtv.com/news/2254239/lady-gaga-meat-dress-vmas/.

Konstantarakos, Myrto. "Which Mapping of the City? *La Haine* (Kassovitz, 1995) and the *cinéma de banlieue*." *French Cinema in the 1990s: Continuity and Difference*, edited by Phil Powrie, 160–71. Oxford: Oxford University Press, 1999.

Le Bret, Phillipine. "La part de gâteau indigeste de Cédric Klapisch." *La Péniche*, April 3, 2011. Accessed December 14, 2014. http://lapeniche.net/la-part-de-gateau-indigeste-de-cedric-klapisch/.

Mariette, Audrey. "Commitment through the Work of Art, Commitment through the Name: The Case of French Directors of "Social Films" in the 1990s-2000s." *International Journal of Politics, Culture, and Society* 23, no. 2 (2010): 157–73.

Paris. Directed by Cédric Klapisch. DVD. Paris: StudioCanal, 2008.

Pellissier, Rémy. "Interview de Cédric Klapisch: le 7e art en péril?" *Le Figaro*, May 14, 2008. Accessed December 13, 2014. http://evene.lefigaro.fr/cinema/actualite/interview-cedric- klapisch-paris-cannes-visions-sociales-1371.php.

Silverman, Kaja. *The Subject of Semiotics*. Oxford: Oxford University Press, 1983.

"Sortie ce mercredi de "Paris" de Cédric Klapisch." *L'Obs*, February 20, 2008. Accessed December 13, 2014. http://tempsreel.nouvelobs.com/culture/20080220.OBS1375/sortie-ce-mercredi-de-paris-de-cedric-klapisch.html.

"Trash Cans Removed after Paris Bombing." *The Spokesman-Review*, August 19, 1995. Accessed October 29, 2015. http://www.spokesman.com/stories/1995/aug/19/trash-cans-removed-after-paris-bombing.

Index

About the Contributors

Anirban Adhya is Associate Professor of Architecture (Urban Design) at Lawrence Technological University in Southfield, Michigan. He is a founding member of the synchRG think tank and an active proponent of the Detroit Studio outreach program. His research focuses on sustainability as ecology with emphasis on well-being and social ethics in places; his scholarship integrates theories and practices of sustainable placemaking and everyday urbanism. His works have been recently published in *Architecture and Sustainability: Critical Perspectives* (ACCO, 2015), *Terrain Vague: Interstices at the Edge of the Pale* (Routledge, 2013), *Urban Design International* (2013), and *The Urban Wisdom of Jane Jacobs* (Routledge, 2012).

Doris Bremm is area coordinator for culture and literature at the Katholische Familienbildungsstätte, Bonn. In her research, she specializes in contemporary literature and intersections between literature and the visual arts and explores contemporary Anglophone literature about visual art as a new way to historicize postmodernism and the postmodern novel. She has published on ekphrastic narratives by Julian Barnes, A.S. Byatt, and Peter Ackroyd in *London in Contemporary British Fiction: The City beyond the City* (Bloomsbury, 2016) and contributed chapters on gender studies and comparative literature to *The English Literature Companion* (Palgrave Macmillan, 2010).

Lauren Curtright is Assistant Professor of English at Perimeter College at Georgia State University. Her doctoral thesis in English at the University of Minnesota explored connections between the Gothic and Modernism across various media, focusing on ways in which writings by, and images of, Edgar Allan Poe have influenced architecture, film, and literature from the antebellum United States to modern Japan. She has published work on

Dorothy Richardson's film writing in *Pilgrimages: The Journal of Dorothy Richardson Studies*, on daguerreotypes of Poe in *Deciphering Poe: Subtexts, Contexts, Subversive Meanings* (Lehigh University Press, 2013), and on racial constructs in her own genealogical record compared to those in Mark Twain's *Pudd'nhead Wilson* in the journal *Letterature d'America.*

Joseph Donica is Assistant Professor of English at Bronx Community College, CUNY. He has published articles and reviews on a wide range of topics including American architecture, 9/11 literature, Arab-American literature, Netflix and the digital future, and disability studies. His latest two articles are "Rethinking Utopia for the Twenty-First Century: The Good Life after Occupy and the Arab Spring" in the collection *The Good Life and the Greater Good in a Global Context* (Lexington, 2015) and "Negative Memory after Katrina: The Persistence of Memoir" in the *Journal of the Midwestern Modern Language Association.* He is currently finishing his first book titled *Inequality's Memory: American Literature after Occupy and the Arab Spring.* He is also coediting a collection titled *Reflections on a Changing Profession: The Future of the English PhD.* He recently participated in an NEH fellowship in D.C. on the history of Islam in the United States.

Lisa FitzGerald is a performance historian whose research interests include environmental theater practice; ecodigital art; urban ecology; and the relationship between art, nature, and technology. From an interdisciplinary background in fine art and site-specific theater, Lisa received her PhD from the National University of Ireland, Galway. She is currently a Carson Fellow at the Rachel Carson Center for Environment and Society in Munich. Her book, *Re-Place: Irish Theatre Environments*, will be published by Peter Lang in Autumn 2017.

Heide Imai holds a PhD in Urban Sociology and is Assistant Professor at the Department of Global Interdisciplinary Studies, GIS, Hosei University, and Research Associate at Keio University, Tokyo. Current publications include *Tokyo Roji: The Diversity and Versatility of Alleyways in a City in Transition* (Routledge, 2016). As a licensed architect, she has practiced in offices in Germany, the Netherlands, and the United Kingdom, and is teaching courses on critical urbanism, urban theory, and so on, as part of her regular field and research trips.

Alexander Kleinschrodt is a musicologist and cultural studies scholar. He currently teaches as assistant lecturer for "Transdisciplinary Theories and Methods" in the Bachelor of Arts program at the Philosophical Faculty, University of Bonn, Germany. From 2012 to 2014, he was research associate in the Department of Musicology/Sound Studies at the University of Bonn. He works in the field of cultural education, connecting the topics music, arts, cultural history, and sustainability and is also active as a journalist for different media, among them the Berlin-based FUTURZWEI Foundation.

Mehdi Kochbati holds a doctoral degree in contemporary American literature from the University of Paris Ouest. His thesis titled "Memory Writing in the Work of Paul Auster: Crossing of Personal and Collective Memory," for which he received a very Honorable distinction, was conducted under the direction of Professor Cornelius Crowley in the Department of English and Anglo-American Studies. He is a researcher at the Anglophone Research Centre (CREA), Section 370 of the CNRS: Human sciences and Humanities at the University of Paris Ouest.

Claudia Mantovan holds a PhD in Sociology. She is a research fellow and teacher in the Department of Philosophy, Sociology, Education and Applied Psychology (FISPPA), University of Padua, Italy. In 2015, she was a visiting fellow at the Centre d'études européennes, Sciences Po, Paris. Her research focuses on immigration, citizenship, participation, coexistence, and conflict in multiethnic neighborhoods; urban segregation of immigrants; and Roma. Her work has been published in several academic journals (such as *Patterns of Prejudice*, *Forum: Qualitative Social Research*, *Déviance et Société*) and edited books. She has also published the books *Quartieri contesi. Convivenza, conflitti e governance nelle zone Stazione di Padova e Mestre* (with E. Ostanel, 2015), *Il ghetto disperso. Pratiche di desegregazione e politiche abitative* (coedited with F. Faiella, 2011), *Le forme della partecipazione politica delle donne. Il Veneto attraverso alcuni casi studio* (coedited with S. Grimaldi and L. Perini, 2011), and *Immigrazione e cittadinanza. Auto-organizzazione e partecipazione dei migranti in Italia* (2007).

Philip D. Plowright is Professor of Architecture (Design, Theory, and History) at Lawrence Technological University in Southfield, Michigan, and editor-in-chief of *Enquiry: The ARCC Journal of Architectural Research*. He holds degrees in studio art, art history, and architecture from the University of Guelph and the University of British Columbia and is currently pursuing advanced research in cognitive linguistics at the Universidad de Castilla-La Mancha. His research centers around issues of meaning, interpretation, and knowledge transfer in socio-formal environments. His recent publications include the book *Revealing Architectural Design: Methods, Frameworks & Tools* (Routledge, 2014) and a chapter in *Architecture and Sustainability: Critical Perspectives* (ACCO, 2015).

Iuliu Ratiu earned a Master of Arts degree in English and Writing from Kent State University (2006) and a PhD in American Literature from SUNY Albany (2012). He was a Marion L. Brittain Postdoctoral Fellow in the Writing and Communication Program at Georgia Tech (2012–2015). Most recently, he was a Stefan Odobleja Fellow at New Europe College, Institute for Advanced Study in Bucharest, Romania (2015–2016), where he worked on a project about the translation of Henry David Thoreau's *Walden* from English to French and from French to Romanian.

Lea Rekow (PhD) is director of Green My Favela, an urban remediation project based in the favelas of Brazil. Her research focuses on transdisciplinary practices for reclaiming degraded land in areas of social exclusion. She is an advisor to the Integrated Media program at Cal Arts and a member of the Institute for Australian Geographers and of New York Women in Film and Television. She is also a sustainability consultant to the GlobalCAD network and was recently advisor to the European Urban IxD program. Previously based in New York, she has held positions as director of Gigantic ArtSpace, media director at the former Center for Peace and Human Security, adjunct professor of media and communications at Pratt Institute, and cultural advisor for Advance at the Australian consulate in New York. She has also held positions as director of the Center for Contemporary Art in Santa Fe and research fellow at the Center for Art and the Environment. She has sat on advisory panels for NYFA, SVA, Parsons, Amnesty International, the MacArthur award, and the LMCC. Lea's work as a writer, filmmaker, researcher, and curator has been supported, reviewed, and presented internationally.

Christopher Schliephake (PhD) is a lecturer and postdoc scholar at the University of Augsburg. His dissertation *Urban Ecologies: City Space, Material Agency, and Environmental Politics in Contemporary Culture* was published in Rowman and Littlefield's *Ecocritical Theory and Practice Series* (Lexington Books, 2015). His research interests include ecocriticism, cultural ecology, cultural memory studies, and classical reception studies. He has dealt with these topics in numerous publications including essays in the peer-reviewed journals *Anglia*, *Amerikastudien/American Studies*, and *Ecozon@*. He recently edited a collection of essays titled *Ecocriticism, Ecology, and the Cultures of Antiquity*.

Katarzyna Szalewska (PhD) is Assistant Professor at the Institute of Polish Philology at the University of Gda sk (Poland). She has written the monograph *Textual Arcade. Reading the City as a Form of Experiencing the Past in a Contemporary Polish Essay* (2012). She has coedited such books as *The Curse of Reality. On Obsessions and Fantasies* (2009), *The North Side of*

Czesław Miłosz (2011), and *Non-places. Spatial Theories in Contemporary Practices of Interpretation* (2014). Her interests include urban anthropology, theory of historiography, relationships between literary studies and cultural geography, and historical research methodology. She has been awarded by the Foundation for Polish Science.

Caitlin Yocco-Locascio is a PhD candidate in the Department of French and Italian at the University of Wisconsin-Madison. She is currently working on her dissertation that examines the connections, limits, and overlaps between the Human and the Posthuman in French literature and film of the twentieth and twenty-first centuries. Her research interests include the contemporary novel, French film, visual culture, and questions of identity.

Karim Wagih Fawzi Youssef is a sessional instructor at Mount Royal University and has a PhD in Environmental Design from the University of Calgary and a master's degree from the University of Montreal. He has worked as a teaching assistant for courses in Community Planning, Sustainable Urban Planning, and Neighborhood Design. He is currently writing a monograph titled *Gated-ness: The Monadic Space of Suburban Neighborhoods*. He has coauthored publications and reports with Dr. Sasha Tsenkova, Professor of Urban Planning and International Development at the University of Calgary.

www.ingramcontent.com/pod-product-compliance
Ingram Content Group UK Ltd.
Pitfield, Milton Keynes, MK11 3LW, UK
UKHW020221290926
15470UKWH00001B/309